APIs AND PROTOCOLS FOR CONVERGENT NETWORK SERVICES

APIs and Protocols for Convergent Network Services

Stephen M. Mueller

McGraw-Hill
New York Chicago San Francisco Lisbon London Madrid
Mexico City Milan New Delhi San Juan Seoul
Singapore Sydney Toronto

Cataloging-in-Publication Data on file with the Library of Congress

McGraw-Hill
A Division of The McGraw·Hill Companies

Copyright © 2002 by The McGraw-Hill Companies, Inc. All rights reserved. Printed in the United States of America. Except as permitted under the United States Copyright Act of 1976, no part of this publication may be reproduced or distributed in any form or by any means, or stored in a data base or retrieval system, without the prior written permission of the publisher.

1 2 3 4 5 6 7 8 9 0   DOC/DOC   0 9 8 7 6 5 4 3 2 1

ISBN 0-07-138880-X

The sponsoring editor for this book was Marjorie Spencer, the editing supervisor was David E. Fogarty, and the production supervisor was Sherri Souffrance. It was set in Vendome by ATLIS Graphics and Design.

Printed and bound by R.R. Donnelley & Sons Company.

This book is printed on recycled, acid-free paper containing a minimum of 50% recycled, de-inked fiber.

McGraw-Hill books are available at special quantity discounts to use as premiums and sales promotions, or for use in corporate training programs. For more information, please write to the Director of Special Sales, Professional Publishing, McGraw-Hill, Two Penn Plaza, New York, NY 10121-2298. Or contact your local bookstore.

Information contained in this work has been obtained by the McGraw-Hill Companies, Inc. ("McGraw-Hill") from sources believed to be reliable. However, neither McGraw-Hill nor its authors guarantee the accuracy or completeness of any information published herein, and neither McGraw-Hill nor its authors shall be responsible for any errors, omissions, or damages arising out of use of this information. This work is published with the understanding that McGraw-Hill and its authors are supplying information, but are not attempting to render engineering or other professional services. If such services are required, the assistance of an appropriate professional should be sought.

CONTENTS

	Preface	xi
Chapter 1	Introduction	1
	1.1 What Is Network Convergence?	2
	1.1.1 Phone Networks	6
	1.1.2 The Internet	7
	1.1.3 Different Approaches to Convergence	11
	1.2 Who's Who in Convergent Networks	13
	1.2.1 ITU-T	13
	1.2.2 IETF	14
	1.2.3 Other Organizations	16
Chapter 2	Software for Managers	19
	2.1 Procedural Programming	20
	2.2 Object-Oriented Programming	22
	2.3 Component-Based Software	29
	2.4 Computer Networks	31
	2.4.1 APIs and Protocols	31
	2.4.2 Text-Based versus Binary Protocols	34
	2.4.3 The OSI Reference Model	39
	2.4.4 The Internet and Its Protocols	43
	2.5 For More Information	45
Chapter 3	The Java Programming Language	47
	3.1 Java's Role in Converged Networks	48
	3.2 The Java Virtual Machine	50
	3.3 The Java Class Libraries	53
	3.4 Java Applets and Security	55
	3.5 Java Servlets	56
	3.6 Java's Event Model	58
	3.7 JavaBeans	61
	3.8 For More Information	63

Chapter 4   XML: The Extensible Markup Language   65

- 4.1 Markup Languages and Structured Data   66
- 4.2 From HTML to XML   68
- 4.3 Defining XML Tags   73
    - 4.3.1 Document Type Definition (DTD)   73
    - 4.3.2 Schemas   78
- 4.4 Displaying XML: CSS and XSL   84
    - 4.4.1 Cascading Stylesheets (CSS)   85
    - 4.4.2 Extensible Stylesheet Language (XSL)   85
- 4.5 Parsing XML: DOM and SAX   88
- 4.6 For More Information   90

Chapter 5   Distributed Computing   91

- 5.1 Introduction   92
- 5.2 Sockets   94
- 5.3 Remote Procedure Calls   96
    - 5.3.1 How Nonremote Procedure Calls Work   96
    - 5.3.2 How Remote Procedure Calls Work   99
- 5.4 Distributed Objects   101
- 5.5 CORBA   103
- 5.6 Java RMI   109
- 5.7 DCOM   114
- 5.8 SOAP   117
    - 5.8.1 Why SOAP?   117
    - 5.8.2 SOAP and XML   119
    - 5.8.3 SOAP and HTTP   124
- 5.9 Comparing Distributed Object Technologies   129
- 5.10 For More Information   130

Chapter 6   Directories   133

- 6.1 Basic Concepts   134
- 6.2 Domain Name System (DNS)   136
    - 6.2.1 The DNS Namespace   137
    - 6.2.2 Nameservers   140
    - 6.2.3 Resource Records   142
- 6.3 X.500   144
- 6.4 LDAP   149
- 6.5 Future Trends   153

Contents

	6.5.1 X.500 and LDAP: Partners or Rivals?	153
	6.5.2 Directory Enabled Network (DEN)	154
	6.6 For More Information	155
Chapter 7	**Telephony for Programmers**	**157**
	7.1 The Voice Network	158
	7.2 Switching: Analog and Digital	160
	7.3 Signaling and Call Processing	163
	7.4 Phone Numbers	165
	7.5 ISDN	166
	7.6 SS7	170
	7.6.1 Circuit-Associated and Common-Channel Signaling	170
	7.6.2 SS7 Network Architecture	171
	7.6.3 SS7 Protocols	173
	7.7 Intelligent Network (IN)	181
	7.7.1 The Road from IN2 to IN0	181
	7.7.2 The Architecture of IN	183
	7.7.3 An Example of an IN Service	186
	7.8 For More Information	188
Chapter 8	**IP Telephony**	**189**
	8.1 What Is IP Telephony?	190
	8.2 QoS for IP Telephony	193
	8.2.1 Why Is QoS Such a Big Deal?	193
	8.2.2 Speech Encoding	194
	8.2.3 RTP and RTCP	195
	8.2.4 INTSERV, DIFFSERV, RSVP, and MPLS	196
	8.3 H.323	197
	8.3.1 H.323 Architecture	197
	8.3.2 H.323 Call Processing	201
	8.4 IP Telephony Gateways	202
	8.4.1 Gateway Architectures	202
	8.4.2 Gateway Protocols	205
	8.5 Routing and Translation for IP Telephony	214
	8.6 For More Information	217
Chapter 9	**Session Initiation Protocol (SIP)**	**219**
	9.1 Background	220
	9.2 SIP Architecture	223

	9.3 SIP Messages	227
	9.3.1 SIP Request Messages	227
	9.3.2 SIP Response Messages	229
	9.4 Session Description Protocol (SDP)	230
	9.5 Call Signaling with SIP	235
	9.5.1 Simple Two-Party Call	235
	9.5.2 Two-Party Call with Redirect Server	243
	9.5.3 Two-Party Call with Proxy Server	246
	9.5.4 Registration	248
	9.5.5 Forked Invitations	249
	9.6 SIP and H.323	253
	9.7 Interworking SIP	256
	9.8 For More Information	258
Chapter 10	Going Further with SIP	259
	10.1 Introduction	260
	10.2 Extending SIP	260
	10.2.1 Adding Request Methods	261
	10.2.2 Adding Headers and Parameters	261
	10.2.3 OPTIONS Requests	262
	10.3 PINT	263
	10.3.1 History	263
	10.3.2 The PINT Milestone Services	264
	10.3.3 The PINT Architecture and Protocol	265
	10.3.4 PINT Applications	268
	10.3.5 SUBSCRIBE, UNSUBSCRIBE, and NOTIFY	272
	10.4 Music on Hold	273
	10.5 Call Forward Busy Line (CFBL)	277
	10.6 SPIRITS	279
	10.7 For More Information	282
Chapter 11	Web-Oriented Service Technology	285
	11.1 Services and Service Logic	286
	11.2 Call Processing Language (CPL)	287
	11.2.1 What It Is and What It's For	287
	11.2.2 A Simple Script: Call Redirection	290
	11.2.3 A More Complicated Script: Call Screening	291
	11.3 SIP CGI and SIP Servlets	293
	11.3.1 SIP CGI	294
	11.3.2 SIP Servlets	298

Contents

	11.4 VoiceXML	303
	11.5 For More Information	309
Chapter 12	**JTAPI**	**311**
	12.1 JTAPI and Computer Telephony Integration (CTI)	312
	12.1.1 What Is CTI?	312
	12.1.2 APIs for CTI	314
	12.1.3 The Service Provider Interface (SPI)	316
	12.2 JTAPI Core	318
	12.2.1 The JTAPI Call Model	319
	12.2.2 Core Object States	322
	12.2.3 A Simple JTAPI Application	325
	12.2.4 JTAPI Event Listeners	327
	12.3 JTAPI Extension Packages	332
	12.3.1 javax.telephony.callcenter	332
	12.3.2 javax.telephony.callcontrol	335
	12.3.3 javax.telephony.media	337
	13.3.4 javax.telephony.mobile	338
	13.3.5 javax.telephony.phone	338
	13.3.6 javax.telephony.privatedata	341
	12.4 For More Information	341
Chapter 13	**JAIN**	**343**
	13.1 Traditional Service Development	344
	13.2 JAIN and Parlay to the Rescue?	347
	13.3 The JAIN API Stack	349
	13.4 JAIN's Development Process	353
	13.5 JAIN Protocol APIs	354
	13.5.1 JAIN TCAP	354
	13.5.2 JAIN SIP	361
	13.5.3 JAIN MGCP	367
	13.5.4 JAIN INAP	368
	13.6 Java Call Control (JCC) API	371
	13.6.1 Core JCP and JCC Classes	371
	13.6.2 JCP and JCC States	373
	13.6.3 JCC Call Processing	377
	13.7 For More Information	382
Chapter 14	**Parlay**	**383**
	14.1 Overview	384

14.2 Parlay's Interface Definition Language . . . 386
    14.2.1 Naming Conventions . . . 386
    14.2.2 Basic Data Types . . . 387
    14.2.3 Arrays . . . 387
    14.2.4 Constructed Types . . . 388
    14.2.5 Method Definitions . . . 390
14.3 Setting Up a Parlay Application . . . 391
    14.3.1 Authentication . . . 394
    14.3.2 Requesting Access to Parlay . . . 399
    14.3.3 Discovering Parlay Services . . . 401
    14.3.4 Subscribing to a Parlay Service . . . 402
14.4 Parlay Call Processing . . . 405
    14.4.1 Registering for Call Events . . . 405
    14.4.2 Handling Call Events . . . 408
    14.4.3 Handle Results from a Request . . . 410
14.5 Parlay Types Used in This Chapter . . . 412
14.6 For More Information . . . 420

Appendix A Acronyms . . . 421

Appendix B Web References . . . 429

Bibliography . . . 431

Index . . . 437

PREFACE

> Another damned, thick, square book! Always scribble, scribble, scribble! Eh, Mr. Gibbon?
> —William Henry, Duke of Gloucester, upon receiving from Edward Gibbon, Volume II of "The Decline and Fall of the Roman Empire"

This book aims to fill a gap in the literature by gathering in one place a survey of at least some of the more important software technologies that lie behind voice and data network convergence. There's a lot of ground to cover and, to be frank, the biggest motive for writing this book was my own inability to keep all this stuff straight in my head. As long as I had to work it out for myself, it seemed worth putting things in a form that might spare others at least some of the effort I had to go through.

Superficial white papers and exhaustive—not to mention exhausting—specifications of convergent network technologies can be found, but it's harder to find anything between these two extremes. Furthermore, much of the relevant material is scattered across the Internet. My goal was to find as many of these informational bits and pieces as I could, correlate them, and put them in a form that is, hopefully, more useful than a mere recitation of extracts from the standards. Some repetition of material from the original sources was unavoidable, but I've tried to pick and choose, highlighting what seems essential, adding commentary, and pointing the reader toward sources of further details.

You'll find lots of examples of code and protocol message exchanges accompanied by running discussions of what's going on. I think this is a better way to present this material than making the reader slog through long catalogs of features and capabilities when they just want to get a first impression of how things work. Despite the many code examples, I must, however, emphasize that this is not a manual for developing convergent network software. The examples should get you to the point where you can see the patterns that lie beneath a particular technology. But they won't make you an expert.

Another gap I've tried to fill in the existing literature is a tendency to concentrate on either the data or the voice side of convergence. A lot of prerequisite knowledge is assumed and readers who are unfamiliar with those topics are often left to their own devices trying to get up to speed. For example, CORBA is frequently mentioned as an element of network service architectures and APIs. Readers who aren't familiar with CORBA may feel they've been asked to put together a puzzle with missing pieces. Perhaps they can make out the shapes of those pieces from the holes they leave but not what's on them. I've tried as best I could to include enough information that one could build a reasonably complete picture, even if some details are only sketched in.

Having said this, two warnings are in order. First, I make no claims to have covered all software technologies that bear on convergent networks. Given the pace at which new technology is being created in this area, I'd probably have gotten myself into a Sisyphean mess if I'd attempted it. Nevertheless, I think that what's presented here is a good sample of the current state of the art. If I've succeeded, the reader should be able to figure out roughly where things not covered fit in the overall framework.

Second, in any discussion of network technologies, convergent or otherwise, it pays to remember the Cold War motto, "Trust, but verify." That goes also for this book. Networks are inherently complex. Many subtle details have to be just right for anything to work at all. With so much to coordinate and so much changing out from under us as networks and standards themselves change, it's all too easy for someone to state—in perfectly good faith—something that is quite simply not true. One of the biggest problems we face working in this area is sorting out facts from guesses. Even the standards themselves—perhaps because they are so new and have yet to undergo thorough review—are often vague. I've done my best to sift through all the evidence I could find and piece together a consistent story. I offer my apologies for all mistakes and misperceptions that have nevertheless found their way in.

I assume most of my audience has a technical background, but much of this book should be useful for others as well. The latter should feel free to skim along whenever there's more detail than they want or need, particularly in the code and protocol walkthroughs. Recognizing that not every reader will come to this book with the same background, Chapter 2 provides a basic introduction to selected topics in software. Likewise, Chapter 7 provides a basic introduction to telephony.

The first half of the book looks at software technologies for convergent networks that have their roots in the data world, in particular, distributed computing and the Internet. The second half turns to the voice world and the intersection of data networking technology with traditional telephony.

Preface

Chapter 1, Introduction, describes the landscape in which network convergence is being played out, laying out the shape of the traditional voice telephone network and that of data networks, especially those based on Internet protocols, plus standards bodies and other organizations that are creating the rules by which the convergence game is played.

Chapter 2, Software for Managers, provides background material on software technology for readers who are not programmers.

Chapter 3, The Java Programming Language, covers Java, a programming language that has had, and continues to have, a significant influence on the design and deployment of convergent network software.

Chapter 4, XML: The Extensible Markup Language, describes XML, a new language that builds on the capabilities of HTML, the language in which most Web pages are now written. It's likely XML will eventually replace HTML as the foundation of the Web. One of XML's biggest strengths is that it can be extended to support applications other than just Web page display, including applications that combine traditional telephony with Web interfaces.

Chapter 5, Distributed Computing, describes techniques for building applications in which cooperating programs run in parallel on a collection of interconnected machines. After introducing some general principles—sockets, remote procedure calls, and distributed objects—this chapter goes on to cover the most important technologies for building distributed, object-based applications: CORBA, Java RMI, DCOM, and SOAP.

Chapter 6, Directories, covers a key element of every network, namely, a way to find services, platforms, and users, and connect them to each other, with particular focus on DNS, X.500, and LDAP.

Chapter 7, Telephony for Programmer provides background material on the traditional voice phone network for readers who are familiar with software, but not with telephony.

Chapter 8, IP Telephony, introduces some basic principles of VoIP technology and emerging architectures and protocols that link IP networks with the PSTN.

Chapter 9, Session Initiation Protocol (SIP), describes an IETF protocol that is increasingly seen as part of the foundation for convergent network services.

Chapter 10, Going Further with SIP, builds on the material in Chapter 9, demonstrating how SIP can be used to build services that equal and exceed those in the traditional phone network, including services that use the PINT and SPIRITS protocols now emerging from the IETF.

Chapter 11, Web-Oriented Service Technology, describes CPL, SIP CGI, SIP servlets, and VoiceXML, all of which are Web-oriented technologies for building and deploying convergent network services:

Chapter 12, JTAPI, describes the Java Telephony API for computer telephony applications. JTAPI was one of the first convergent service APIs. Though its emphasis is on enterprise network applications, such as call centers, it has strongly influenced public network service APIs, such as JAIN and Parlay. Even if you aren't particularly interested in call centers and such, you should read this chapter because it lays the foundation for JAIN.

Chapter 13, JAIN, describes the Java APIs for Integrated Networks that can be used to build convergent network services.

Chapter 14, Parlay, describes a language-independent API with strong ties to JAIN. Like JAIN, Parlay also supports convergent network service development, while adding features to support third-party programming, features that have since been incorporated into JAIN as well.

One final note, before diving into the material at hand. As this is being written, the networking industry has undergone a veritable meltdown in the financial markets. One might wonder where all this convergence stuff is headed or whether it will even get anywhere. My reply is that, if you sincerely believe voice and data networking aren't going to be one of the fundamental technologies of our economy in the future, then by all means shift your attention elsewhere. I, for one, believe otherwise.

There will be some inevitable slowing down from the frantic pace set in the late 1990s, but that won't necessarily be such a bad thing. A lot of just plain awful software was written under the justification that it had to be that way when you "do things on Internet time." Would you want to fly in a plane or call a 911 system whose software was developed on Internet time? (I rest my case.) Perhaps we can now take another look at such hoary chestnuts as "A job worth doing is worth doing well." One can only hope.

Acknowledgments

I'd like to thank my colleagues at SBC Technology Resources who, over the years, have built a culture that is at once challenging and supportive. I'd especially like to thank Anil Bhandari, Alex Huang, Satish Parolkar, Tim Schroeder, and Mark Wuthnow, who provided feedback of immeasurable value to me. My brother, Mike Mueller, provided me with invaluable background material on SS7 from his years of experience in that field. I'd also like to thank my colleagues who contributed—knowingly or not—to the material in this book through any number of hallway conversations over the years. Among them, I'd particularly like to note Tom Adams, Will Chorley, Phil Cunetto, Jim Doherty, Dave Harber, John Lemay, John Palmore, Richard Payton, Marco Schneider, and Jeff Scruggs. Finally, I'd like to thank my managers (yes, you read that right), Jeff Johnson and Monte Cely, who had the patience to let me to pursue this to the end, even as it grew larger than we ever expected. I should also thank them for making it clear that patience has its limits. There's nothing like a deadline, even if it's somewhat flexible, to get an author to the last page and the final revision.

—STEPHEN M. MUELLER

CHAPTER 1

Introduction

Birds of a feather will gather together.

—*Robert Burton*
(Anatomy of Melancholy)

1.1 What Is Network Convergence?

Networks connect things. Trading networks connect buyers and sellers. Kinship networks connect family members. Old boy networks connect old boys. Phone networks—wireline and wireless—connect callers. Radio and TV networks connect broadcasters and viewers. The Internet connects computers.

Getting connected is easy if you stay within a given network's boundaries. Connecting from one network to another or sharing information between networks is another matter. Methods range from the merely adequate to the nonexistent. That's not surprising because networks often develop in isolation from one another. Thus, each network develops its own *protocols* (defined sets of messages), for making and maintaining connections and for exchanging information among its members. If networks are to communicate with each other across their normal boundaries, they must agree on protocols that reconcile their differences.

Protocols that allow one network to talk to another are nothing new. They're used to place calls from one country's phone network to another's. They're used in calls from wireless to wireline phones and vice versa. But these are all phone networks and thus have many more similarities than differences. Trying to connect traditional voice and data networks, which differ greatly from each other, is a lot more complicated. That's what network convergence is all about. And that's why it's taking a lot longer to do than some had predicted (perhaps without giving enough thought to what had to be done).

Network convergence is often thought to be more or less the same as *Voice over IP (VoIP)*, a service in which phone calls are carried entirely or in part on the Internet or a private IP network rather than just on the traditional *Public Switched Telephone Network (PSTN)*.[1] VoIP is an impor-

---

[1] The traditional voice network and its constituent networks go by several names. In this book, *PSTN* refers to all the world's interconnected traditional (i.e., non-Internet) voice networks. The PSTN can be viewed either as a single network or, more accurately, as a collection of networks, each having strictly defined points of interconnection. These could be where one carrier's network meets another's, where one country's network meets another's, or where a wireline meets a wireless network. Some definitions omit things like corporate voice networks from the PSTN, since they aren't, strictly speaking, "public." Others split out things like wireless voice networks or ISDN. While it can be useful to draw such distinctions for some purposes, when most people talk about the PSTN, they implicitly assume these are included. To escape this quandary, some have proposed the terms *Switched Circuit Network (SCN)* and *Global Switched Telephone Network (GSTN)*. Both are plausible alternatives. We could use any of these, but we have to pick one, so *PSTN* it is.

tant part of network convergence. But there's a lot more to convergence than just VoIP.

Besides traveling across the Internet, calls can stay on the PSTN but record information about themselves on the Internet, or they can use information from the Internet to decide how they should be handled. Calls in the PSTN can be launched from the Internet. Phones can be used to access Web pages. Information about a subscriber's location can be shared among wireline and wireless networks. Equipment attached to home and office networks can interact with public networks, like the Internet and PSTN. The technology used in public networks can be applied to private networks that blend traditional telephony and Internet protocols. All these and more fall within the scope of network convergence.

In a seminal 1996 paper [66], David Messerschmitt identified the extent to which telecommunications, as exemplified by the PSTN, and computing, as exemplified by the Internet, were becoming increasingly intertwined. He proposed the model shown in Fig. 1.1, in which *net-*

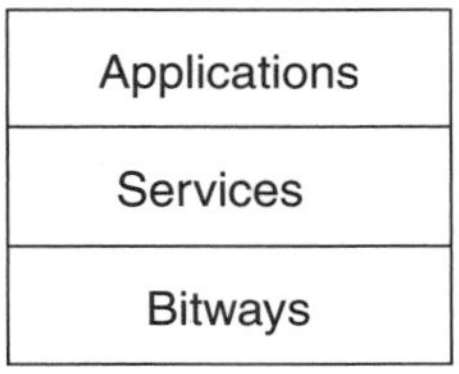

Figure 1.1
A simple model of network applications.

Applications
Services
Bitways

worked applications—for example, e-mail, voice telephony, Web browsing, file transfer, or video conferencing—use supporting *services* of the computing or communications infrastructure. These services can be used to build all sorts of applications. They include such things as media transport, file management, encryption, and key distribution. For an application to be networked, there has to be a way to get bits from one place to another, a function that's provided by *bitways*. Many discussions of convergence focus almost exclusively on bitway technologies and far less on services. This book takes the opposite approach and deals almost exclusively with services and their relation to applications.

Messerschmitt goes on to group network applications into four categories, as shown in Fig. 1.2. Two of these pertain to ways in which users interact with an application:

- In *user-to-server* applications, users connect to remote systems to access or interact with information on those systems.

Figure 1.2
Types of network applications.

	Immediate	Deferred
User-to-server	Video on demand, Web browsing	File transfer
User-to-user	Voice telephony, Video conferencing	Email, Voicemail

- In *user-to-user* applications, users interact with each other directly. Remote systems assist, but they recede into the background, becoming at times almost invisible.

The other two categories relate to the temporal aspect of a user's interaction with an application:

- With *immediate* applications, users and servers interact in real-time. Delays can be critical.
- With *deferred* applications, time can pass without adversely affecting service. In fact, as with voicemail, the fact that time can pass between one operation (leaving a message) and another related operation (retrieving a message) is crucial to the usefulness of many deferred applications.

Network applications can be built in two ways, as shown in Fig. 1.3:

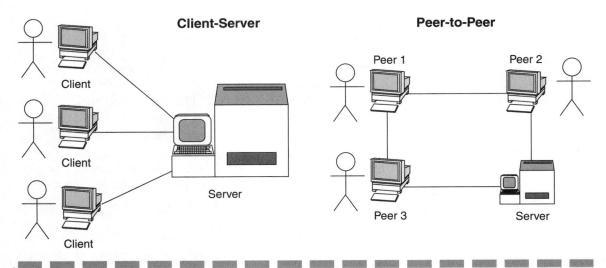

Figure 1.3
Client-server and peer-to-peer architectures.

- A *client-server* architecture connects a user at a *client terminal* to a *server* that offers information or services. User-to-server applications are always supported with a client-server architecture, as shown in the left half of the figure. User-to-user applications may be supported by a client-server architecture, as shown in the relation of Peer 2 and Peer 3 in the right half. A client-server architecture is especially suited to deferred user-to-user applications. The server acts as a buffer that can hold data to be sent from one user to another, regardless of whether the recipient is ready to receive or not at the time it is sent.
- A *peer-to-peer* architecture connects users at *peer terminals* directly to each other. User-to-user applications may be supported by a peer-to-peer architecture as shown in the relation of Peer 1 and Peer 2.

A final way in which network applications may be categorized is according to their mode of deployment, as shown in Fig. 1.4:

- *Vertically integrated* applications use a separate infrastructure with tightly coupled bitways and services for each application.
- *Horizontally integrated* applications use:
  - One or more integrated bitways
  - A set of shared services available to all applications

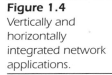
Figure 1.4
Vertically and horizontally integrated network applications.

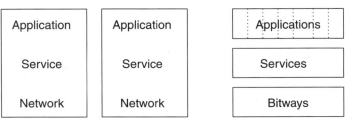

An advantage of horizontal integration is that one bitway can support many applications. At the same time, each application can use as many bitways of as many kinds as it needs, rather than trying to force a single network to do everything on its behalf. Horizontal integration also encourages *open interfaces* at the boundaries between bitways and services, and between services and applications. Open interfaces have several notable characteristics:

- Freely available specifications, as opposed to closed interfaces with characteristics that are not made known publicly or made known only in part
- A variety of implementations, as opposed to products available from only a single vendor
- Wide acceptance among users and infrastructure suppliers, precisely because specifications and implementations are so easy to obtain

To a great extent, the traditional PSTN was vertically integrated. The Internet, on the other hand, is horizontally integrated. The PSTN excelled at immediate user-to-user applications, while the Internet was better at user-to-server applications (with deferred user-to-user applications occupying a middle ground). Convergence is bringing the benefits of horizontal integration to the PSTN while opening the way to new kinds of user-to-user applications that take advantage of the unique capabilities of the Internet and other networks using Internet protocols. The next two sections take a closer look at these networks to clarify their differences and how they might benefit from convergence.

1.1.1 Phone Networks

Traditional voice networks, as exemplified by the PSTN, limit the extent to which they connect with other networks. Even a seemingly trivial feature like passing caller ID information across phone networks from one end of a call to another is by no means guaranteed. Because of this, the constituent networks of the PSTN are sometimes said to be *siloed* apart from each other.

Changes and additions to PSTN protocols are reviewed carefully to make sure they don't break the PSTN's member networks. And because so many networks could be affected by a breakdown, representatives of all or most of them participate in the evaluation. Not surprisingly, PSTN protocols develop slowly, and new ones are introduced seldom and with great care.

Large centralized network elements hold most of the PSTN's intelligence. These network elements communicate among themselves using a variety of complex protocols. The sources of this complexity are many.

Some of it is just the inherent challenge of connecting any two phones in the world, while crossing corporate, national, and technological boundaries. Some of it results from having to deal with more than a hundred years of incremental changes. Some of it results from compromises made to satisfy the often competing agendas of government bodies, network operators, and equipment vendors. And some of it may even have resulted from the temptation to be clever just for its own sake.

Another characteristic of the PSTN is that new service providers face high barriers to entry. Among the most daunting of these are the financial barriers. Phone networks require huge investments in equipment and wiring. Once such a network has been deployed, it's hard for someone else to duplicate it. And unless a new entrant can duplicate or improve on what's already in place, there's little incentive for customers to abandon their incumbent network operator. Until recently, potential newcomers in some areas were blocked by government regulations. In many countries, the phone company was a monopoly—often run by the government—and it was simply illegal to set up competing networks.

Despite its underlying complexity, the PSTN is easy to use—at least for its original purposes and for some new features, such as Caller ID. It reaches an extremely high percentage of its potential customers. It is extremely stable and reliable. To a great extent, these virtues result from its maturity. To put it simply, the PSTN has had a long time to improve and extend its reach.

1.1.2 The Internet

As mentioned, change within the PSTN has, until recently, been a leisurely affair. Two developments have started to change that. First, the traditional PSTN monopolies—private and government-run—are being spun off, broken apart, and opened up, thus introducing a new element of competition. While the size and scope of that competition is debated by some, it's been enough to start tipping the PSTN culture ever so gradually in favor of innovation over stability.

A second development that's changing the network balance of power is the Internet and other *IP networks*, i.e., those networks that use the suite of protocols founded on the *Internet Protocol (IP)*, the Internet being the largest such public network. (Private networks built on Internet protocols are often called *intranets*.) The success of IP networks and the Internet—with regard to technology and economics—has forced traditional network operators to take it seriously as a potential competitor, as a potential partner, and as a model for the their own networks' evolution.

Where the PSTN traditionally has been characterized by centralized intelligence and dumb terminals (phones), IP networks have always distributed intelligence across all attached elements, those at the center and those at the edges. Because its intelligence is distributed, the Internet presents low barriers to entry for new service providers. Just attach a server to the Internet and you can start offering services. (Getting somebody to use and pay for those services is an entirely different matter, a lesson that's all too familiar to many upstart dot-coms.)

Barriers to entry for new IP protocols are comparably low. For one thing, it's just a lot easier to design a new IP protocol, because they're generally simpler than those in the PSTN (which is not to say that they are simple). Anyone can create a new IP protocol and use it on the Internet. All you need is two machines that speak the protocol to each other and you're on your way. The trick is getting enough other people to use the protocol to make it useful and attractive. While there are standards bodies that that oversee protocols for IP networks, they tend to use natural selection at least as much as official decree in picking and choosing among new protocols.

Many of the differences between the Internet and the PSTN grew out of their differing histories and purposes. The PSTN, as a regulated network, has had to meet legally mandated standards of performance. These led its operators to stress stability and reliability over openness and simplicity. Nothing could be done that might compromise network performance. Every contingency had to be accounted for, with a consequent increase in the complexity of phone network protocols, equipment, and operating procedures. In the not too distant past, it led to prohibitions against connecting equipment not directly under the control of network operators, a significant obstacle to innovation by outside parties until fairly recently.[2]

---

[2] Among the more hilarious Bell System high-jinks of this era was the Hush-A-Phone ban of the mid-1950s. Hush-A-Phone was a plastic gizmo that snapped onto a phone's mouthpiece to block office noise. The old Bell System operated under a system of tariffs that prohibited "foreign attachments" to its equipment. The Bell System owned everything up to and including phones. In their eyes, that made the Hush-A-Phone a foreign attachment. The FCC saw it their way and banned Hush-A-Phone. Hush-A-Phone appealed to the courts and won. Some believe this was the first crack in the Bell System that led to its eventual breakup in 1984. As silly as the Hush-A-Phone ban might seem now, it wasn't the silliest example of a bygone era. In North Carolina, customers could not even put plastic covers on their phone books. Again, the courts thought otherwise.

Introduction

By contrast, there's been a lot more room for experimentation on the Internet. Like the PSTN, the Internet was designed to be highly reliable. But where PSTN reliability includes time and quality measures—number of seconds to get dial tone after going offhook, voice clarity, and the like—Internet reliability boils down to a simple guarantee that data sent from one place will—somehow and eventually—get to its destination. Some Internet protocols relax even these requirements, sending data on its way with no guarantee it will ever arrive, or that it will arrive in the same order it was sent. Beyond this, just about anything goes.

The disadvantage of the Internet and other networks that use IP protocols is that it's hard to build applications with them that meet strict real-time performance requirements. The PSTN uses *circuit switching*, while the Internet uses *packet switching* to route data from one place to another. Circuit switching is fundamentally more suited to real-time applications, like making voice calls (Fig. 1.5).

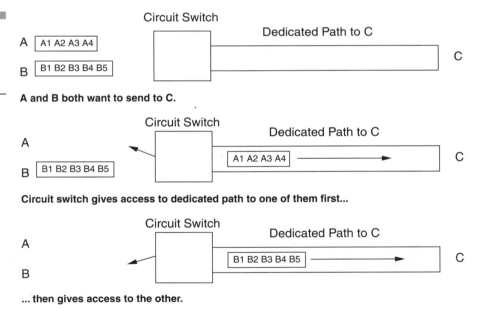

Figure 1.5
Circuit switching provides dedicated access to transmission paths.

To see why this is so, consider the difference between circuit and packet switching in the context of a highway system. You could run a "circuit-switched highway" by offering it to drivers on a one-at-a-time reservation basis. If somebody wanted to go from one place to another, they would put in a request to use the highway serving that route. If it's free, off they would go, with the whole road to themselves. If not, they're denied access until it's available. Obviously, once they're on the road, they get very good performance. The drawback is that much of the highway's potential capacity is wasted since only one user is allowed on at a time.

A "packet-switched highway" would let anyone on the road at any time until there's no more room left for new traffic (Fig. 1.6). None of

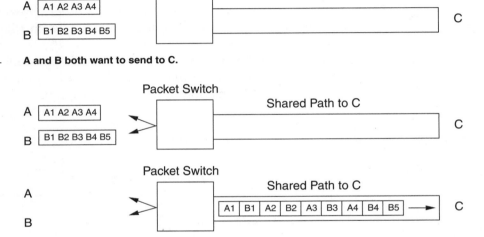

Figure 1.6
Packet switching provides shared access to transmission paths.

the road's capacity is wasted, but performance drops as traffic increases. Rather than a few persons getting guaranteed good performance, everybody gets whatever performance can be sustained under existing conditions, which may range from excellent to intolerable.

The PSTN works like a circuit-switched highway. Once you're on, you get excellent performance. If all available circuits are busy, the network plays a tone to tell you that you've been denied access and will have to try again later. With the packet-switched Internet, you may never be denied access, but the access you get may not be worth much. When things get too crowded, portions of your conversation may be delayed, arrive out of order, or never arrive at all. Up to a point, these problems can be handled on the receiving end of a VoIP system by ignoring missing packets and packets that arrive out of order. If there aren't too many of these, quality won't suffer much. But there's only so far you can go with this method. The balancing act with VoIP is figuring out how to offer something acceptably close to the reserved highway performance of the PSTN under the Internet's free-for-all rules.

The impetus for convergent network technology is the realization that the PSTN and Internet each do some things well and other things not so well. So, rather than try to turn the Internet into a duplicate of the PSTN and vice versa, the focus is on blending the two networks. Ideally, the result will behave as a single integrated network that retains the advantages of both, while at the same time offering new features that integrate their functions. As a vision, network convergence is being driven by market forces. As a reality, it's being driven by technology. The technology that's most important for achieving the goals of network convergence is software.

I make this claim because, when you get right down to it, data transmission, no matter how fast or sophisticated, is just plumbing. All it does is get things from point A to point B. Admittedly, without it you have no services. But if that's all you have, then all you have is the digital equivalent of a water company. A company in the data transmission business must decide whether that's all it wants to be or whether it thinks it can do more with that data to make it more valuable to potential customers (and, hence, more profitable to the company). Since data services are software, it stands to reason that a company that wants to offer profitable data services is a company that's going to be heavily invested in software, either as a developer or as an informed partner with and purchaser of software from companies that specialize in converged networks and their services.

1.1.3 Different Approaches to Convergence

Now that we've examined some of the characteristics of the PSTN and the Internet (and other IP networks) in a bit more detail, let's return to convergence and what exactly it can do. I'll use a simple two-part model of a communication network as the basis of this discussion. The first and most basic function of a communication network is *transport*, moving data from one place to another.[3] The other essential function of a communication network is *control*, which gets data to the correct destination and issues the signals needed by services associated with that data.

---

[3] Don't confuse *transport* as used here with the layer 4 transport protocols of the OSI network model described in Sec. 2.4.3.

When connecting the PSTN with an IP network such as the Internet, there are several ways to combine their control and transport functions. The most obvious is to connect both the control and transport functions of each network. When the IP network under consideration is the Internet, this is the same as Voice over the Internet service. A phone call may start in either the PSTN or an IP network, travel to a *gateway* where it crosses over to the other network—there to continue its journey—and finally end up at a terminal on that network.[4] The key element is the gateway, which connects the transport and control functions of the two networks by converting data and messages from PSTN to Internet form, and vice versa. I cover the technology needed for this in Chap. 8.

The PSTN and an IP network can also be tied together by leaving transport alone and connecting only the control function. This is how the *PSTN/Internet Interworking (PINT)* and *Service in the PSTN/IN Requesting Internet Service (SPIRITS)* working groups of the *Internet Engineering Task Force (IETF)* have approached convergence.[5] The voice part of a PINT- or SPIRITS-enabled call stays entirely within the PSTN, but information about that call may pass from the Internet to the PSTN (SPIRITS) or vice versa (PINT).[6] We'll look at those two protocols in more detail in Chap. 10.

PINT and SPIRITS exchange control information between an IP network and the PSTN while a call is in progress. A simpler form of control interworking is also possible in which users access the Internet to update information about their PSTN services. Thus, users might change their speed calling codes by filling out a form at a phone company Web site. Software at the Web site would then interact with the necessary PSTN network elements to update that data. This is in contrast to the tradi-

---

[4]This is just one of many possible scenarios. A call could go from PSTN terminal to IP terminal; it could go from PSTN terminal to PSTN terminal via an IP network (entering the IP network by one gateway and leaving by another); it could go from IP terminal to IP terminal via the PSTN; and so forth.

[5]Working group names are often all lowercase, for example, *pint* or *spirits*. It's not uncommon for the protocol issuing from such a group to have the same name in uppercase, for example, *PINT* or *SPIRITS*. But it's not uncommon to see the group name given in uppercase. One may also see constructs such as *Pint* and *Spirits*. So how is one to know what is being referred to? Context.

[6]One could also leave the voice stream entirely within the Internet's transport function and send control messages back and forth between the Internet and PSTN. While one can imagine applications operating in this fashion (for example, an Internet phone call getting caller ID information from PSTN databases), little work has been done yet along these lines.

tional means of updating PSTN services by interacting with IVR dialogs and pressing keys on the phone. For sophisticated services, this can be so cumbersome it renders them all but unusable. Giving users Web access to their service data promises to make possible many services which are otherwise impractical. The software technology that might support this is the topic of the first half of this book.

1.2 Who's Who in Convergent Networks

1.2.1 ITU-T

One of the most important organizations laying the foundations for convergent networks is the *International Telecommunications Union—Telecommunications Sector (ITU-T)*. The ITU was established in the 1860s as the *International Telephone and Telegraph Consultative Committee (CCITT)*[7] to standardize telegraph networks. It's now an agency of the United Nations,[8] where it oversees global standards for radio and phone communication. Any government that belongs to the UN can belong to the ITU as a *member state*. Others—commercial network operators, equipment vendors, and other organizations—can join as *sector members* in one or more of the three *Sectors* of the ITU:

- The *ITU-R (Radiocommunication) Sector* was formed in 1993 out of its predecessors, the *International Radio Consultative Committee (CCIR)* and the *International Frequency Registration Board (IFRB)*. The ITU-R continues its oversight of international standards for the allocation and use of radio spectrum.

- The *ITU-T (Telecommunications) Sector* was formed in 1993 from the CCITT. The ITU-T oversees international standards for telephony.

- The *ITU-D (Development) Sector* helps countries to implement and expand their communications infrastructures.

---

[7]The abbreviation comes from the French translation of the name.

[8]Historically, most phone companies were government-run; thus, the ITU's placement within the UN.

Of these three, the ITU-T is the most important with regards to network convergence standards. The ITU-T is composed of *study groups*. Each study group handles standards for a subject area such as transmission, operations, or switching and signaling. The number of study groups varies according to current need. For the period 1997-2000, there were fourteen study groups. Each study group is divided into *working parties* that divide their work into topics called *questions*.

Members of the ITU-T submit proposals for new and modified standards to the study groups. These may eventually become *ITU-T draft standards* that are published in the reports of their study group. When a draft standard is mature enough, it's sent to all ITU-T members for final comments and consideration for approval by its study group. Once approved, it becomes an *ITU-T Recommendation*.[9]

1.2.2 IETF

The other big player in convergent networks is the *Internet Engineering Task Force*, which oversees many (but not all) Internet standards. Unlike the ITU, the IETF has no official status or set membership. Its authority flows from a general consensus in the Internet community to abide by the decisions it makes. This method works because the Internet had its roots in a tightly knit group of researchers. Consensus was a natural way to reach harmony in this environment. From that consensus came the success of today's Internet. New participants find it hard to argue with that success, so they're willing to keep playing—for the most part—by the original rules.[10]

The IETF has eight primary *study areas* that are broken up into *working groups*. Each working group concentrates on a subject. Most of their work is done online using e-mail. Anybody can participate in an IETF working group. Decisions are made by group consensus, not by vote, as in the ITU.

---

[9]*Recommendation*, not *Standard*, is used to preserve the sensibilities of member states who, for reasons of sovereignty and national dignity, don't appreciate having standards dictated to them. Member states are within their rights to ignore a "Recommendation," but in practice they don't.

[10]When compared with their early years, the Internet and IETF may now seem much less harmonious, with consensus harder to reach and corporate politicking much more in evidence. While true, the Internet and the IETF are still far more open and consensus-based than not.

Introduction

Anybody can submit a proposal, known as an *Internet Draft*, to the IETF. These are published without review on the IETF's Web site (http://www.ietf.org), but publication carries with it no implied endorsement by the IETF. Internet drafts have a lifespan of six months. After that time, they are removed from the IETF's Web site.

The only official documents of the IETF are *Requests for Comments (RFCs)*. Every RFC has a unique number and is stored permanently at the IETF Web site. Not all RFCs are standards. An *informational RFC* provides background information the IETF feels would be useful to readers. An *experimental RFC* describes a protocol or system of widespread interest that is being tried out. *Standards track RFCs* are the IETF's standards. They have three levels, as described in RFC 2026:

- A *Proposed Standard* is "generally stable, has resolved known design choices, is believed to be well understood, has received significant community review, and appears to enjoy enough community interest to be considered valuable. However, further experience might result in a change or even retraction of the specification before it advances."

- A *Draft Standard* is the next stage in the IETF standards track. A Proposed Standard moves to this level when it has "at least two independent and interoperable implementations from different code bases ... for which sufficient successful operational experience has been obtained. ... Elevation to Draft Standard is a major advance in status, indicating a strong belief that the specification is mature and will be useful."

- A *Standard* is the final stage. Few IETF standards reach this level. Most go no further than the Draft Standard level. A Standard is a Draft Standard "for which significant implementation and successful operational experience has been obtained. ... [It] is characterized by a high degree of technical maturity and by a generally held belief that the specified protocol or service provides significant benefit to the Internet community." Besides their regular RFC number, Standards also receive an STD number. Standards define the lowest common denominator protocols of the Internet and include things such as the specifications for TCP and IP (RFC 0793/STD 0007 and RFC 0791/STD 0005, respectively) and SMTP, the Simple Mail Transfer Protocol (RFC 0821/STD 0010).

RFC 2026 also states that some RFCs are categorized as *Best Current Practices (BCPs)*. These are "designed to be a way to standardize practices and the results of community deliberations." BCPs specify the normal operating procedures and assumptions under which the IETF operates. RFC 2026 is itself a BCP.

1.2.3 Other Organizations

While the ITU-T and IETF are the dominant standards-setting bodies in the world of convergent networks, many other organizations also play an important role. Among these are the following:

- The *World Wide Web Consortium (W3C)* is an organization of companies that develops open standards to ensure the coherent and unified evolution of the Web. The W3C oversees HTTP, HTML, and XML.
- The *European Telecommunications Standards Institute (ETSI)* develops standards for the European Union and makes contributions to the ITU-T. Of particular interest is its *Telecommunications and Internet Protocol Harmonization Over Networks (TIPHON)* project that is defining VoIP architecture and protocol requirements.
- The *ATM Forum* is an industry consortium that promotes the growth of Asynchronous Transfer Mode (ATM) technology by encouraging equipment and service interoperability. They have done extensive work on quality-of-service issues.
- The *Parlay Group* is an industry consortium that is developing an open object-oriented API for network services.
- *T1* develops U.S. standards for wireline communications, including methods for interworking with IP networks. T1 is sponsored by the *Alliance for Telecommunications Industry Solutions (ATIS)* and is accredited by the *American National Standards Institute (ANSI)*.
- The *Telecommunication Industry Association (TIA)* develops U.S. standards for wireless communications, including methods for interworking with IP networks.
- The *Institute of Electrical and Electronics Engineers (IEEE)* has standardized protocols for local area networks (LANs).
- The *3rd Generation Partnership Project (3GPP)* is defining standards for the next generation of *GSM (Global System for Mobile communication)* wireless networks, including *General Packet Radio Service (GPRS)* and *Enhanced Data rates for GSM Evolution (EDGE)*. They are working closely with the Parlay Group.

Introduction

- The *Object Management Group (OMG)* is standardizing the *Common Object Request Broker Architecture (CORBA)* to provide distributed object technology in a multiplatform, multilanguage, and multiprotocol environment.
- The *Enterprise Computer Telephony Forum (ECTF)* is developing standards and guidelines for computer telephony.
- The *PAMforum* is defining APIs for *Presence and Availability Management (PAM)*. They are working closely with the Parlay Group.
- The *Open Systems Gateway Initiative (OSGi)*[11] is defining APIs that can be used to support services that are loaded and run on a services gateway, such as a set top box, cable modem, DSL modem, PC, or dedicated residential gateway.

---

[11] The lowercase *i* at the end is correct.

CHAPTER 2

Software for Managers

All programs are dull.

—Stan Kelly-Bootle

This chapter is aimed at readers who are not programmers. It presents an overview of several areas in software technology that pertain to convergent networks. The subject matter is admittedly dry, and there's no getting around that. Hopefully, those who stick with it will find it's been worth their while further on.

The first topic will be methods of designing and writing software. Of these, object-oriented programming is the dominant model for software design today. However, I'll cover procedural programming first, since it's a necessary prerequisite to an understanding of the object-oriented approach and its benefits. The next topic will be a brief overview of component-based software, an important application of object-oriented programming. The final section introduces some basic principles of computer networking.

2.1 Procedural Programming

Traditional software design—*procedural programming*—grew out of developers' experience with languages like Fortran, COBOL, and C. In procedural programming, a problem is first broken into separate tasks. These tasks are broken into subtasks, which are broken up still further, and so on, until it's apparent there's no point in going further. For each subtask, a separate piece of software, called a *procedure*, is written, followed by a *main program*, or *main procedure*, that calls those procedures in the order needed to carry out the primary task.

To see how this works, let's say we are going to write a simple program that manages bank accounts. The basic actions to be supported are getting account balances, making deposits, withdrawing funds, and transferring funds from one account to another. Each of these actions will be a procedure. They all need to authenticate the person trying to perform them, so authentication will be another procedure. Deposits and withdrawals update the current balance, so updating the balance is another procedure. Transferring funds is just making a withdrawal from one account and making a matching deposit to another, so the transfer procedure can use the other two. The last thing the program needs is a procedure to retrieve the current balance of an account. Everything else in the program will be procedures for interacting with users of the program, which can be skipped for now. At this point, there are no more procedures to be found.[1]

---

[1] Well, there may be, but this will do for now.

Figure 2.1 shows the design. The main program calls procedures for input and output (get... and display...), authentication, making transfers and deposits, getting balances, and withdrawing funds (and some others not listed that are represented by the box containing only "..."). The transfer procedure uses the deposit and withdraw procedures. The withdraw procedure uses the procedure for getting a balance, and both the deposit and withdraw procedures use the procedure for updating a balance.

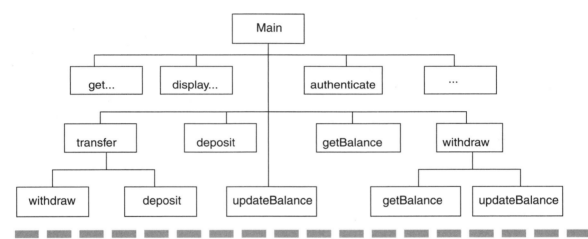

Figure 2.1
Procedural diagrams show who calls what in a procedural design.

To do anything, procedures must be given data on which to work and they must make their results available to the rest of the program. There are several ways to do this. A piece of data may be *global,* which means it can be used anywhere in a program. Any procedure can access all of a program's global data. Any changes a procedure makes to global data are visible throughout the program. While using global data is easy, it can lead to subtle problems if one part of a program changes global data and another part of the program expects it to be left unchanged. When that happens, it can be hard to find what caused the problem.

If a piece of data isn't global, then it's said to be *local* to the part of a program in which it appears (for example, in the main procedure or in one of the other procedures) and can be used only in that part of the program. Local data can be passed from one procedure to another as a *parameter,* or *argument,* of a procedure call.[2] In most programming languages, a procedure call with arguments looks something like this:

```
foo (bar);
```

where `foo` is the name of the procedure being called and `bar` is the data passed to `foo` as a argument.

A third way to pass data between a procedure and the code that calls it is to declare the procedure as a *function*. A call to a function can be used in expressions, just like any other piece of data. The function uses data passed to it in arguments to calculate a result and returns that result for use in an expression. Here's an example of a function call in an expression:

```
i = round (3.9876) + 6;
```

After this code is executed, the value of `i` is 10 (the value of the function `round` applied to the argument 3.9876, which is 4, plus 6).

2.2 Object-Oriented Programming

Procedural programming's emphasis on procedures as the basic element of software design reduces the importance of data, which can end up just sloshing about from one procedure to another. This leads to bugs that happen when data changes unexpectedly as it passes through all those procedures. Other problems result when data is used in more than one way (for example, representing speed in miles per hour in some places and in kilometers per hour elsewhere). Object-oriented programming, as exemplified in languages like C++ and Java, turns the procedural model upside-down, using data as the starting point for program design.

The core concept in object-oriented programming is, of course, *objects*. Coming up with an iron-clad definition of objects leads quickly into some challenging philosophical problems. I'll just assume readers have an intuitive sense of what I mean when I talk about objects in the physical world, and that I can use that intuition to talk about objects in the software world.

---

[2]While one can draw subtle distinctions between these two terms, in practice they are used almost interchangeably. I call them *arguments*.

Just as a physical machine is a collection of interacting parts—physical objects, like gears, pulleys, and levers—an object-oriented program is a collection of interacting data parts—software objects. These software objects may represent physical objects (for example, people, books, automobiles). They may represent things that, while not tangible, are nonetheless real in some sense (for example, accounts, sales, debts). They may be things that are seen only inside the program, software "gears" and "pulleys" that connect objects to each other (for example, hash tables, arrays, and lists).

Like physical objects, software objects have *properties*, facts that describe them or things that arise from their behavior. Properties are things like account numbers, titles, weights, temperatures, colors, speeds, ages, genders, numbers of dollars owed, and so on. Properties distinguish one object from another and help you keep track of the interesting characteristics of an object (where what is interesting about an object depends on what you are trying to do).

Besides properties, software objects have *methods*, things you can tell an object to do, such as, report the current value of one of its properties, change the value of a property, or carry out some task. Methods are procedures—procedures that can be applied only to the objects with which they are associated. Using the machine analogy, methods roughly correspond to the knobs, dials, and buttons that control a machine's operation.[3]

A set of objects that shares the same properties and methods forms a *class*. The individual objects are called *instances* of the class. For example, if we were writing a program dealing with means of transportation, we might group all objects whose properties include having four wheels and an internal combustion engine in a Car class and all objects having two wheels and pedals in a Bicycle class.

You could go farther and move all properties and methods common to cars and bicycles into a *superclass* called VehicleWithTires. That way, a programmer who wants to create a new class that represents a different sort of vehicle with tires—say a Bus class—won't have to rewrite the common code shared by all vehicles with tires. In object-oriented

---

[3]Methods are sometimes called *messages*. Rather than saying that we invoke, or call, a method to make an object do something, we might say that we *send a message* to the object that tells it what to do. But this is just a different name for the same thing. Unless you're dealing with distributed objects (about which I'll say more later), there really aren't any messages sent at all. While this message passing talk has some theoretical charms, it causes much confusion, because many people assume (not surprisingly) that when you say you are sending a message, you are indeed sending a message.

terminology, this is called *inheritance* (so called because the corresponding class diagrams look like family trees.) A class like Car is *derived from*, or is a *child* of, the *parent class* VehicleWithTires. An object that is an instance of Car is also an instance of VehicleWithTires and similarly for instances of Bicycle and Bus. Taken together, a parent class and its child classes form a *class hierarchy*, as shown in Fig. 2.2.

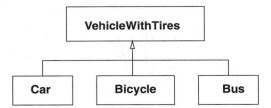

Figure 2.2
Inheritance from the parent VehicleWithTires class forms a class hierarchy.

The bulk of an object-oriented program is code that defines the properties and methods of the classes it uses. Classes are not objects. They're templates that describe what instances of those classes will look like and how they will behave. In order to do anything, an object-oriented program must also have code that creates instances of its classes—the program's objects. To carry out its tasks, the program calls the methods of these objects.

To show how this is done, see Fig. 2.3, which is a simple Java implementation of the VehicleWithTires class and a few of its subclasses. Instances of VehicleWithTires have a data element, currentSpeed, in which the program stores the current speed of a vehicle in miles per hour. You could store it as kilometers per hour, but it makes no difference to users of this class since they have no direct access to currentSpeed. This introduces another of the key ideas of object-oriented programming: *encapsulation*.

Normally, all an object's users can see are its methods. They have no direct access to its data. Instead, they get to an object's data through its methods. The data is "encapsulated" within the object and access to the data is controlled by the methods. With VehicleWithTires objects, all users see are methods to set and obtain the speed in either miles per hour or kilometers. These methods take care of all necessary conversions inside themselves. From a user's perspective, the properties of a VehicleWithTires object are mph and kph, and these properties are accessed through the methods getMph, setMph, getKph, and setKph.

Encapsulation makes it easier to change one part of a program without affecting other parts as an unexpected side effect. Encapsulation also

```
public class VehicleWithTires {
   protected float currentSpeed; // in mph
   // Methods to set and read the current speed
   public void setMph (float speed) { currentSpeed = speed; }
   public void setKph (float speed) { currentSpeed = speed * 0.6; }
   public float getKph () { return mphToKph (currentSpeed); }
   public float getKph () { return currentSpeed; }
   private float mphToKph (float mph) { return mph * 1.6; }

   // Method to initialize a VehicleWithTires
   VehicleWithTires () { currentSpeed = 0.0; }
}
public class Car extends VehicleWithTires {
   // Properties only a Car can have
   float loanBalance;

   // Methods to get and set Car properties
   public float getLoanBalance () { return loanBalance; }
   public void reduceLoanBalance (float amt) { loanBalance -= amt; }

   // Method to initialize a Car
   Car (float initLoan) { loanBalance = initLoan; }
}
public class Bicycle extends VehicleWithTires {
   public float getFps () { return CurrentSpeed * 5280 / 3600; }
   public void setFps (float fps) { currentSpeed = fps * 3600 / 5280; }
}
public class VehicleSystem {
   static public void main (String [] args) {
      Car car1 = new Car (1000.0);
      Car car2 = new Car (2000.0);
      Bicycle bike1 = new Bicycle();
      Bicycle bike2 = new Bicycle();
      car1.setMph (60);
      car2.setKph (100);
      car1.reduceLoanBalance (10.0);
      bike1.setMph (10);
      bike2.setFps (bike1.getFps());
   }
}
```

Figure 2.3
Java code for `VehicleWithTires` and selected subclasses.

makes it easier to put different faces on the same data to fit different needs. In the example, you could change the representation of currentSpeed from mph to kph without having to change any code written by users of the VehicleWithTires class and its child classes (assuming you make all the necessary changes in methods like getMph and getKph). In a procedural program, if you were to change the interpretation of currentSpeed, you would also have to find and change every use of `currentSpeed` to fit. It's all too easy in those circumstances to miss some of the necessary changes or to make incorrect changes.

The collection of methods that an object makes available to its users is called its *interface*. In Java and C++, interface methods are easily found because they are labeled with the keyword `public`. Not surprisingly, interface methods are often referred to as *public methods*.

An object may have other methods that support its internal mechanisms, but these are not part of its interface. Noninterface methods can be used only by the object itself or by a limited set of authorized objects. As far as most users are concerned, it's as if the noninterface methods don't even exist. In Java and C++, noninterface methods are labeled with the keywords `protected` and `private`. In Fig. 2.3, `mphToKph` is a noninterface method called by the public method `getKph` as part of the internal workings of `VehicleWithTires`.

`Car` adds another property, `loanBalance`, besides the `currentSpeed` property it inherits from `VehicleWithTires`. Users of `Car` objects gain access to this property through the `getLoanBalance` and `reduceLoanBalance` methods. `Bicycle` adds a different property, `fps` (feet per second). But it adds this property not as a separate data item, but rather as a new interface method that presents `currentSpeed` in a different way. The last class in Fig. 2.3, `VehicleSystem`, is where the real work (such as it is) gets done. It creates some instances of `Car` and `Bicycle` and then invokes various methods in those objects' interfaces.

Turning to the banking problem we tackled procedurally in Sec. 2.1, an object-oriented solution would first identify the kinds of objects with which the program will be working.[4] These determine the classes in the program. A plausible list would include `Account`, `AccountOwner`, `AccountId`, `OwnerId`, and the `Bank` itself. There are two ways the program could handle account balances: It could use a `Balance` class, or it could just use decimal numbers. (Making the choice of which to do is where science and engineering give way to at least some art.) Digging down a bit further, it appears that `AccountId` and `OwnerId` share many characteristics, so they should be child classes of a parent `Id` class. Finally, it may not be a good idea to make the `Account` class do everything on its own. So it splits into separate child classes for `CheckingAccount` and `SavingsAccount`.

---

[4] A process known as *object-oriented analysis and design (OOAD)*, that is an entire art unto itself.

The result is the *Unified Modeling Language (UML)* static class diagram shown in Fig. 2.4. UML is a standardized notation for designing and describing an object-oriented system that's been developed under the auspices of the Object Management Group (OMG). UML is actually a collection of notations, each covering different aspects of a software system. Because UML is intended to describe many kinds of systems, not all its notations may be needed for every project. Some, such as the static class diagrams discussed here, will be used on just about any conceivable object-oriented system. Others, such as statecharts and the *Object Constraint Language (OCL)*, provide ways to give more information about the runtime behavior of a system. Still other notations describe the deployment architecture of a distributed system. Obviously, if a system isn't distributed, deployment diagrams will be irrelevant. The decision whether to use things such as statecharts and OCL depends, in part, on the level of rigor demanded of a project.

Each box in a UML static class diagram represents a class, the name of which appears in the top section of the box. The middle section lists the data items contained by instances of that class. Note that some of these data items are themselves objects (for example, the `id` property in `Account` and `AccountOwner`). The bottom section of each box lists methods. To keep the diagram from getting too cluttered, I've shown only a few examples of these. A hollow arrow leading from one class to another shows inheritance, with parent classes at the arrow end of the line.

A class diagram doesn't show objects. Individual objects can be created and destroyed throughout the life of a running program, so any picture of the objects in a system is just a snapshot of its state at some point in time. For example, when the banking system starts up, it might have a single bank object with two account objects and a single owner object as shown in Fig. 2.5. The arrows in this diagram don't show inheritance, but rather the fact that one object has another as part of its internal data.

Object-oriented programming seems to fit more naturally with how people think of the world. That makes it easier for programmers and their clients to converse with each other about the subject matter of a program. Consider the banking classes in Fig. 2.4 and compare them with the procedural diagram in Fig. 2.1. The class diagram shows a collection of reasonably well-understood business objects and their properties. The procedural diagram shows little more than a collection of activities and an efficient way to organize them as procedures. Which do you think more closely models a banker's view of his business?

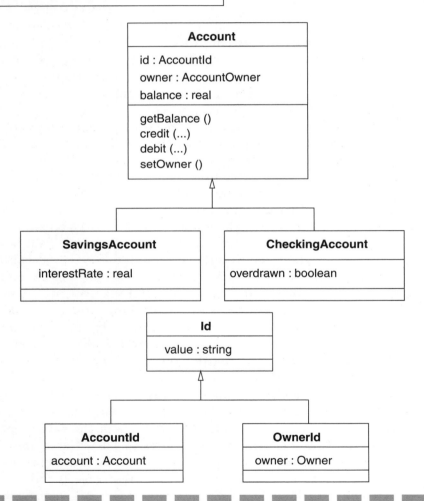

Figure 2.4
UML static model of the classes in an object-oriented bank system.

Software for Managers

Figure 2.5
Dynamic object diagrams are a snapshot of a program's state at some point in time.

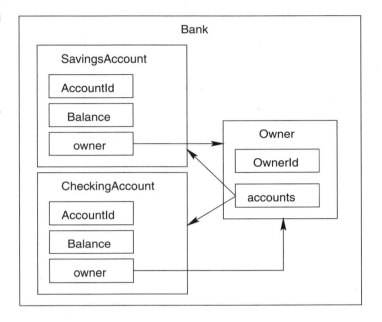

2.3 Component-Based Software

Software components are much simpler than they're often made out to be. They're nothing more than reusable software building blocks, prewritten chunks of code that can be combined with custom-written code to make an application. Components free programmers from having to write their own code for problems that have already been solved elsewhere (and likely solved better than most programmers would do on their own).

In a very loose sense, the elements of a procedure library are software components. Under that interpretation, just about any C program is an example of component-based software, because it almost certainly uses one or more of the procedures in C's standard library. Components of that library handle common tasks like input and output (for example, `printf` and `getchar`) and manipulating strings of characters (for example, `strcpy` and `strstr`).

In the object-oriented world, components are more narrowly defined. An object-oriented component is a class that provides a well-defined interface that exposes three crucial aspects of itself to the programs in which it is used:

- A component class exposes its *attributes* or *properties*, data about an instance of the class that would be of interest to a user of the component. The *attribute interface* lets programs find out the value of a particular attribute. It may also allow programs to modify the value of an attribute.

- A component class exposes its public *actions* or *methods*. These define the things that a program is allowed to do with the component. Among these methods are those that implement the attribute interface of the component.

- A component class provides an *event* interface, by which it can notify other parts of a program when an instance of it has changed.

Component classes must typically be written according to rules that allow programming tools to determine their attributes, actions, and events. These tools then present components to programmers, often in a visually oriented environment, in which programs are constructed by joining icons that represent individual components and their event inputs and outputs.

A common use of object-oriented components is in tools that help programmers design graphical user interfaces (GUIs). The visual elements of such GUIs include things like windows, text areas, input boxes, radio buttons, sliders, knobs, and scrollbars. Users of a GUI-building tool typically see a blank work area and a palette of user interface elements that can be added to the GUI under construction. They drag and drop palette elements to their desired position in the GUI and set the properties of those elements—their colors, fonts, default input values, and so on. Behind the scenes, the tool generates a program that represents each interface element with an instance of a component class that implements it. (These classes are sometimes called *widgets*.)

Component-based tools can do more than just help GUI developers. They can also provide components for activities like working with a database or interacting with phone equipment. Just as with GUI development, programmers have a blank work area and palettes of components. They drag and drop these components into the work area and set their properties. Furthermore, components may have inputs and outputs, and the outputs of one component can be linked to the inputs of another by connecting the two in the work area. Connecting components in this way sets up chains of actions that lead from one component to another.

Programmers needn't restrict themselves to the components that come with their chosen tools. They can also import components obtained from third parties, assuming these components follow the programming rules that allow the tool to discover things about it, such as its properties, inputs, and outputs. One such set of rules is defined by JavaBeans, a framework for building components with the Java programming language (see Sec. 3.7 for more on JavaBeans).

2.4 Computer Networks

While computers, on their own, are a significant technological force, their influence is increased many times when they're connected to each other in public and private networks. While computer networks have been around in some form or another for much of computing's history, it's only been recently that they got much visibility. Arguably, this occurred in the late 1980s and early 1990s as LANs and the Internet reached critical mass. The rapid growth of computer networks, coupled with the parallel evolution of voice networks, provided the medium within which network convergence took root.

This section provides a brief overview of some basic concepts of computer networking, with a focus on those that have influenced network convergence. The first topic is a discussion of APIs and protocols. Both are important in computer networking, and both are sometimes confused with each other. Next, we'll turn to the OSI reference model for computer networks, a widely accepted tool for categorizing and understanding network protocols. Using the OSI model as a reference point, we'll finish this section with a (very) brief look at the Internet and its protocols.

2.4.1 APIs and Protocols

An *application programming interface (API)* describes a packaged collection of prewritten procedures (or classes and methods in the object-oriented world) that programmers can incorporate into their own software. It defines how programmers see the resources of a particular computing system (which may be a single machine or a collection of machines cooperating together). A *protocol* describes a set of messages that may be exchanged between two or more parties (people, machines, programs, etc.) to carry out a specified set of tasks. APIs and protocols resemble each other in several ways.

- A protocol specifies messages that communicating parties can exchange. An API specifies the procedures a software developer can use in a program.
- A protocol specifies all the data items that must appear in each of its messages. An API specifies all the data items (arguments) that must appear in calls to each of its procedures.
- A protocol specifies the order (if any) in which its messages must be exchanged. An API specifies the order (if any) in which its procedures must be called.
- A protocol specifies replies that may be returned after sending one of its messages. An API specifies values that may be returned after calling one of its procedures.
- A protocol specifies the effects of its messages on the machines that receive them. An API specifies the effects of its procedures on the program in which they are called.

To highlight the similarities and differences between APIs and protocols, consider a simple system that does nothing but print text strings. Here's some C code that defines an API for this system:

```
typedef char* PRINTER_ID;
typedef int BOOLEAN;
const int SUCCESS = 0;
const int NO_SUCH_PRINTER = 1;
const int OTHER_ERROR = 2;
/*
   Print the supplied string on the named printer.
   Return SUCCESS if the call succeeds,
   NO_SUCH_PRINTER if the named printer is not known,
   and OTHER_ERROR for all other errors.
*/
int print_string (PRINTER_ID printer, char* string);
```

As you can see, an API contains only the code a programmer needs in order to call its procedures, plus some extra information (represented by C language comments here) to describe those procedures. Here's a program fragment using the API just defined:

```
int result;
result = print_string ("AdminPrinter", "The quick brown fox . . .");
```

Protocols can be defined in several ways, but all are just ways to specify the protocol's messages and what characters will appear in what order in those messages. For simple protocols, a plain English description may be enough. Here's an English description of a protocol for the printing system:

> A print request message consists of the string `PRINTSTRING`, one or more spaces, the name of a printer in double quotes, one or more spaces, and the string to be printed in double quotes, in that order.[5]
>
> A system that receives the `PRINTSTRING` message will try to print the supplied string on the indicated printer. It will respond with one of the following messages:
>
> `STATUS OK` String printed successfully.
> `STATUS NO_SUCH_PRINTER` String not printed, printer not known.
> `STATUS OTHER_ERROR` String not printed, other error.
>
> A machine using this protocol might send the message:
>
> `PRINTSTRING "AdminPrinter" "The quick brown fox . . ."`
>
> and get back the response:
>
> `STATUS OK`

if the request succeeds.

In some sense, every API defines a protocol, since it describes how a programmer can communicate with the software described by that API. Likewise, the software that implements a protocol often has a published API. At times, a protocol implementation and its API become so closely intertwined that it's nearly impossible to separate the two.[6] Perhaps the occasional confusion of APIs and protocols is not so surprising after all.

Another source of confusion is the fact that protocols and object-oriented APIs are both often documented with what are known as *message sequence charts (MSCs), ladder diagrams,* or *ping-pong diagrams*. These diagrams show examples of how an API or protocol will behave in different circumstances, by showing all participants in a scenario (objects in an

---

[5]This example, simple as it is, shows why English descriptions are seldom adequate on their own as a protocol definition. It might seem we've given a reasonably complete description of the `PRINTSTRING` message. However, it leaves several questions unanswered:

- What character set should we use in these messages? ASCII? EBCDIC? Unicode?
- Is a double quote a single character, or is it two single quotes?
- What should we do if the string to be printed includes a double quote?

[6]To some extent, this has happened with LDAP and its C language API, as described in the IETF's RFC 1823.

object-oriented diagram) and the messages they exchange (or methods they call) as the scenario runs its course. Participants appear at the top of the diagram, each in its own box, with a line extending down from it. A message or method call from one participant to another appears as an arrow labeled with the message or method name going from the sender to the receiver. Arrows near the top of the diagram represent messages and procedure calls that occur before those farther down.

Fig. 2.6 shows a ladder diagram for a scenario using the printer API. Fig. 2.7 shows a diagram for the corresponding scenario using the printer

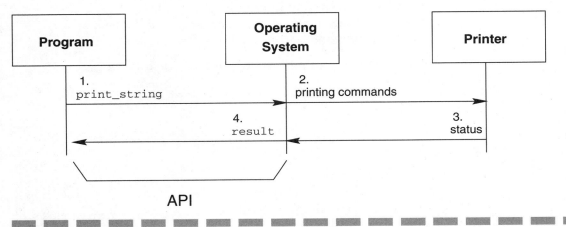

Figure 2.6
Ladder diagram: An example of the printer API in use.

protocol. The numbering of the arrows in these diagrams shows the order in which events occurs. Not all ladder diagrams are numbered in this way. However, even without numbers, the order of events can easily be determined by following the arrows as they lead from one to another, top to bottom.

2.4.2 Text-Based versus Binary Protocols

It's long been a point of discussion in the networking community whether protocols should be text-based or binary. Text-based protocols use printable characters, so their messages can be made readable. One frequently used technique for making text-based protocols easy to read

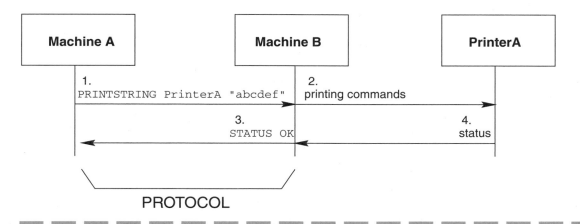

Figure 2.7
Ladder diagram: An example of the printer protocol in use.

is to use keywords that precede values. Another common practice is to separate fields in a message with a carriage return and linefeed characters, so that each will appear on its own line. This isn't necessary, but it does make it easier for humans to see what those messages contain. Text-based protocols are sometimes called (with some exaggeration) *self-describing,* because it's often possible to decipher their rough meaning without much documentation.

Binary protocols encode information as strings of bits, the meaning of which is determined by where they appear in a message. The example protocol in the previous section was text-based. The following is a binary equivalent of that protocol.

A print request message consists of the following fields:

- *A 4-bit code indicating this is a request. The code for a request is 0.*
- *A 4-bit code indicating what action is being requested. The code for a print request is 2.*
- *An 8-bit integer containing the number of characters in the printer name.*
- *An ASCII string containing the printer name.*
- *An 8-bit integer containing the number of characters in the string to be printed.*
- *The ASCII string to be printed.*

A system that receives this message will try to print the supplied string on the indicated printer. It will respond with a message containing the following fields:

- *A 4-bit code indicating this is a response to a request. The code for a response is 3.*
- *A 4-bit code that contains the code of the request to which this is responding. For a print response, this will be 2.*
- *A 4-bit status code containing the result of the print request:*

    0 String printed successfully.

    1 String not printed, printer not known.

    2 String not printed, other error.

Instead of sending a text message like:

```
PRINTSTRING "prt1" "abc, def"
```

the system would send the following (hexadecimal) bits:

```
020370727431086162632C20646566
```

Instead of getting back the response:

```
STATUS OK
```

if the request succeeds, the systems would get back the following:

```
3200
```

The IETF, and thus the Internet community as a whole, tends to favor text-based protocols. The ITU, and thus the traditional voice and data networking community, has tended to favor binary protocols.

An argument made in favor of binary protocols is that they're more efficient because they use up less bandwidth. In the example above, the text-based print request takes up 29 characters. The binary version uses only 15 characters (an ASCII text character is eight bits, a hexadecimal digit is four bits). The text-based response is nine characters and the binary response is only two characters.

Advocates of text-based protocols make the following arguments in favor of their preference:

- Engineering all protocols for bandwidth efficiency is no longer as important as it used to be, given today's network speeds.

Software for Managers

- It's much more difficult to write software to create and interpret binary messages. This raises the cost of software that uses binary protocols and greatly increases the likelihood of errors.
- Because it's harder for those with limited resources to get into the binary protocol game, fewer people experiment with them, thus, there's less innovation. Furthermore, less experimentation in the early stages of development may leave design flaws undiscovered until it's much more expensive to fix them (or even too expensive, in which case everybody just has to live with them).
- It's more difficult to develop services that use binary protocols, because their messages aren't easily interpreted on the fly. One needs to use another piece of software—often expensive—to print them in a human-readable format. The other alternative is laboriously to decode messages by hand. Either way, the result is fewer developers writing software that uses these protocols.
- Many protocols exchange only a few relatively short messages, thus rendering the network efficiencies of a binary protocol less compelling.
- Because of the way some binary protocols are encoded, there may be so much overhead that messages end up nearly as big as, and sometimes even bigger than, their text-based equivalents.

The last point bears clarification. Many binary protocols, SS7 among them, are defined with ISO's *Abstract Syntax Notation 1 (ASN.1)*. ASN.1 is a language for describing the information in a protocol's messages. By itself, ASN.1 doesn't say what those messages will look like "on the wire." It just specifies things such as "the first field of a message is an integer" or "the second field will be either a string or another integer depending on whether the first field's value is 0 or 1."

The following is an example of a simple ASN.1 definition:

```
Example DEFINITIONS ::= BEGIN
  TransactionID ::= INTEGER
  Request ::= PrintableString
  RequestMessage ::= SEQUENCE {
    TransactionID,
    Request
  }
  ResponseMessage ::= SEQUENCE{
    TransactionID,
    result INTEGER {success(0), failure(1)}
  }
END
```

ASN.1 definitions are translated to strings of bits with *encoding rules*. Two sets of rules are commonly used: the *Basic Encoding Rules (BER)* and the *Packed Encoding Rules (PER)*. Even a partial description of these would take many pages. Instead, I'll go through just enough of BER to give some feel of how it works. ASN.1 has several built-in types, such as `BOOLEAN`, `INTEGER`, `BITSTRING`, `OCTETSTRING`, `PrintableString`, and `REAL`. ASN.1 also lets protocol designers define new types, just as they can in programming languages such as Java and C, using type constructors like `SEQUENCE` (roughly like a C `struct`) and `CHOICE` (roughly like a `union` with a discriminant in C++). BER defines a binary encoding of the type, length, and value of each field in an ASN.1 definition.

For example, the boolean value `TRUE` is encoded as follows:

```
0101FF
```

The first `01` indicates the type (`BOOLEAN`). The next `01` is the length of the value to follow (one byte). The final `FF` is the value itself. A `PrintableString` whose value is `status` would be:

```
1306737461747573
```

where type is `13`, length is 6 bytes, and `737461747573` is the string in ASCII characters.

You may have noticed that, so far, the length has always been a single byte. That's not a problem with boolean values since their length is always one. But strings can be any length. For a long enough string, 1 byte isn't going to be big enough to hold the length value. In BER, that maximum length is 127, the largest number that can be represented with 7 bits (you'll see shortly why not all 8 bits get used). BER has a clever way around this problem.

If the length is more than 127, the length field can be extended to more than 1 byte by using the first byte to say how many bytes will follow with the actual length. To make sure the software that parses the message can tell what kind of length field it's working with, the first bit of the first byte is reserved to indicate whether this is a 1-byte length (first bit is zero) or a multiple-byte length (first bit is one). That's why 1-byte lengths only go up to 127.

So, if we were to transmit a string of 456 characters, starting with `status`, BER would give the following representation:

```
138201C8737461747573 . . .
```

The type code is still `13`. The first byte of the length has a 1 in the first bit, so the length will occupy more than 1 byte, 2 to be specific. The decimal length value 456 is encoded as `01C8` in the next 2 bytes. Then come 456 bytes of character data, of which only the first six are shown.

This is still simple stuff compared to some of what's possible with ASN.1 and BER. Writing software to generate or parse BER-encoded data is a nontrivial task. Just handling the different kinds of length representations is a fair piece of work (speaking from personal experience) and there's a third form I didn't even cover. Commercial packages can be purchased, but they're not inexpensive. Furthermore, once you start working with messages that combine complex types like SEQUENCE, the type and length information added by BER can end up as many extra bytes.[7] PER greatly reduces the length of messages. The price you pay is that PER is even more complicated than BER and the messages even harder to decipher without software to help out.

While there's been no final resolution of the text-based versus binary protocol debate, the growing tendency is to use binary protocols where you're working closer to the hardware and to use text-based protocols at the higher levels where you're closer to user applications. That still leaves a large middle ground where the debate is carried forward to the delight and edification of all.

The cultural differences between the IETF and ITU have become especially clear at the border of the Internet and the PSTN, where their respective service protocols meet. On the Internet, these protocols are generally text-based. On the PSTN, they are usually binary. The contrast will become even more clear in later chapters, when we look at how the two communities have addressed IP telephony.

2.4.3 The OSI Reference Model

Conceptually, there's not much to computer networking. You have two computers. You connect them with a wire or some similar transmission medium. Then you have them exchange data with each other according to some protocol. As anyone who has tried to make a dialup connection to the Internet or to a computer at work will realize, the concept is simple. The reality is not.

The reason for this situation is the same as the reason programs have bugs: computers don't deal well with unanticipated situations. They are perhaps the ultimate expression of the neurotic sensibility, going to

---

[7]In one application on which I worked, as much as 30-50 percent of every message was overhead.

pieces at the slightest deviation from routine. When there's just one machine running a single program, things are bad enough. When you have two machines trying to talk to each other, each likely to be using its own program to set up and maintain the connection, you're just asking for trouble. Each new machine and program you add into this mix just compounds the problems. What's surprising is not that it can sometimes be hard to set up a network connection, but that it sometimes works on the first try.

The sheer complexity of computer networking—all the details that have to be handled flawlessly—naturally leads one to look for ways to simplify things. As with many computing problems, a divide-and-conquer method—chopping a big problem into smaller, relatively independent pieces with clearly defined points of interaction—has proved useful. In computer networking, thinking along these lines crystallized in the early 1980s with the *ISO Open Systems Interconnection (OSI) Reference Model* (see Fig. 2.8).

The OSI model divides a computer network connection into seven layers, each representing a separate aspect of the computer networking problem. Each layer solves only the issues defined for it, assuming layers below it will provide services on which it can depend. Though not all parts of the OSI seven layer model have stood the test of time, it remains a common point of reference in discussions of computer networking software.

Physical layer   Defines how bits are to be sent from one machine to another on a link that connects them. Examples of physical layer issues are how many volts represent a 1 and how many represent a 0, how many fractions of a second a bit lasts, and the physical characteristics of connectors used to attach a computer to a link (number of pins, uses of pins, and so forth). Every machine in a network has to support the physical layer. Examples include *RS232C* and *Ethernet*.

Datalink layer   Ensures that bits travel reliably from one machine to another across a physical layer link that connects them. Every machine in a network has to support the datalink layer. Examples include the various flavors of *LAP (Link Access Protocol)*, including *LAPB* and *LAPD*.

Network layer   Exchanges packets between adjacent machines in a network. It uses the datalink layer to ensure reliable transmission of the bits in those packets. One of its primary tasks is figuring out the next machine to which packets should be sent to ensure they arrive at their ultimate destination (*routing*). The network layer also makes sure

Software for Managers

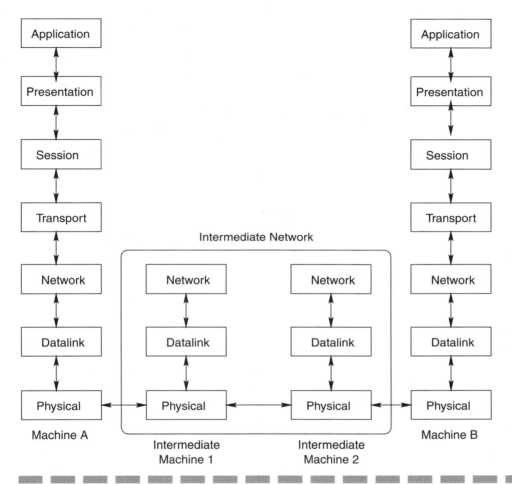

Figure 2.8
The ISO OSI seven layer reference model.

machines don't flood each other with more packets than they can handle (*flow control*). Every machine in a network has to support the network layer. Examples include *X.25* and *IP (Internet Protocol)*.

Transport layer Provides a logical channel for reliably exchanging data between programs that are executing at different endpoints on a network. It uses the network layer to ensure data gets to the right destination. The transport layer makes sure data gets there in the correct order and that no data is lost en route. The transport layer is typically implemented only on machines at the endpoints of a network. For most programmers developing network-enabled software, this is the level at which they set up and work with network connections. The details of the physical, datalink, and network layers are typically hidden from them within transport connections. The Internet *Transmission Control Protocol (TCP)* is a transport protocol. Because it doesn't provide reliable transmission, the Internet *User Datagram Protocol (UDP)* may or may not be considered a transport layer protocol, depending on how strictly one adheres to the OSI definitions. In common practice, UDP is considered a transport protocol, since it runs on top of a network protocol (viz., IP).[8]

Session layer Provides services that support a session composed of multiple transport layer streams, all supporting a single application running between two machines. In practice, this layer never caught on. If it were implemented, the session layer would typically appear only on machines at the endpoints of a network. Not surprisingly, there are no well-known examples of session layer protocols, though the ISO 8327 standard did touch on it.

Presentation layer Defines the format in which data is exchanged (for example, the number of bits in an integer, encrypted or not). This is another orphan layer, though perhaps not to the same extent as the session layer. If it were implemented, the presentation layer would typically appear only on machines at the endpoints of a network. In practice, many functions of the presentation layer have been absorbed into what is now called "middleware" (see Chap. 5). Some people put cryptographic protocols and standards in the presentation layer. These would include such things as the *Data Encryption Standard (DES)* and *Rivest-Shamir-Adelman (RSA)* public key cryptography.

Application layer Defines protocols that can be used to build applications (but not the applications themselves). Examples would include the Internet *File Transfer Protocol (FTP)* and *Simple Mail Transfer Protocol (SMTP)*. The application layer may itself be layered for applications in which more than one application protocol is used. The application layer is typically implemented only on network endpoints.

---

[8]Some things fit the OSI model well. Others not so well. What's important here is not a precise classification of protocols, but rather a general sense of where they fit into the model and how they relate to other protocols. With UDP, what's important is that it is an end-to-end protocol that runs on top of a network layer protocol. Together, these trump the fact that UDP lacks end-to-end reliability mechanisms when deciding where to put it in the OSI model.

2.4.4 The Internet and Its Protocols

While attempts have been made to apply the full OSI model to the Internet, a four layer approach, as illustrated in Fig. 2.9, works better. The

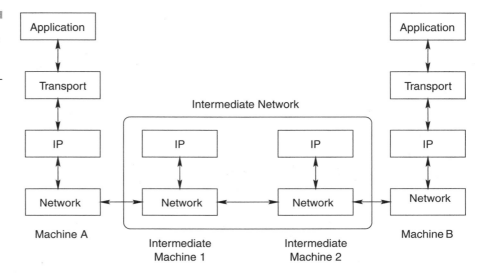

Figure 2.9
A four layer model of the protocols used in the Internet.

network layer at the bottom contains the protocols used in the network(s) on which the everything else runs. These are not Internet protocols as such. Rather they are the infrastructure of the Internet. Those protocols may themselves be layered (probably with a physical and a datalink layer), but that's not visible or relevant in this model. Ethernet would be an example of one implementation of this layer.

The next layer up, the *Internet Protocol (IP) layer*, is the first specifically defined in the Internet architecture. The IP layer is based on the IP protocol. In the context of the OSI model, IP is a network layer protocol. IP also lets different networks, each with its own physical and datalink protocols, connect together as a single logical *internetwork*. The *Internet* (with a capital "I") is a specific instance of an IP-based internetwork.

The next layer up—the *transport* or *end-to-end layer*—provides a logical channel for exchanging data between programs executing at different endpoints on an IP network. Two protocols dominate the Internet's transport layer:

Transmission Control Protocol (TCP) Provides a reliable channel between two machines. Messages sent via TCP are guaranteed to arrive at their destination. They are further guaranteed to arrive in the same same order they were sent. The combination of TCP and IP is often called *TCP/IP*. If only TCP is referred to, it may be assumed that IP lies underneath it.

User Datagram Protocol (UDP) Provides an unreliable channel between two machines. Messages, called *datagrams*, sent via UDP are neither guaranteed to arrive at their destination nor to arrive in the same order they were sent. This may seem useless, but it turns out to be handy for applications where it's more important that data travel quickly than that all of it get there or in the right order (Internet telephony, for example). Applications that use UDP are also free to add their own reliability mechanisms if so desired. When combined with IP, it's often called *UDP/IP*. If only UDP is referred to, it may be assumed that IP lies underneath it.[9]

The *application layer* contains the Internet application protocols. This layer contains the vast majority of Internet protocols. I already referred to a couple of them—FTP and SMTP—in the previous section. As in the OSI model, Internet application protocols provide services on which applications may be built. Sometimes it happens that a protocol and the application built on it have the same name (for example, FTP), which can be confusing. Another, and more subtle, source of confusion is the flexibility of Internet protocol layering, since an application protocol can work through TCP or UDP or it can bypass the transport layer entirely and talk directly to the IP layer.

UNICAST, MULTICAST AND BROADCAST Unicast, multicast, and broadcast are three different modes of sending messages. While they aren't specific to the Internet, the terms appear often in discussions of Internet protocols. A *unicast* message goes to a single receiving host. A *multicast* message goes to a defined set of hosts. A *broadcast* message goes to every host on a defined network (which may be part of another network such as the Internet). The applications of these to convergent network services will become evident as we go on.

---

[9]Note that the Internet transport layer, because it includes the unreliable UDP, is more flexible than the OSI definition of a transport layer.

2.5 For More Information

Java: An Introduction to Computer Science and Programming [86] provides an introduction to programming in general, and object-orientation using the Java programming language (see Chap. 3) in particular. *The Absolute Beginner's Guide to Programming* [81] is highly recommended as an elementary introduction. Information on UML may be found at the OMG's Web site (http://www.omg.org). Interest in UML has exploded since its introduction several years ago. There has been a corresponding explosion in the number of books on that topic. Fowler and Scott's *UML Distilled* [18] is an especially good introduction. *Component Software: Beyond Object-Oriented Programming* [88] discusses component software, comparing various technologies in a comprehensive, though at times pedantic, style. Andrew Tanenbaum's *Computer Networks*, now in its third edition, [89] is a classic on computer networking. Larry Peterson's *Computer Networks: A System Approach* [82] is a worthy successor. For more on ASN.1, [8] is about as painless an introduction as is possible.

CHAPTER 3

The Java Programming Language

I feel impelled to speak today in a language that in a sense is new.

—*Dwight D. Eisenhower*

3.1 Java's Role in Converged Networks

Java is an object-oriented programming language developed by Sun Microsystems in the early 1990s. Java's creators designed it specifically for writing programs that run on equipment attached to a network. Besides the core Java language, the Java standard also specifies a large collection of software libraries that support everything from enterprise-class Web servers to cellphones and smartcards. Java's designers were guided by three central ideas:

- *Java programs should be portable.* One should be able to send executable Java code from one machine to run on another. Java programs should not depend on or assume specific hardware or operating systems. A Java program should run without change on any device that supports Java technology, an essential feature in networks, where there is no way to know in advance what sorts of devices will be connected to and working with each other.

- *Java programs should be secure.* Sharing information across networks creates benefits, but it also creates risks. Since network-attached devices can exchange Java programs among themselves, there must be a way to ensure they can be executed without undue risk of adverse consequences to their recipients.

- *Java programs should be lightweight and distributable.* If Java programs are to be sent across networks, they cannot be too large, or they will use up too much bandwidth and slow down overall system performance while platforms are waiting for executable code to arrive.

When Java was first introduced in the mid-1990s, some people predicted that Java-enabled Web browsers would replace Windows as the standard operating environment for desktop PCs. They were wrong. As it turned out, Java came to be far more significant as an enterprise computing technology. That makes sense, because enterprise software runs across networks on a variety of platforms, precisely the environment Java's designers targeted. Remember also that, as computing depends more and more on networks, the desktop PC is just one device among many, and not always the most important one at that. Devices of all kinds—desk phones, cellphones, PDAs, even household appliances—are now, or soon will be, network-enabled.

The Java Programming Language

The sheer numbers of devices that connect to networks will force programmers to write software that runs on many more platforms than just the PC. Learning to write software for even one platform is a major undertaking. Faced with the choice of learning a bit about many platforms or learning just one platform well, most developers prefer to specialize. Hence, the popularity of Microsoft Windows. Java lets programmers apply a single set of skills to many platforms, thus increasing the potential market for their work and reducing their chances of betting on the wrong horse.

Besides the Java language itself, several Java-based technologies warrant mention because of their relevance to the convergent network computing environment.

PersonalJava PersonalJava is a collection of Java classes that have been optimized for writing application that run on network-aware consumer devices, like set-top boxes, smartphones, game consoles, PDAs, televisions, and so forth. PersonalJava has features designed specifically for devices with limited resources. Among these is a version of the Java user interface API that's been tuned for small displays and limited input facilities (for example, devices having no keyboard or mouse). Since PersonalJava is a subset of Java's full set of class libraries, applications written in PersonalJava can also run on devices that support the full Java environment. For example, a developer might build an electronic TV guide that runs on PersonalJava-enabled set-top boxes. The same application could also run on a Java-enabled handheld remote control, a PC, or even a smartphone. A particular benefit PersonalJava brings to the world of network-enabled devices is Java's security model (see Sec. 3.4). This will let users load new applications into their devices, without worrying too much about the consequences. Trusted application writers will sign their code with a certificate issued by a third party who vouches for their trustworthiness. Only applications that have been properly authenticated in this manner could run on such a device.

EmbeddedJava PersonalJava is sometimes confused with EmbeddedJava. Like PersonalJava, EmbeddedJava defines a subset of the Java class libraries. However, EmbeddedJava's target is special-purpose processors embedded in equipment that may or may not be connected to a network. Embedded applications are usually invisible to consumers, though the processors on which they run are often built into consumer devices. PersonalJava, on the other hand, aims directly at consumer applications. Furthermore, EmbeddedJava assumes either a character-based display or no display at all, while PersonalJava generally assumes at least some sort of display.

JavaCard Java Card is a collection of Java classes for writing smart card applications. While smart cards have not had much success in the United States, they are popular in Europe and Asia. Smart card vendors see Java Card as a means to encourage developers to write smart card applications, an activity that was formerly the province of a handful of specialists.

Enterprise Java Enterprise Java provides the framework of classes needed to build distributed enterprise applications. Among its features are the *Java Database Connectivity (JDBC)* API for database access, Enterprise JavaBean components, servlets, and XML support. Also known as *Java 2 Enterprise Edition (J2EE)*.

Jini Jini is a Java-based distributed processing technology that lets all sorts of devices—not just computers—cooperate with each other on networks.

JTAPI JTAPI, the Java Telephony API is a framework for developing Java-based CTI applications. JTAPI is covered in Chap. 12.

JAIN JAIN, the Java APIs for Integrated Networks, is an emerging framework for developing communication applications that can run within many kinds of networks. JAIN is covered in Chap. 13.

Parlay Parlay is a set of APIs being developed by the Parlay Group. To some extent, its goals overlap with those of JAIN, though Parlay supports languages other than Java. Recognizing the potential for duplicated effort, the designers of JAIN and Parlay are now working closely with each other to ensure both efforts are compatible. Parlay is covered in Chap. 14.

3.2 The Java Virtual Machine

The *Java Virtual Machine (JVM)* is the means for achieving Java's design goals. The JVM specification defines a processor that could, in principle, be implemented as hardware, much like an Intel Pentium or Motorola PowerPC. In practice, the JVM is a program called an *interpreter* that emulates a hardware JVM's behavior.[1] The machine language for the

---

[1] Actually, the JVM has been implemented in hardware. To date, price of these so-called "Java chips" has rendered them less than competitive with other processors.

The Java Programming Language

JVM is called *byte code*. An implementation of the JVM reads Java byte code and executes it by invoking the specific commands of the hardware on which it is running.

Java compilers translate Java programs into byte code. Because every implementation of the JVM uses the same byte code language, the compiled version of a Java program will run on any hardware that has a JVM implemented for it (Fig. 3.1). From Java's point of view, the code

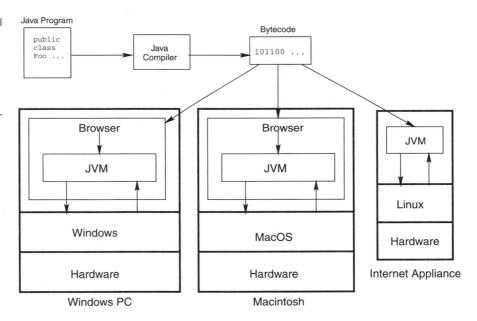

Figure 3.1
Java compilers produce byte code executables that can be executed on any machine with a JVM.

runs on only one platform: the JVM. Some refer to this as "write once, run anywhere." A Java program written on a Windows PC and compiled into byte code there can run on any Macintosh or a Linux PC with a JVM. A program written for a set-top box can run on a cellphone, if they both have JVMs and the program uses capabilities common to both those devices.

There are several ways an implementation of the JVM can be made available to users. It can be provided as a separate program that is invoked whenever Java code is to be executed. It can be built into programs that will be executing Java programs as part of their normal activities (e.g., in a Web browser). Finally, it could be built directly into the operating system as a service ("execute Java") just like any other (e.g., "list directory").[2]

Besides making code portable from one machine to another, the JVM has another feature, *automatic garbage collection,* that simplifies Java programming when compared with C++, a language Java resembles in many respects. In C++, programmers request bits and pieces of memory as needed by their programs. Since memory is a limited resource, programs have to let the system know when they are done using it. A common—perhaps the most common—bug in C++ programs is for a program to free up memory before the program is finished with it. When the program tries to use that memory later, it may find that something else has gotten hold of it and and written something there in the meantime. This sort of accidental memory sharing usually leads to odd program behavior and crashes. Just like C++ programmers, Java programmers also obtain memory as needed. But Java programmers don't have to worry about freeing it themselves, since the JVM figures out on its own when memory is no longer needed and can be returned to the system.

Another common—but more insidious—problem in C++ programs is for memory to be allocated by a program but never freed. The longer the program runs, the more memory it allocates out of a steadily diminishing memory pool, a so-called *memory leak.* Eventually, all available memory is used up and the program—or even the entire machine—crashes. In the best case, memory leaks are discovered and fixed during testing. But they may not occur until the program has been released on an unsuspecting world and run long enough in actual use to expose the problem. Memory leaks aren't as troublesome in Java programs, because the JVM frees memory on its own.[3]

---

[2] As this was being written, Microsoft had just announced they will not include a JVM in their Windows XP operating system. Some think this means Java programs will no longer run on Windows. This is most emphatically not the case. Microsoft is not the only supplier of a JVM for Windows. Purveyors of Java-based software need only supply one of these other JVMs to users. The situation is analogous to playing streaming audio in a format Microsoft doesn't support. Users download or are otherwise provided with software that understands the format. The only difference is that, in one case Microsoft provides the software, in the other somebody else provides it.

[3] Of course, a program may run out of space because it simply needs more memory than is available on a given machine. But there's no way to handle that, other than rewriting the program to use less memory or adding more physical memory.

Garbage collection is a controversial topic in the world of convergent networks. Traditionally, phone networks have been required to adhere to strict measures of performance, many of them time-based. Meeting these requirements means that every aspect of call

3.3 The Java Class Libraries

Another benefit Java offers to programmers is its collection of prewritten classes, the *Java class libraries*. These support common programming tasks, like creating and interacting with a GUI or setting up and communicating over a network connection. The Java class libraries also have APIs for more specialized functions, like telephony and security. Because of the Java class libraries, Java programmers need learn only one set of APIs, no matter where they plan to develop or deploy their software. This was an important contributor to Java's rapid growth in popularity. (See Fig. 3.2.)

Java is not the only language with a bundled code library. The C language's standard libraries are every bit as much a part of that language as the Java class libraries are for Java. However, when compared with the Java libraries, the C libraries have a much more limited scope: basic file operations, input and output, simple string manipulation, fundamental interactions with the operating system for memory and process management, and some common mathematical functions.

C++, the object-oriented successor to C, inherited C's libraries and added a few of its own, but again these did not cover much territory: files, I/O, and a bit of math. It was often noted that C++ needed a larger set of standard class libraries, but these were long in coming, and some of the results ended up being more expressions of their designers' technical brilliance than tools for the journeyman programmer (the Standard Template Library being a particularly egregious example of this

---

processing must cooperate if the poor performance of one piece is not to jeopardize the performance of the whole. If not implemented properly, garbage collection can occur whenever a system decides it's necessary. If something more important is going on at the same time, that won't matter, since space has to be freed up if the system is to continue working. The potential for garbage collection to occur at the worst possible time should be obvious. Careful design of the garbage collection software can offset much of this. The open question is whether that's enough. Experts still disagree over the answer.

Another factor to keep in mind is that software can be partitioned so that the parts most sensitive to garbage collection's performance issues can often be kept separate from it. For example, a service that takes data from a phone call and does something with it for off-line storage doesn't have to be part of call processing. Call processing can hand the data off to the service and then continue on its own. The service can process the data independently of what the call does after that. If it encounters garbage collection, no harm is done to call processing.

Figure 3.2
Some members of the Java class libraries. Classes are grouped into *packages* of related classes. Package names are in bold type here.

java.awt	java.io	java.net
Color	File	DatagramPacket
Component	InputStream	DatagramSocket
Button	FileInputStream	InetAddress
Canvas	StringBufferInputStream	PasswordAuthentication
Checkbox	OutputStream	ServerSocket
Container	FileOutputStream	Socket
ScrollPane	StringOutputStream	URL
Window	StreamTokenizer	URLDecoder
Dialog	. . .	URLStreamHandler
Label		. . .
Scrollbar		
TextComponent	**java.util**	**java.lang**
TextArea		Character
TextField	Calendar	Math
Cursor	Date	Process
Font	Dictionary	String
Image	Hashtable	StringBuffer
Polygon	Random	Thread
. . .	StringTokenizer	. . .
	TimeZone	
	. . .	

tendency). Commercial vendors have filled the gap with products of their own. While useful, these are not standardized.[4]

Perhaps the most successful cross-platform class libraries before Java were those for the Smalltalk language. Unfortunately, the Smalltalk libraries were crippled by the fact that there were competing versions of Smalltalk, each with its own set of class libraries. While similar, these libraries were just different enough that programs written for one vendor's Smalltalk would not run with another vendor's version.[5]

Java's class libraries build on more than a decade of industry and academic experience with object-oriented programming, incorporating many of the best practices accumulated over that time. Consequently, many programmers find Java's libraries much easier to learn and use

---

[4] Windows programmers might argue that Microsoft's C++ libraries are an exception. However, those libraries are of no use to a C++ programmer who doesn't develop for Windows.

[5] It's roughly the same as would happen if there were different versions of the JVM, each with slightly different sets of byte code instructions. Java portability would then be lost. This was the basis for much of Sun's disagreement with Microsoft's handling of Java.

than their predecessors, another reason for the language's popularity. Java's network programming classes are a good example of this. Using just the classes `Socket` and `ServerSocket` and a few of their methods, it's possible to write a simple client-server application. The `java.net` package includes these and other classes for every sort of network programming task, from decoding URLs to authenticating users who request a network connection. Other packages in the Java class libraries offer comparable sets of classes for other problem domains.

3.4 Java Applets and Security

Another of Java's key ideas is that a JVM can be incorporated into a Web browser. The browser can then download Java programs from the Internet in the same way it downloads other Web content. Once downloaded, these Java programs, called *applets*, run on the browser's built-in JVM.[6] This is a powerful model for software distribution, but it has inherent risks. An applet with free rein on the system into which it is loaded could wreak havoc in short order. Furthermore, an otherwise trusted applet might be altered by a bad guy without the original author's knowledge.

The JVM includes features that help to protect users from malicious programs. The most basic of these is the *Byte Code Verifier*. Before the JVM will run a Java program, it makes several checks of the program's byte code. The first thing it does is to ensure that all the byte code is in the proper format. If it isn't, the JVM refuses to run the program, on the assumption it may have been tampered with. Next it tests the code

---

Microsoft provided a Java programming tool of its own, *Visual J++*, that supported special instructions in the Microsoft JVM that weren't a part of the standard JVM. Programs that used these instructions could run on Windows PCs but nowhere else. Microsoft's argument was that Java is just another programming language, and they added the nonstandard instructions so programmers could take advantage of Windows features that would otherwise be inaccessible to Java programmers. Sun's counterargument was that portability was part of the essence of Java, so that what Microsoft was offering wasn't really Java at all.

[6]While browser-executed applets got a lot of attention, they aren't nearly as common as Sun originally expected. Their biggest drawback was the time it took to download them on a modem connection. Most Java programs today are either standalone applications or *servlets* that run within Web servers. However, the increasing availability of home broadband Internet access could change that.

to make sure it doesn't try to access memory outside its address space or manipulate data of one type as though it were of another type.

The second part of Java security is the *Java Applet Class Loader*. As a Java program runs, it creates and uses instances of various classes. The class loader makes sure a program doesn't try to bypass Java's built-in security by substituting its own versions of classes from the standard Java class libraries. The standard Java classes are designed to behave themselves. There's no telling what a replacement might do, so the class loader prevents it from happening.

The part of Java that enforces security is the *Java Security Manager*. A Java program can't do anything to the system on which it runs, or any other system, without going through the security manager. Whenever a Java program is about to do something risky (say, writing to a file or opening a network connection to another host), it first asks permission from the security manager. If the program doesn't have explicit permission to carry out the operation, the security manager stops it in its tracks. Every host that runs Java programs can assign permissions for the security manager to use in deciding what's allowed. If no special permissions are given, a default set is used that's very restrictive. For example, under normal circumstances, the security manager won't let applets read or write files on the system where they're running or open network connections to any system other than the one from which they were downloaded.

As with any security technology, the JVM provides no guarantees of absolute safety. However, attacks are less likely and the JVM's security features are subject to extensive external review, further reducing the chance of exploitable holes in its security system.

3.5 Java Servlets

A Java applet is a Java program that can be loaded and run by a Web browser. A Java *servlet* is a Java program that can be loaded and run by a Web server. Servlets can be used for the same things as other *server-side programming* technologies: interacting with users through Web pages that contain forms to be filled in and generating dynamic Web pages with content that varies according to the circumstances under which it is retrieved.

One of the most common server-side technologies is the *Common Gateway Interface (CGI)*. A programmer using CGI first writes a program to generate and process the desired Web page (Perl is an especially popular language for this). This is stored in a location known by the Web server to hold the CGI programs for which it is responsible. This location and the program name are then incorporated into the URL provided by the user in his Web browser, for example:

```
http://www.someserver.com/cgi-programs/someprogram.pl
```

The server invokes the indicated CGI program (`someprogram.pl`), which builds a Web page and sends it to the user as shown in Fig. 3.3.

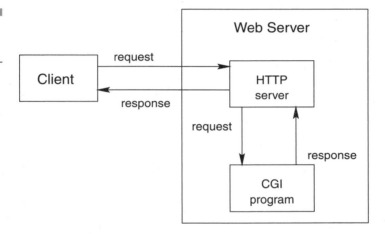

Figure 3.3
Common Gateway Interface (CGI).

That Web page may contain fields that invoke other CGI programs, thus providing some interactivity.

While CGI is an adequate solution, it isn't ideal. CGI emerged in the early days of Web technology and began to show some weaknesses as demands on Web servers increased. One of the biggest disadvantages of CGI programs is that an entirely new process must be created by the Web server to run it every time a user links to it. Starting new processes is a fairly time-consuming and expensive activity. For heavily visited Web sites with lots of dynamic or interactive content, this can be a drain on capacity. Furthermore, because a CGI program is run anew every time it is invoked, any startup or cleanup code will be executed as well. Like starting a new process, this sort of work (for example, opening and closing database connections) can be time consuming and expensive.

Java servlets offset these disadvantages and bring several advantages of their own. A Java servlet does not run as a separate process. Rather, it runs as a subtask of the server, just as an applet runs as a subtask of a browser. Unlike an applet, which typically loads and runs and then disappears when the user navigates to another Web page, a servlet once loaded stays loaded. That means that a servlet can be loaded, execute its startup code, and then handle invocations by users who navigate to it. If the servlet opens a database connection during its startup phase, that connection can stay open to be used every time the servlet is invoked.

Java servlets do not all have to be loaded at once. Only some, or even none, of them may be loaded when the server starts. Servlets are then loaded as requests arrive for them. It is even possible to add new servlets to an already running server, thus extending its capabilities, because of this on-demand loading. The first request for a given servlet will take a bit longer because of the extra time it takes to load, but requests after that will run quickly because the servlet is already present.

Another advantage of servlets is the Java language. To the extent that Java is portable, Java servlets are portable from one server to another. Java programmers also benefit from the features of Java—object-orientation, garbage collection, the Java class libraries, and so forth. Some of the most popular languages for CGI scripts—Perl in particular—are notorious for the ease with which it is possible to write a program that even its original author will later find unreadable.

3.6 Java's Event Model

Most telecom software is inherently event driven. Something happens and one or more network elements must see that it happened and respond appropriately: a subscriber goes offhook, digits are dialed, feature codes are entered, one switch seizes a trunk to another switch, and so on. One of Java's greatest advantages for network programming is its well-developed model for writing event-driven programs. This model is used extensively in Java APIs for communication services (for example, JTAPI and JAIN, which are described in Chaps. 12 and 13).

In Java, any object can generate an *event*. This event is an object that contains information, such as why it was generated, who generated it, or relevant values at the time it occurred. An object can generate just one kind of event or it can generate many different kinds of events. Different kinds of events are represented by objects that are instances of different event classes.

The Java Programming Language

Other objects can register themselves with an event-generating object to let it know that they want to be informed when it generates events. These objects are called *event listeners,* or just *listeners.* If an object generates many kinds of events, its listeners can register for all or only some of them. Event listeners must provide a method for an event-generator to call when it produces an event of a specific sort.

Figure 3.4 shows the pattern of classes of the Java event model applied to a simple telephony application. Event classes derive from the built-in `java.util.EventObject` class. The main purpose of

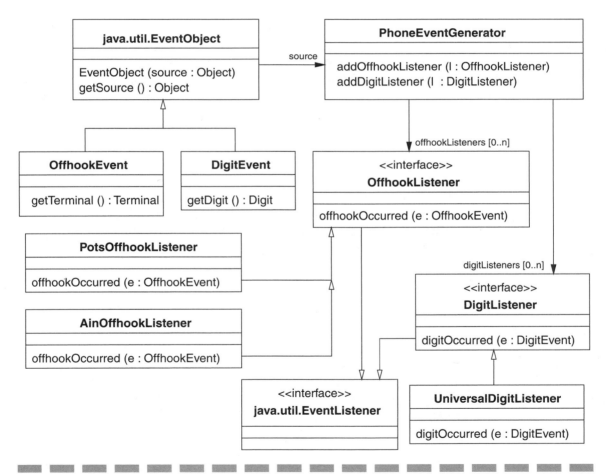

Figure 3.4
An application of the Java event model pattern to telephony.

`java.util.EventObject` is to let event listeners find the object that generated an event by using its `getSource` method. There are two different kinds of event classes, representing the two different kinds of events that will be generated: `OffhookEvent` and `DigitEvent`. These represent a user going offhook and a user dialing a digit, respectively.

Events are generated by event sources, instances of `PhoneEventGenerator` in this example. Event-generating objects must offer their listeners methods with which to register their interest in any or all of the events they might produce (`addOffhookListener` and `addDigitListener` here). Event-generating objects must also keep track of objects that have registered with them (the `offhookListeners` and `digitListeners` elements of `PhoneEventGenerator`).[7]

Java provides the `java.util.EventListener` interface from which to derive listener classes. The `java.util.EventListener` class has no methods, so the author of an event-generating class must provide interface classes that derive from `java.util.EventListener`. These interface classes define methods that listeners should use to be informed of events. Listeners implement these interfaces and provide code that does whatever it is they want done when the desired event occurs.

In Fig. 3.4, the two listener interfaces that derive from `java.util.EventListener` are `OffhookListener` and `DigitListener`. These provide `offhookOccurred` and `digitOccurred` methods. The `PotsOffhookListener` class implements the `OffhookListener` interface for an application that supports a POTS phone line, while `AinOffhookListener` implements it for an application that supports a phone line with AIN offhook detection features. We can assume digit detection is handled the same way no matter what, so there's only one class, `UniversalDigitListener`, that needs to implement `DigitListener`.

Figure 3.5 shows a scenario that uses these event classes. Two different instances of `PotsOffhookListener` first register themselves with an instance of `PhoneEventGenerator`. Sooner or later the phone that is monitored by the event generator goes offhook. The event generator creates a new instance of `OffhookEvent` and calls the `offhookOccurred` method in all objects that registered for that event.[8] Once they get the event, it's up to them to do something with it.

---

[7] It is customary for event-generating objects to also offer methods with which listeners can de-register themselves. We've not shown those here in order to reduce clutter.

[8] It might be more appropriate to create a separate instance of `OffhookEvent` for each listener.

The Java Programming Language

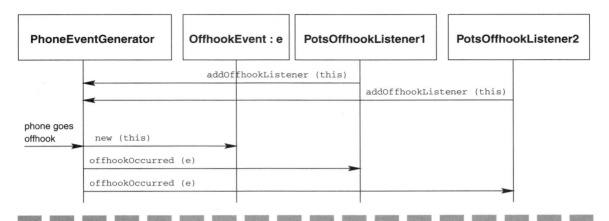

Figure 3.5
An example of event registration and notification.

3.7 JavaBeans

JavaBeans is Sun's framework for writing Java component software. JavaBeans are often used for visual user-interface elements, such as scrollbars and buttons. They can also be used for nonvisual program elements. JavaBeans free programmers from having to implement complex "plumbing" code over and over again. Among the tasks a bean can handle on behalf of a programmer are:

- Managing pools of available objects in a running system, so that enough are on hand to meet demand without having so many as to waste system resources.
- Adding new classes and objects into a running system.
- Spreading objects and classes across a distributed system.
- Finding objects in a distributed system.
- Managing simultaneous requests to access the same items of information.
- Ensuring that objects persist across reboots and system failures.

Any class that follows the JavaBean coding rules is a JavaBean. By following those rules, a bean class makes its properties, events, and methods known to bean programming tools (sometimes called *beanboxes*). Instances of JavaBean classes are called *beans*.

Any method declared `public` is visible to bean tools. Bean properties are defined by a pair of methods of the following form:

- `public <type> getProperty()`
- `public void setProperty (<type> val)`

These methods define a bean property named `property` whose type is `<type>`.[9] For example, if a class has public methods `getBalance` and `setBalance`, then a bean tool would decide that `balance` is a property of that class. Besides these regular properties, beans can also have indexed properties, bound properties, and constrained properties. An *indexed property* has multiple elements, like an array, each of which can be set or retrieved individually. A *bound property* sends a notification event whenever its value changes. A *constrained property* also sends a notification event when its value changes, but the change can be vetoed by objects that are listening for that event. Bean events are handled by the standard Java event model as described in Sec. 3.6.

Enterprise JavaBeans (EJB) extend the component model of JavaBeans to support business logic and data in server-side enterprise programs. It's common practice in client-server systems to simplify client programs by moving business-specific functions to the server (either directly or via an intervening system), so-called *thin client* programming. The problem for the server is that business-specific code gets mixed in with code for handling network connections, authentication, transactions, databases, and the like. EJBs simplify this by providing a separate place to put business code apart from all the low-level infrastructure code.

Unlike regular JavaBeans that can run on any JVM, EJBs run inside a special environment, called an *EJB container*. The EJB container runs on an *EJB server*. The EJB container and server provide the support services needed by an EJB—transaction monitoring, network connection management, security and naming services, and so on—thus freeing the bean programmer from having to worry about all that in his own code.

There are two kinds of Enterprise JavaBeans. *Session beans* provide client programs with services provided by the server on which they run. *Entity beans* represent server data objects, such as orders or customer accounts. Changes to an entity bean are persistent. This means that:

---

[9]The property name is not capitalized when it stands on its own, but it is capitalized in the `get` and `set` methods. This pattern is followed for all JavaBeans properties.

- When one client program changes data in an entity bean, the next client program to use that bean will see the results of those changes.
- An entity bean will preserve its state across system crashes.

3.8 For More Information

There are many excellent and some not-so-excellent books on Java. Niemeyer and Knudsen's *Learning Java* [67] is a good introduction. Sun's own Java Web site, `http://java.sun.com`, has an excellent series of online tutorials that cover everything from the basics to advanced topics. Sun's site is also the best place to find information on the most recent additions to the Java class libraries and other emerging Java technologies.

For the working programmer, the online documentation for the Java class libraries that comes with the Java SDK plus the *Java Developer's Almanac* [7] are essential. The standard reference for language lawyers is Gosling, Joy, and Steele's *Java Language Specification* [19].

CHAPTER 4

XML:
The Extensible
Markup Language

A place for everything and everything in its place.
—*Isabella Mary Beeton*
(The Book of Household
Management, 1861)

4.1 Markup Languages and Structured Data

XML is the successor to HTML (HyperText Markup Language), one of the core technologies of the World Wide Web. Both XML and HTML are *markup languages*. Markup languages are nothing more than a set of *tags*—strings like `title` or `phoneNumber`—that can be added to Web pages and other documents to mark their *structure*. Like objects in object-oriented programming, it's easier to explain what structure is by example than by trying to give a strict definition. Consider the following:

```
John Doe 1234 Main Street Anytown WI 55555
```

Looking at this, you can probably tell it's an address. Even though it has no punctuation, there's a good chance you can figure out the pieces of which it's composed: name, street, city, state, ZIP code. In so doing, you've added structure to what is otherwise just a string of characters. So, for a person who knows the system, this is structured data.

People are good at finding structure in data, even when there's not much to work with. They're so good at it, they sometimes find structure when it's not even there.[1] Computers, on the other hand, need careful programming if they are to use the structure in a given batch of data. While you could write a program to interpret the example address, it could easily be confused by even minor changes, since there are so many ways to represent an address. Could it handle addresses with and without punctuation? What about five-digit and nine-digit ZIP codes? What about state abbreviations versus full state names? What about addresses that are broken across lines? Or what about this:

```
1234 Main Street Anytown WI 55555 John Doe
```

While this violates the normal rules for writing addresses, most people could draw the reasonable conclusion that it's the address of a person whose name is John Doe. Would (or even should) the program be able to handle something like that? Odd things do turn up, and it would be best to have a program that could deal with them gracefully.

---

[1] An example would be Pyramidology, which claims the Egyptian pyramids are a sort of giant petrified encyclopedia in which are embedded all sorts of mathematical and other information.

XML: The Extensible Markup Language

The problem here is that, from the computer's (rather dullwitted) point of view, the addresses it gets are just strings of unstructured characters. Humans are good at making something of an unstructured mess. Computers aren't.

The solution is not to leave everything up to humans. Nor is it to try to make computers as good as people are at discovering structure on their own (at least not if you want a solution within the next few decades, if even that). The solution is to provide humans a way to help the computer, without making them do too much work. That's where markup languages come into play.

A markup language is a tool people can use to indicate structure in data they want a computer to process. Returning to the address example, consider this:

```
<address>
   <name>John Doe</name>
   <street>1234 Main Street</street>
   <city>Anytown</city>
   <state>WI</state>
   <zip>55555</zip>
</address>
```

This is, in fact, an example of how an address could be marked up with XML. Even if you don't know any XML at all, you can probably see what's going on here (the human ability to discover structure at work again).

There are now labels—`<address>` and `</address>`)—called *tags* that surround the entire address. Each piece of the address is also marked with its own tags so there can be no doubt what's what. Using these tags, a program that works with addresses doesn't have to know much about how addresses are laid out. It just needs to know that it will find an address between `<address>` and `</address>`, a name between `<name>` and `</name>`, a street between `<street>` and `</street>`, and so on. You could even give it something like this:

```
<address>
   <state>WI</state>
   <city>Anytown</city>
   <street>1234 Main Street</street>
   <zip>55555</zip>
   <name>John Doe</name>
</address>
```

and a program that understands the tagging rules will be able to work out its meaning. Without tags, this would be:

```
WI Anytown 1234 Main Street 55555 John Doe
```

which even a human might have trouble deciphering.

The advantages of using tags to structure data should be obvious, especially when it's data to be shared among different systems. Even more sophisticated applications of XML described elsewhere in this book go even farther and provide tools to support applications as diverse as distributed object programming (see Sec. 5.8) and scripting languages for phone services (see Sec. 11.2). But before we look at those those, we must first dig deeper into XML.

4.2 From HTML to XML

HTML and XML are both applications of the *Standard Generalized Markup Language (SGML)*. SGML is a *metalanguage*, a language that is used to define markup languages. Because HTML and XML follow SGML's rules, HTML and XML documents share many similarities. One of these is the way their tags look. SGML says that tags must come in pairs—a start tag and a corresponding end tag. The two tags surround some *content*:[2]

```
<tag>some content</tag>
```

If `<tag>` is a start tag, then its corresponding end tag has the same name with a slash in front: `</tag>`. Tags come in two flavors, semantic tags and formatting tags. They represent two different approaches to structuring data. *Semantic tags* describe the content they surround, for example, a title, a heading, a phone number, or a quotation:[3]

```
<title>Moby Dick</title>
<heading>Chapter 1</heading>
<phonenumber>999-555-1234</phonenumber>
<quotation>I have not yet begun to fight.</quotation>
```

Semantic tags add lots of structure information to a document. *Formatting tags* specify the desired appearance of the content they surround, for example, bold, italic, or underlined:

---

[2]This is also called the *value* of the tag.

[3]These and subsequent examples of formatting tags are not samples of either XML or HTML. They serve only to demonstrate the principles of markup languages in general, and SGML-based markup languages in particular.

XML: The Extensible Markup Language

```
Some <bold>bold</bold> text
Some <italic>italic</italic> text
Some <underline>underlined</underline> text
```

Formatting tags add a minimum of structuring information to a document. A formatting program (like a Web browser) reads files that contain markup tags of either kind and figures out how best to display them based on those tags.

Formatting programs display content that is surrounded by semantic tags in some reasonable fashion. For example, titles typically use characters that are larger and/or bolder than those in the body of the document. With formatting tags, formatting programs can display the content as specified or, if that's not possible, use the best available substitute. For example, if a word is tagged as italic, but italic characters aren't available, it could be underlined instead. Formatting programs must also deal with tags they don't recognize. A common solution is to treat content surrounded by unrecognized tags as though that content has no tags at all.

Markup languages solve the problem of how to display information from the Web when there are so many different kinds of devices—each with its own features and limitations—on which it might be viewed. HTML reduces the possible ways of formatting data to just a few and then lets the Web browser on each device figure out how best to deal with these on its own. By reducing the ways information can be formatted, HTML makes it possible to write Web pages that can be viewed on many different devices.

Figure 4.1 is a simple HTML document. The entire document is surrounded by `html` tags. These mark everything between them as HTML content. The document has two parts, a head and a body, marked by corresponding tags. Within the head is a tagged title. The body starts with a level one heading, identified by the `h1` tag. Following that are three paragraphs of text separated by paragraph tags (`<p>`).

Notice that the first paragraph of the body has no starting `<p>` tag and none of the paragraphs has a closing `</p>` tag. HTML tries to make the author's job easier by relaxing SGML's rule that a `<tag>` must always have a matching `</tag>`. While HTML formatting programs can usually figure out the right thing to do when end tags are left out, it sometimes makes for unexpected results. It also makes HTML formatting programs much more complicated. In retrospect, this is one of those things that seemed like a good idea at the time, but has probably turned out to be more trouble than it's worth. XML

```
<html>
   <head>
      <title>A simple HTML document</title>
   </head>
   <body>
      <h1>Level 1 heading</h1>
         It was Augustus who first, under colour of this law applied legal
         inquiry to libelous writings provoked, as he had been, by the
         licentious freedom with which Cassius Severus had defamed men and
         women of distinction in his insulting satires. Soon afterwards,
         Tiberius, when consulted by Pompeius Macer, the praetor, as to
         whether prosecutions for treason should be revived, replied that
         the laws must be enforced. He too had been exasperated by the
         publication of verses of uncertain authorship, pointed at his
         cruelty, his arrogance, and his dissensions with his mother.
      <p>
         It will not be uninteresting.
      <p>
         Tacitus, <i>The Annals<i>
   </body>
</html>
```

Figure 4.1
A simple HTML document.

does away with this and insists that start tags always be matched with an end tag.[4]

All the tags discussed so far are semantic tags. The last tag in the Fig. 4.1, `<i>`, is a formatting tag that suggests using italic type or the best substitute if italics aren't available. The HTML text in Fig. 4.1 might be formatted as shown in Fig. 4.2 (the actual formatting would depend on the program used to display it).

Although HTML is a powerful tool, the Web is outgrowing it. One of HTML's biggest problems is its limited set of tags. The number of HTML tags is undeniably large; some critics say there are already far too many. (That's part of the reason Web browsers have become such large

---

[4]There is one exception to this rule. If an XML start tag and end tag have nothing between them:

`<tag></tag>`

the pair may be replaced with:

`<tag/>`

Figure 4.2
Formatted version of the the HTML in Fig. 4.1.

> **A simple HTML document**
>
> **Level 1 heading**
>
> It was Augustus who first, under colour of this law applied legal inquiry to libelous writings provoked, as he had been, by the licentious freedom with which Cassius Severus had defamed men and women of distinction in his insulting satires. Soon afterwards, Tiberius, when consulted by Pompeius Macer, the praetor, as to whether prosecutions for treason should be revived, replied that the laws must be enforced. He too had been exasperated by the publication of verses of uncertain authorship, pointed at his cruelty, his arrogance, and his dissensions with his mother.
>
> It will not be uninteresting.
>
> Tacitus, *The Annals*

and complex pieces of software.) But even if you don't think there are too many HTML tags as it is, getting the World Wide Web Consortium (or W3C, the standards body responsible for HTML) to approve new tags is a slow and uncertain process. There are some workarounds when the tags you need don't exist, but they aren't convenient. For example, to display mathematical formulas or music, one must typeset the information with something other than HTML and then convert the result into an image file that can be embedded in an HTML document, a tedious process at best.

Another disadvantage of HTML is that it is almost entirely a formatting language. Even its semantic tags are little more than high-level formatting tags. There's much useful information embedded in Web pages, but it's hard for other programs to extract it. For example, consider a Web-based phone directory. A reasonable way to display the entries in such a directory would be as follows:

Name	Phone Number	Type
Wilkins Micawber	555-4321	Home
Harold Skimpole	555-1234	Work
Thomas Gradgrind	555-6666	Home

The HTML for this describes a three-column table with four rows. Each row in the table is marked by a `<tr>` (table row) tag, each column by a `<td>` (table data) tag:

```
<table BORDER COLS=2>
    <tr>
       <td>NAME</td>
       <td>PHONE NUMBER</td>
       <td>TYPE</td>
    </tr>
    <tr>
       <td>Wilkins Micawber</td>
       <td>555-4321</td>
       <td>Home</td>
    </tr>
    <tr>
       <td>Harold Skimpole</td>
       <td>555-1234</td>
       <td>Work</td>
    </tr>
    <tr>
       <td>Thomas Gradgrind</td>
       <td>555-6666</td>
       <td>Home</td>
    </tr>
</table>
```

While this is perfectly serviceable if all you want to do is display the data, what if you wanted to take this information and add it to another address book, perhaps the one in your e-mail program or the one in your PDA? A program to do that would have to know that the data is formatted as a table. It would have to know that the first column holds the name and the second column holds the phone number. And it would have to know that the first row can be ignored since it is just column headings. You could write such a program. But what if you change the display from a table to a list? For example:

Wilkins Micawber 555-4321 Home

Harold Skimpole 555-1234 Work

Thomas Gradgrind 555-6666 Home

The HTML now looks like this (`<dl>` is a definition list, `<dt>` is a definition term, and `<dd>` is a "definition definition"):

```
<dl>
    <dt> Wilkins Micawber</dt>
    <dd> 555-4321Home</dd>
    <dt> Harold Skimpole</dt>
    <dd> 555-1234Work</dd>
    <dt> Thomas Gradgrind</dt>
    <dd> 555-6666 Home</dd>
</dl>
```

The program will need many changes to handle that. Or suppose you leave the display as a table but add a new column for address, between the name and phone number? That will also lead to many changes in the program.

Could you perhaps mark content by what it means, instead of what it should look like? Here's how that might work with the address book:

```xml
<addressbook>
   <owner>David Copperfield</owner>
   <entry>
      <name>Wilkins Micawber</name>
      <phone type="Home">555-4321</phone>
   </entry>
   <entry>
      <name>Harold Skimpole</name>
      <phone type="Work">555-1234</phone>
   </entry>
   <entry>
      <name>Thomas Gradgrind</name>
      <phone type="Home">555-6666</phone>
   </entry>
</addressbook>
```

Even without explaining the new tags that have been introduced, it's much easier to see what sort of information you're dealing with here. This is, in fact, an example of XML. So where do these new tags come from, how does a browser know they're being used correctly, and how would an XML address book like this be displayed? The next few sections tackle those questions.

4.3 Defining XML Tags

4.3.1 Document Type Definition (DTD)

Recall that SGML is a metalanguage, a language for creating other languages—markup languages, in this case. While XML, like HTML, is an application of SGML and thus a language in its own right, it is an application of SGML that is itself a metalanguage. XML is, in fact, a simplified version of SGML. SGML's advantage is that it's flexible. This makes it possible to create new markup languages tailored for specific problems, like the address book markup language in Sec. 4.2. SGML's disadvantage is that it's complicated. XML's designers knew that HTML wasn't powerful enough to support everything people wanted to do on the Web. They knew that SGML was up to the task, but they also knew it would never catch on because it was so hard to use. Their solution? A simplified version of SGML for the Web, now known as XML.

Both XML and SGML define new tags in a *Document Type Definition (DTD)*. The DTD determines whether a document is *valid*, i.e., whether it follows the tagging rules set forth in its associated DTD. An XML document can be *well-formed*, i.e., it may follow all the syntactic rules of XML, without being valid. An invalid document may have no associated DTD and thus have no tagging rules at all, or it may violate one or more of the rules in its DTD.

EXAMPLE: AN ADDRESS BOOK DTD To see how DTDs work, let's continue with the address book example. The first thing it needs is a tag for the address book. Then it needs tags for entries, names, and phone numbers. There also has to be a way to say that an address book may have zero or more entries and that each entry has a name and a phone number, in that order, and nothing else. Figure 4.3 shows a DTD that does all this.

Figure 4.3
A DTD for an address book.

```
<!ELEMENT addressbook (owner, entry?)>
<!ELEMENT owner (#PCDATA)>
<!ELEMENT entry (name, phone)>
<!ELEMENT name (#PCDATA)>
<!ELEMENT phone (#PCDATA)>
<!ATTLIST phone type (Work | Home) "Home">
```

In XML (and SGML), a start tag, its corresponding end tag, and everything in between are called an *element*. The text that appears between the start and end tags is the *value* associated with that tag (sometimes called its *content*). An ELEMENT definition in a DTD creates a new tag. The first line of the DTD in Fig. 4.3 creates the <addressbook> tag and says that between the start and end tags of an address book there must be one <owner> element and zero or more <entry> elements:

```
<!ELEMENT addressbook (owner, entry?)>
```

Under the rules of this DTD, an address book that has entries but no owner is invalid. Since nothing else in this DTD contains an address book element, XML calls <addressbook> the *root element* of the DTD.

XML: The Extensible Markup Language

The second line creates the `<owner>` tag and says that it may contain only character data (or as SGML calls it, "parsed character data," which is why `#PCDATA` has that P in it):

```
<!ELEMENT owner (#PCDATA)>
```

The third line creates the `<entry>` tag and says that an entry consists of a `<name>` and a `<phone>` element, in that order:

```
<!ELEMENT entry (name, phone)>
```

The fourth and fifth lines create the `<name>` and a `<phone>` tags and say that both may contain only character data:

```
<!ELEMENT name (#PCDATA)>
<!ELEMENT phone(#PCDATA)>
```

A start tag may have *attributes* of the form `name="value"`. Attributes provide extra information about the element in which they appear.[5] In the XML address book, the `type` attribute indicates whether a phone number is for home or work.[6] The last line of Fig. 4.3 defines the `type` attribute and associates it with the `<phone>` tag. This attribute may take one of two values: `Home` or `Work`. If omitted, `type` takes the default value `Home`:

```
<!ATTLIST phone type (Work | Home) "Home">
```

DOCUMENT TYPE DECLARATIONS Assuming you have a file, say *addresses.xml*, containing an address book marked with the tags just defined, how would an XML-enabled browser, on any other XML tool, know that it is an XML file, the contents of which are to be interpreted with some particular DTD? The first thing you need to do is start the file with an XML header line. This XML header tells the XML software that what follows is content marked with XML that uses features from XML version 1.0:

```
<?xml version="1.0"?>
```
[7]

---

[5] The quotes around an attribute value are mandatory in XML. HTML lets you leave them off.

[6] It can be hard to decide whether a given piece of data should be represented as an attribute or as the value of an element. Either way may often be suitable.

[7] This is an XML *processing instruction*. Processing instructions take the form

```
<?application-name instruction-list?>
```

and pass information to the indicated application.

After the XML header line, there is a `!DOCTYPE` *document type declaration* that specifies the DTD that will be used to interpret XML tags in the rest of the document. The general format of a document type declaration is:

```
<!DOCTYPE root-element dtd-definition>
```

where `root-element` is the root element of the DTD and `dtd-definition` is the DTD itself. There are severals ways to provide a `dtd-definition`. You can put the entire text of the DTD right in the `!DOCTYPE` declaration:

```
<!DOCTYPE addressbook [
<!ELEMENT addressbook (owner, entry?)>
<!ELEMENT owner (#PCDATA)>
<!ELEMENT entry (name, phone)>
<!ELEMENT name (#PCDATA)>
<!ELEMENT phone (#PCDATA)>
<!ATTLIST phone type (Work | Home) "Home">
]>
```

You can name a local file that contains the DTD:

```
<!DOCTYPE addressbook SYSTEM "addressbook.dtd">
```

Or you can use a URL to name a remote file that contains the DTD:

```
<!DOCTYPE addressbook SYSTEM
    "http://somesystem.somewhere.com/dtds/addressbook.dtd">
```

PUBLIC AND PRIVATE DTDS The `SYSTEM` keyword in a `!DOCTYPE` declaration means you are using a *private DTD*, i.e., one you created for your own use. DTDs can also be declared as `PUBLIC`. *Public DTDs* may be approved by a standards organization, or they may be agreed to by an ad hoc group with a shared interest in the subject matter of the DTD. A public DTD declaration has both a DTD name and an optional DTD URL. The DTD name is used by an XML-enabled program to find the DTD in some location known to that program. If the DTD can't be found there, the program uses the DTD URL to find it:

```
<!DOCTYPE root-element PUBLIC "DTD-name" "optional-DTD-URL">
```

DTD names have an odd format. If ISO has approved the DTD, its name starts with the string "ISO." If a standards group other than ISO approved it, the name starts with a plus sign (+). If it's not been approved by a standards group, the name starts with a hyphen (−). Following the approval marker is a double slash (//), the DTD owner's name, another double slash, the type of document described by the DTD, another double slash, and finally, an ISO 639 language identifier (for example, "EN" for English). HTML has a DTD of its own. Some HTML tools add HTML's DTD declaration to the files they generate:

```
<!DOCTYPE HTML PUBLIC "-//W3C//DTD HTML//EN">
```

The World Wide Web Consortium (W3C) is the owner of this DTD. The W3C is not a recognized standards group (at least not in the eyes of those who make the rules for this sort of thing), so the name starts with a hyphen. There is no DTD URL, since it's assumed any browser reading an HTML document already knows HTML and has no need to find its DTD.

Other public DTDs have been or are being defined to handle things like chemical data (Chemical Markup Language, or CML) and mathematical formulas (Math Markup Language, or MML). Another likely subject for public DTDs is the sort of personal information one might store on a PDA: address books, calendars, to-do lists, and memos. These would make it easier to offer standardized network services for keeping track of such information and synchronizing it with mobile devices like PDAs and cellphones.

Various organizations are defining public DTDs they can use to exchange commercial and other data among themselves.[8] For example, the news industry, under the auspices of the International Press Telecommunications Council (IPTC) has defined NewsML, an XML-based markup language for structuring multimedia news reports. NewsML will let software do things like see the difference between the headline and the body of a story or distinguish between an English and a French version of the same story. Similar efforts are under way to standardize XML frameworks for the exchange of invoices, customer data, catalogs, receipts, real estate listings, and other information. Of particular interest with regard to convergent networks are Call Processing Language (CPL) and VoiceXML, both of which are covered at more length in Chap. 11.

---

[8]Of course, there is the matter of who's in charge. Anybody who wants to can create a DTD and release it to the world. If two DTDs cover the same subject matter, you've got a problem. Which one should you use? In some cases, that's slowed the adoption of XML when there was no obvious choice. Recognizing this as a problem, more and more joint working groups are being set up so there will be a consensus in their areas of interest.

4.3.2 Schemas

Schemas are one of the newest features of XML. In one sense, it's easy to say what schemas are: They're a better way to do the same things as DTDs. I won't describe all the ways schemas improve on DTDs because some of them are, frankly, esoteric. There are, however, two obvious advantages of schemas over DTDs:

- DTDs are written in their own language. Schemas are written in XML. To work with DTDs, you need to learn two different languages—XML and the DTD language—and use two different parsers. To work with schemas, you need only know how schemas use XML. And since schemas are written in XML, they can be parsed by any XML parser.
- Even more importantly, schemas let you provide more information about the structure of data that goes into your documents. XML tools can use this information to run a lot more checks on a document before it gets used. For example, in a DTD that defines a ZIP code tag, all you can say is that the content surrounded by such a tag is a sequence of characters. With schemas, you can say that a valid ZIP code must contain exactly five characters, and that each character must be a numeric digit (ignoring nine-digit ZIP codes, to keep things simple).

There was no practical way the DTD language could be modified to remedy its shortcomings. This was the incentive for developing XML schemas. As time goes on, schemas will supplant DTDs as the preferred means for defining XML documents. Several public DTDs have already been or are being converted to schemas (for example, DocBook, an XML markup language for books and articles).

Schemas have a much richer set of basic data types than are provided by DTDs. The author of a DTD is pretty much limited to saying that the value of a tag is a string of characters (`#PCDATA`). Schemas let the designer say that a tag value must be a string, a number—even a specific type of number, such as an integer or a decimal number, a date, a time, a URI, an interval of time, and so on. Besides these predefined types, designers can create new types of their own.

As previously noted, XML schemas are written in XML. That means XML schemas have their own set of tags and attributes. In fact, there is a DTD that defines these tags. Because schemas do the same thing as DTDs, there is also an XML schema for XML schemas. There are about

30 tags and attributes that can be used in schemas, but I'll focus on just a few of these by working through portions of an example taken, with some modifications, from the Primer section of the XML Schema specification (which may be found at `http://www.w3c.org`).

Figure 4.4 shows a purchase order tagged according to the XML markup language, defined by the schema in Fig. 4.5. According to that schema, content marked by the `<purchaseOrder>` tag is of type `PurchaseOrderType`:

```
<xsd:element name="purchaseOrder" type="PurchaseOrderType">
```

Figure 4.4
Purchase order: po.xml

```
<?xml version="1.0"?>
<purchaseOrder orderDate="1999-10-20">
   <shipTo>
      <name>Alice Smith</name>
      <street>123 Maple Street</street>
      <city>Mill Valley</city>
      <state>CA</state>
      <zip>90952</zip>
   <shipTo>
   <items>
      <item partNum="872-AA">
         <productName>Lawnmower</productName>
         <quantity>1</quantity>
         <price>148.95</price>
      </item>
      <item partNum="926-AA">
         <productName>Baby Monitor</productName>
         <quantity>1</quantity>
         <price>39.98</price>
         <shipDate>1999-05-21</shipDate>
      </item>
   </items>
</purchaseOrder>
```

In schemas, `<xsd:element>` is one of the special schema tags. By convention, all these tags begin with the string `xsd:` (presumably, this stands for "XML schema document"). The first line of the schema establishes the use of this `xsd:` prefix convention:

```
<xsd:schema xmlns:xsd="http://www.w3.org/2000/08/XMLSchema">
```

This line sets up an XML *namespace*. It says that any tag with the prefix `xsd:` should be interpreted as following the rules of XML schemas as documented by the W3C in August of the year 2000. The URL `http://www.w3.org/2000/08/XMLSchema` in this line doesn't have to lead to a document or a Web page at all. It's just a string that's known to

```
<xsd:schema xmlns:xsd="http://www.w3.org/2000/08/XMLSchema">
<xsd:element name="purchaseOrder" type="PurchaseOrderType"/>
<xsd:complexType name="PurchaseOrderType">
  <xsd:sequence>
    <xsd:element name="shipTo" type="USAddress"/>
    <xsd:element name="items" type="Items"/>
  </xsd:sequence>
  <xsd:attribute name="orderDate" type="xsd:date"/>
</xsd:complexType>
<xsd:complexType name="USAddress">
  <xsd:sequence>
    <xsd:element name="name" type="xsd:string"/>
    <xsd:element name="street" type="xsd:string"/>
    <xsd:element name="city" type="xsd:string"/>
    <xsd:element name="state" type="xsd:string"/>
    <xsd:element name="zip" type="xsd:decimal"/>
  </xsd:sequence>
</xsd:complexType>
<xsd:complexType name="Items">
  <xsd:sequence>
    <xsd:element name="item" minOccurs="1" maxOccurs="unbounded">
      <xsd:complexType>
        <xsd:sequence>
          <xsd:element name="productName" type="xsd:string"/>
          <xsd:element name="quantity">
            <xsd:simpleType>
              <xsd:restriction base="xsd:positiveInteger">
                <xsd:minInclusive value="1"/>
                <xsd:maxInclusive value="100"/>
              </xsd:restriction>
            </xsd:simpleType>
          </xsd:element>
          <xsd:element name="price" type="xsd:decimal"/>
          <xsd:element name="shipDate" type="xsd:date" minOccurs="0"/>
        </xsd:sequence>
        <xsd:attribute name="partNum" type="SKU"/>
      </xsd:complexType>
    </xsd:element>
  </xsd:sequence>
</xsd:complexType>
<xsd:simpleType name="SKU">
  <xsd:restriction base="xsd:string">
    <xsd:pattern value="\d{3}-[A-Z]{2}"/>
  </xsd:restriction>
</xsd:simpleType>
</xsd:schema>
```

Figure 4.5
Purchase order schema: po.xsd.

be globally unique. Since the W3C controls all its URLs, it knew it could use this URL as the required globally unique string, without having to worry that some other organization might try to use it to set up a competing namespace.

XML tools use the namespace definition to interpret tags that begin with `xsd:` and tell them apart from other tags. With namespaces, two tags whose names would otherwise clash with each other can be told apart, for example, `xsd:element` and `chemistry:element` (assuming you've also set up a `chemistry:` namespace).

The prefix didn't have to be `xsd:`. The following line at the beginning of the schema would let you use `blurfle:element` instead of `xsd:element` and it would still mean the same thing (`xsd:schema` keeps the `xsd:` prefix, because `blurfle:` isn't set up until this statement has been processed):

```
<xsd:schema xmlns:blurfle="http://www.w3.org/2000/08/XMLSchema">
```

The important thing is not the prefix string; it's the string with which it's equated.[9] XML tools that work with schemas should know to look for a namespace declaration like this to find out what version of XML schemas is being used. It's the W3C's job to let XML tool vendors know about all valid XML schema namespaces (so far there's only `http://www.w3.org/2000/08/XMLSchema`). It's the XML tool vendors' job to keep track of and interpret those namespaces correctly.

So, having put this excursion into XML namespaces behind us, it's back to `<xsd:element>`. In an XML schema, this tag does the same things that an `<!ELEMENT...` definition does in a DTD. It defines a tag, gives it a name (`purchaseOrder`), and says what the type of its content will be (`PurchaseOrderType`). XML has no built-in type for purchase orders, so you have to create a definition from scratch. `PurchaseOrderType` is what's known as a *complex type*. Complex type definitions are tagged by `<xsd:complexType>`:

```
<xsd:complexType name="PurchaseOrderType">
...
</xsd:complexType>
```

Elements that appear between the start and end tags of an instance of a complex type are tagged with `xsd:element`. The entire collection

---

[9]This string, by the way, doesn't even have to be a URL. It just has to be globally unique. But URLs are an easy way to create globally unique names (assuming you've registered an Internet domain name like `w3c.org` for yourself out of which to construct them).

of elements in the complex type is marked by the `xsd:sequence` tag. `PurchaseOrderType` has two elements defined for it: `<shipTo>` and `<items>`. The content of `<shipTo>` is of type `USAddress`. The content of `<items>` is of type `Items`. Both of these are themselves complex types. Note also the shorthand notation `/>`. When a tag has no content, rather than using both the start and end tags `<tag></tag>`, one may use `<tag/>` by itself instead:

```
<xsd:sequence>
   <xsd:element name="shipTo" type="USAddress"/>
   <xsd:element name="items" type="Items"/>
</xsd:sequence>
```

Following the definition of the elements of `PurchaseOrderType` is an `xsd:attribute` definition that sets up an attribute named `orderDate` of type `xsd:date` for the `<purchaseOrder>` tag:

```
<xsd:attribute name="orderDate" type="xsd:date"/>
```

The `xsd:date` is a *built-in* or *simple type*. The XML schema specification defines more than 40 simple types. Among these are `byte`, `string`, `integer`, `positiveInteger`, `negativeInteger`, `float`, `boolean`, `time`, `timeDuration`, `date`, `month`, `year`, and `recurringDate` (the `xsd:` prefixes have been omitted). The interested reader can find complete details on these types in the W3C's specification for XML schemas.

Moving on, there are definitions of the complex types `USAddress` and `Items`. The structure of these is the same as that of `PurchaseOrderType`, a sequence of element definitions followed by attribute definitions. All the elements of `USAddress` have simple types:

```
<xsd:complexType name="USAddress">
  <xsd:sequence>
    <xsd:element name="name" type="xsd:string"/>
    <xsd:element name="street" type="xsd:string"/>
    <xsd:element name="city" type="xsd:string"/>
    <xsd:element name="state" type="xsd:string"/>
    <xsd:element name="zip" type="xsd:decimal"/>
  </xsd:sequence>
</xsd:complexType>
```

`Items` is a bit more interesting. Like the other complex types, it is a sequence of elements. The declaration of the first of these, `item`, includes two attributes not seen until now: `minOccurs` and `maxOccurs`. `minOccurs` says that an instance of `Items` must have at least one

XML: The Extensible Markup Language

`<item>`} element. `maxOccurs` says there is no limit on the number of `<item>` elements that may appear in an instance of `Items`. If these attributes don't appear in an element declaration, they both take a default value of 1:

```
<xsd:complexType name="Items">
  <xsd:sequence>
    <xsd:element name="item" minOccurs="1" maxOccurs="unbounded">
      <xsd:complexType>
        <xsd:sequence>
```

Note that `item`'s declaration doesn't have a `type` attribute. Its type is defined right after its declaration as an `xsd:complexType`. Like other complex types, this one is a sequence of element declarations. `productName` is a string:

```
<xsd:element name="productName" type="xsd:string"/>
```

`quantity`'s declaration introduces another feature of XML schemas, the ability to create new types from the simple built-in types. Once defined, these new types are treated just like simple types. In the example below, the newly defined type is based on the simple type `xsd:positiveInteger`. There are several ways to create a new type from a simple type, among them restricting the values of the new type to some subset of the original type. Here, the schema restricts a `quantity` to be a `positiveInteger` that's between 1 and 100, inclusive:

```
<xsd:element name="quantity">
  <xsd:simpleType>
    <xsd:restriction base="xsd:positiveInteger">
      <xsd:minInclusive value="1"/>
      <xsd:maxInclusive value="100"/>
    </xsd:restriction>
  </xsd:simpleType>
</xsd:element>
```

The `price` element has no new surprises. It's a decimal number:

```
<xsd:element name="price" type="xsd:decimal"/>
```

`shipDate` is a date that may or may not be present:

```
<xsd:element name="shipDate" type="xsd:date" minOccurs="0"/>
      </xsd:sequence>
```

Finally, `item` elements have a `partNum` attribute whose type is `SKU`. Everything else in the schema just ends tags that were started previously:

```
            <xsd:attribute name="partNum" type="SKU"/>
        </xsd:complexType>
    </xsd:element>
  </xsd:sequence>
</xsd:complexType>
```

The `SKU` type shows another way that simple types may be created. Here the schema restricts instances of `SKU` to `string`s that match the pattern in the `<xsd:pattern>` tag:

```
<xsd:simpleType name="SKU">
  <xsd:restriction base="xsd:string">
    <xsd:pattern value="\d{3}-[A-Z]{2}"/>
  </xsd:restriction>
</xsd:simpleType>
```

This example should give you some sense of what can be done with schemas. There are many other ways to create and use types in XML schemas. The interested reader should refer to the Web sites at http://www.w3c.org and http://www.xml.org for more information.

4.4 Displaying XML: CSS and XSL[10]

A DTD defines tags and specifies where they may or must appear in an XML document. But it says nothing about how to display marked up XML documents. Browser makers are not likely to add address book tags to their software; imagine trying to persuade them to add more esoteric tags like `<atom>` or `<molecule>` (to give just a few examples from the Chemical Markup Language). Because XML is not a markup language, but rather a way to define new markup languages, there are an infinite number of possible XML tags. So how are XML tags formatted?

The solution is similar to the way those tags are defined in the first place. Display definitions go in a separate file and you tell the browser (or other display program) where to find them. The W3C has created two different languages for writing these display instructions: *Cascading Style Sheets (CSS)* and the *Extensible Stylesheet Language (XSL)*. CSS can be used with both HTML and XML. It is simpler but less powerful than XSL. Most modern browsers support CSS in some fashion. XSL can

---

[10]This section gives some more detail on how XML works, but it is not essential to the material in the rest of the book and may be skipped.

be used only with XML and SGML. Because XSL is so new, some browsers give it only partial support, and others give it no support at all.

4.4.1 Cascading Stylesheets (CSS)

Cascading Stylesheets are fairly simple. For every tag you create, you also specify the formatting to be applied to its content. CSS style definitions can be embedded right in an HTML or XML file, or they can be put in a separate file that's referred to by the HTML or XML file that uses them. Here's an example of a CSS style definition for address book names that displays them in 14-point, bold, red font:

```
<STYLE TYPE="text/css">
<!--
name { color: red; font-size: 14pt; font-weight: bold }
-->
</STYLE>
```

Tags inside other tags inherit the styles already being used by the outer tags (hence the "cascading" part), unless the inner tags redefine something on their own. For example, suppose you modify the address book DTD so that a name is now made of two separate elements, a last name and a first name. The following CSS definition would cause first names to be 14 point, plain, and red, while last names would be 16 point, bold, and red:

```
<STYLE TYPE="text/css">
<!--
name { color: red; font-size: 14pt; font-weight: bold }
firstname { font-weight: normal }
lastname { font-size: 16pt }
-->
</STYLE>
```

In this stylesheet, first names inherit color and font size from name's style, but override its font weight. Last names inherit color and font weight, but override font size.

4.4.2 Extensible Stylesheet Language (XSL)

One drawback of CSS is that information must be presented exactly in the order in which it appears in an XML document. Furthermore, no data can be omitted from the display. But it's often useful to present the same data in different ways or to show only some of the data in a file.

For example, an XML address book could include a lot more than just names and phone numbers. It could hold just about any kind of information about a person. But you don't necessarily want to show all that information in the same order every time you display it. Sometimes you'd like to see home phone numbers first. Other times you might want to see e-mail addresses first. XSL lets you generate many different views of the same XML file.

XSL is related to *DSSSL (Document Style Semantics and Specification Language)*, the formatting and display language for SGML. DSSSL is based on Scheme, which is based on the LISP programming language. All this makes DSSSL both powerful and complex. XSL is a lot simpler than DSSSL, but it still offers much of DSSSL's power.

XSL can format XML, just like CSS. It can also rearrange the elements of an XML document before they are displayed. XSL can translate an XML document to another format like HTML or PostScript. If you have the appropriate XSL translator, you can view an XML document in just about any format. XSL can even translate from one form of XML to another.

Figure 4.6 shows some XSL code that converts an XML address to HTML. The first thing to point out is that the first line of this code marks this as an XML file, i.e., XSL is itself an application of XML:

```
<?xml version="1.0"?>
```

Figure 4.6
XSL code to transform XML address book to HTML.

```
<?xml version="1.0"?>
<xsl:stylesheet xmlns:xsl="http://www.w3.org/TR/WD-xsl">
<xsl:template match="/addressbook">
    <HTML>
    <HEAD>
    <TITLE>Address Book for
    <xsl:value-of select="owner"/>
    </TITLE>
    <BODY>
    <TABLE BORDER COLS=2>
    <TR>
        <TD>NAME</TD>
        <TD>PHONE NUMBER</TD>
    </TR>
    <xsl:for-each select="entry"/>
        <TR>
            <TD><xsl:value-of select="name"></TD>
            <TD><xsl:value-of select="phone"></TD>
    </xsl:for-each>
    </TABLE>
    </BODY>
    </HTML>
</xsl:template>
</xsl:stylesheet>
```

The stylesheet is marked with the `<xsl:stylesheet>` tag:

```
<xsl:stylesheet xmlns:xsl="http://www.w3.org/TR/WD-xsl">
```

The contents of the stylesheet are a number of templates, each marked by `<xsl:template>` tags. Each of these tags has a `match` attribute. The XSL processor scans an XML file looking for tags that correspond to `match` attributes. When it sees a matching tag, it outputs whatever follows in the template.

The first template in Fig. 4.6 handles matches of the `<addressbook>` tag:

```
<xsl:template match="/addressbook">
```

The next few lines are just regular HTML to be output once the `<addressbook>` tag is found:

```
<HTML>
<HEAD>
<title>Address Book for
```

The next line is an XSL element that instructs the XSL processor to look in the `<addressbook>` element for an element tagged as `<owner>` and output its value:

```
<xsl:value-of select="owner"/>
```

Then there's some more HTML to write out:

```
</TITLE>
<BODY>
<TABLE BORDER COLS=2>
<TR>
    <TD>NAME</TD>
    <TD>PHONE NUMBER</TD>
</TR>
```

Now things get interesting. An `<addressbook>` element may contain any number of `<entry>` elements. The formatting program doesn't know how many there will be, but it should output each one in the same way. The `<xsl:for-each>` tag scans through all the elements in the current element, the tag for which is indicated by the `select` attribute and outputs the `<xsl:for-each>` contents for each such element:

```
<xsl:for-each select="entry"/>
```

Each entry is a row of an HTML table that contains a name and a phone number. Here you see the `<xsl:value-of>` tag used to extract these from the XML:

```
<TR>
  <TD><xsl:value-of select="name"></TD>
  <TD><xsl:value-of select="phone"></TD>
</xsl:for-each>
```

Finally, the formatting program outputs some HTML to finish things off and close any XSL tags that are still open:

```
    </TABLE>
    </BODY>
    </HTML>
</xsl:template>
</xsl:stylesheet>
```

4.5 Parsing XML: DOM and SAX[11]

A *well-formed* XML document is one that adheres to the rules of XML but not necessarily to those of a DTD or schema. A *valid* XML document is one that:

1. Is well formed
2. Has an associated DTD or schema
3. Adheres to the rules of that DTD or schema

A document may be well formed but not valid. However, if it is valid, it must be well formed.

Because XML is much stricter than HTML with regard to SGML's syntactic rules, it's possible to write an *XML parser*, a program that can analyze any XML document to determine its validity and its structure from its DTD. Another program can use such a parser to help it process an XML document: formatting it for display, extracting interesting bits of information to send to another program, and so on. In fact, XSL processors use the output of an XML parser to do their work.

There are two kinds of XML parsers. The first reads an XML document and fires off an event whenever it encounters something interest-

---

[11]This section gives some more detail on how XML works, but it is not essential to the material in the rest of the book and may be skipped.

ing, like a start tag, an end tag, or an attribute.[12] An XML processor that uses an event-generating parser listens for these events and responds to the ones it finds interesting. There's no official way to do event-driven XML parsing, but the *Simple API for XML (SAX)*, a joint effort by a group of XML programmers, is the de facto standard.

The other way to parse XML is called the *Document Object Model (DOM)*. Unlike SAX, DOM is a W3C standard. Rather than generating events and sending them to another program, DOM builds an "object model" of an XML document, a data structure that contains the entire document in a form that makes it possible for another program to extract information from it, using a standard DOM API. Visually, the DOM data structure looks like a tree. Figure 4.7 shows the XML address book example displayed as a DOM tree.

---

[12]Section 3.6 gives background information on events and their handling.

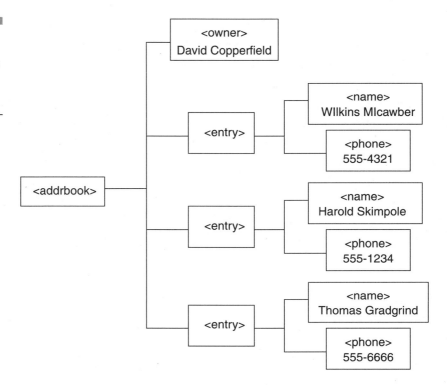

Figure 4.7
A DOM tree corresponding to the XML address book (attributes not shown).

There are advantages and disadvantages to both SAX and DOM. SAX is best when you're working with large documents. SAX passes the pieces of a document to your program one event at a time. DOM gives you the entire document as a tree. DOM's advantage is that it maps well to object-oriented languages, since its output is an object made of other objects. DOM is also better at handling complex transformations where data in one part of an XML document affects the data processing elsewhere. With SAX, programmers have to keep track of this themselves and understand how the order of events might affect the program. With DOM, programmers get everything at once. If they need to look at a bit of data elsewhere in the document before deciding how to proceed, they just go and get it from the document's DOM tree.

4.6 For More Information

This chapter does not even scratch the surface of what XML can do. The most up-to-date information can be found at the Web sites of the W3C (http://www.w3c.org) and the XML Web site of the Organization for the Advancement of Structured Information Standards (OASIS): http://www.xml.org. Another good source of information is http://www.xml.com, which has numerous tutorials and white papers.

CHAPTER 5

Distributed Computing

I propose getting rid of conventional armaments and
replacing them with reasonably priced hydrogen bombs
that would be distributed equally throughout the world.
—Idi Amin, former Ugandan dictator

5.1 Introduction

Distributed computing is the science (or perhaps I should say the art) of building applications with code and data spread across many machines, all connected to each other on a network. Both the telephone network and the Internet can be viewed as distributed systems. There are many reasons you might want to distribute a system across a network, including:

- Running several pieces of a program on several machines at the same time may get the job done faster than if you run them on a single machine. It may also be less costly to run them on several slow and cheap machines than on a single fast and expensive machine.
- If many people spread across a wide area need to use a lot of data that's concentrated in one place, it's often easier to take programs to the data than vice versa. This is also true when many different bits of data at a distant location are used to compute a single value.
- Redundancy—many machines all doing the same thing—protects against component failures. If one machine goes down, others may be able to take over for it.
- As systems change, data and services may be moved from one machine to another. It's generally easier to have one server keep track of where everything is than it is to have every client keep track of it separately. Client programs can go through this server to get at distributed resources. That reduces the number of things that must be changed when pieces of a system move about.
- The applications you're trying to implement (for example, call processing or e-mail) may be inherently distributed.

At its most basic level, distributed computing is nothing more than passing data between processes that run on different computers, i.e., defining and implementing a protocol. While getting data from one machine to another is all you need to do to write a distributed application, it's not easy making sure the right data is sent or that it's interpreted correctly by the receiver. Rather than make all programmers handle this themselves, the solutions to these problems are provided in *middleware*. Middleware is off-the-shelf software that connects two otherwise separate applications. It's often called *plumbing*, because it connects the endpoints of an application and passes data between them. Common types of middleware include the following:

Sockets These provide a programming interface to the two endpoints of a network connection that simplifies the underlying details of that connection.

Database access systems These sit between programs that want to interact with a database and the files that form the database. They translate requests from client programs into operations on the database and send back results in a format that client programs understand.

Transaction processing monitors These oversee *transactions*. A transaction is a piece of work that either fails or succeeds. A transaction can have many steps. For the transaction to succeed, each step must succeed. If any step fails, the entire transaction fails. When a transaction fails, the system needs to go back to the state that it was in before the transaction started. This is called *rollback,* and any changes the transaction made up to the point of failure are said to be "rolled back." Transaction processing monitors make sure transactions finish without errors. If an error occurs, the monitor takes care of rolling back changes. A familiar example of a transaction is making a withdrawal from a bank account. If such a transaction were to fail after your balance had been reduced but before you actually got your money, you'd certainly want the change to your balance rolled back. In this case, a transaction processing monitor would see that your balance is set back to its original (unreduced) value.

Remote procedure calls (RPC) A tool that programs on one computer can use to invoke services on another computer. With RPC, requests for remote services look like normal, nonremote procedure calls to the author of the requesting program.

Object managers Software that supports *distributed objects*. With an object manager, clients can request the services of a software object on another machine without knowing anything about where that object resides. An object manager gets these requests as *remote method calls,* the object-oriented equivalent of an RPC, forwards them to the appropriate servers, and returns the results.[1]

We'll now look at some of these types of middleware in more detail.

---

[1] Some refer to these as *Object Request Brokers (ORBs)*. Strictly speaking, an ORB is CORBA's version of an object manager. To avoid confusion, we speak of "object managers" when we mean it in the general sense, and "ORBs" only when referring specifically to CORBA's object manager.

5.2 Sockets

It's open to debate whether sockets should be classified as middleware at all. Nevertheless, it's worth discussing them, simply because they are such an important foundation for distributed computing and are so universally available. Any platform that supports TCP/IP most likely supports sockets also as its basic network programming interface.

A socket is an endpoint in a communication path between two machines. When a programmer wants to set up a connection to a remote machine, he opens a socket on his local machine. The remote machine must open a corresponding socket to accept the connection. Once these two sockets are opened, messages between the local and remote machines may be exchanged through them.

Every socket has a two-part address. The first part is the *Internet address* of the machine on which the socket resides, for example, 64.208.32.100, or the equivalent name, www.google.com.[2] The second part of the socket address is a port number. Every port has an associated application that handles messages coming in through it. This may or may not be the same application that opened the socket for that port. Port numbers range from 0 to 65,535 (though many are already assigned to standard applications, like FTP and Web servers).

Sockets come in three flavors:

Stream sockets These provide reliable data transport using the TCP part of TCP/IP. Everything that's sent is guaranteed to arrive at its destination and to get there in the order in which it was sent.

Datagram sockets These provide a means to exchange data packets (called *datagrams*), with no guarantee they ever arrive or that they arrive in the correct order. The means for doing this is the *User Datagram Protocol (UDP)*, which, like TCP, sits on top of IP. Why on earth would anyone use such an unreliable service? Speed. Because there isn't much checking back and forth, datagrams that do arrive get there in a hurry, and the sending and receiving sides don't have to do much work loading and unloading them. This is important for applications like VoIP, where it's better to get most of your data quickly than all of it slowly. These applications just ignore data that never

---

[2] Or more accurately, for those who care about such things, the first part is the Internet address of the network interface through which the socket is accessed on the machine on which it resides.

arrives and throw away data that arrives out of order. Datagrams are also used in applications that find out what's attached to a network. There can be many machines on a network, and setting up stream sockets is a lot of work, especially if all you want to say is something on the order of "Who are you?" and "I'm so and so."

Raw sockets These get you to the lowest-level protocols, for example, plain IP without TCP or UDP. It's used to develop and test new protocols.

To put all this on a more solid footing, here's a fragment of a Java program for a stream server socket that listens on port 2000 for a client to connect to it. When a connection is made, the server reads the message it got and sends a reply:

```
ServerSocket server = new ServerSocket (2000);
while (true) {
   // Wait for connection from client.
   Socket client = server.accept();
   //Set up input and output streams on connection.
   DataInputStream instream =
      new DataInputStream(client.getInputStream());
   DataOutputStream outstream =
      new DataOutputStream (client.getOutputStream());
   // Get message on instream, send reply on outstream.
   String msg = instream.readUTF();
   outstream.writeUTF("Goodbye");
   instream.flush();
}
```

The corresponding client opens a socket to the server, sends it a message, and gets back a reply:

```
// Open socket to server.
Socket server = new Socket ("www.simpleserver.com", 2000);
//Setup input and output streams on socket.
DataInputStream instream =
   newDataInputStream (server.getInputStream());
DataOutputStream outstream =
   newDataOutputStream (server.getOutputStream());
// Send message on outstream, get reply on instream.
outstream.writeUTF ("Hello");
outstream.flush();
String msg = instream.readUTF ();
```

As you can see, using sockets to exchange messages between machines isn't difficult. However, sockets don't make it any easier to process those messages. They just make it easier to shuttle them from one place to another. The rest of this chapter discusses middleware technologies that provide services that go beyond what sockets do.

5.3 Remote Procedure Calls

A remote procedure call (RPC) is a way for a program on one computer to request services on another computer. Though implemented as an exchange of messages between two machines, RPC systems make this look like normal nonremote procedure calls to the author of a program requesting a remote service. To see how remote and nonremote procedure calls differ, consider two different ways of accomplishing a workplace task.

Let's say you have to put together a report that incorporates the results of some sophisticated statistical analysis. You could write the entire report on your own, including the statistical calculations. The stage at which you do the calculations would be the rough equivalent of a nonremote procedure call. You break off from the main task of report writing for a while, make the calculations ("invoke a procedure"), then return to the main task to incorporate the products of that work ("return results").

But what if you don't have the expertise to make the calculations yourself? Presumably there are people in your organization who can do it. So you provide them with the necessary input data and ask them to do it for you ("invoke a remote procedure"). Once they finish ("return results"), you use what they provide in your report.

5.3.1 How Nonremote Procedure Calls Work

Before looking at how remote procedure calls work, let's first see how nonremote procedure calls work. While it's running, a program's memory is split into two parts: code and data. The code part holds the program's instructions. The data part holds all the program's variables and constants. Every instruction and data element has its own unique address in memory. As it runs, the program uses these addresses to find data and keep track of what instruction to execute next.

A compiler for a high-level language, like Java or C++, translates the data names and statements of a program into something that can be interpreted as addresses when it runs. For example, here's a simple fragment of a C program:

```
int i;
i = 1;
goto label1;
i = 999;
label1: i = i + 2;
```

It could be translated by a compiler into low-level machine language as follows (the numbers at the beginning of each line represent the address at which that instruction or data item will reside in memory):[3]

```
00 : STORE40, 1
01 : JUMP 03
02 : STORE 40, 999
03 : ADD 40, 2
04 : HALT
...
40 : <storage for i>
```

As it runs, a program usually executes instructions in order—i.e., it starts at 00, then goes on to 01, 02, and so on—unless it encounters an instruction like JUMP that takes it somewhere else.

The example program would execute the instructions at addresses 00, 01, 03, and 04, in that order, and then stop running. The current value of the integer i is stored at address 40. The instruction at address 00 puts the value 1 into address 40. The instruction at address 03 adds 2 to whatever is already stored at address 40. The instruction at address 02 would store 999 at address 40, except that the program never gets to that instruction, having jumped around it in instruction 01.

This is all pretty straightforward. Things get more tricky when you add procedure calls. With procedures, a program not only has to jump to the address at the beginning of the called procedure, it also has to get back to the point where it first made the call and continue from there when the procedure is done. If the call includes any values to be passed to the procedure, those must be transferred somehow. And if the procedure returns values, there must be a way to get them back to the calling program.

Here's another fragment of a C program. It sets a variable i to 1 and calls the procedure foo using i's current value as an argument. Then foo adds 1 to whatever value has been passed to it and returns that as a result. Finally, the program sets the variable j to the returned value:

```
int i, j;
i = 1;
j = foo (i);
...
int foo (int n){
   n = n + 1;
   return n;
}
```

---

[3]The machine language used here is not that of an actual machine (at least, not any one I know of), but it is similar to most actual machine languages.

Here's the imaginary machine's code for that:

```
00 : STORE 40, 1
01 : STORE 80, 04
02 : STORE 82, *40
03 : JUMP 20
04 : STORE 41, *81
05 : HALT
...
20 : ADD 82, 1
21 : STORE 81, *82
22 : JUMP, *80
...
40 : <storage for i>
41 : <storage for j>
...
80 : <storage for foo's return address>
81 : <storage for foo's return value>
82 : <storage for foo's n>
```

Step by step, here's what happens:

1. The instruction at address 00 sets i equal to 1, just as in the previous example.
2. The instruction at address 02 stores 04 at address 80. The 04 is the *return address* for foo, i.e., the address of the instruction immediately following the call to foo at address 03. Once foo is done, it jumps to this address, so the program can continue where it left off when it called foo. The compiler makes sure both foo and any code that calls foo know that return addresses are stored at address 80.
3. Since foo has a value passed to it as an argument, that must also be set up before the call. The instruction at address 02 does this, storing what's now in address 40 (that's what the *40 syntax is all about) at address 82.
4. The call to foo occurs at address 03, with a jump to address 20, the first instruction of foo.
5. Since foo's local variable n has already been set to the same value as i, all the instruction at address 20 has to do is add 1 to that value.
6. The instruction at address 21 puts the new value of n into address 81, the location for values returned from foo.
7. The instruction at address 22 jumps to whatever address is stored at address 80. Since that was set to 04 before the program called foo, it jumps to 04.
8. The instruction at address 04 puts the return value from foo (stored at address 82) into the storage that's been set aside for j (at address 41).

While there are variations on this pattern in practice, it should be enough to give some idea of how procedure calls work in a nondistributed program.

5.3.2 How Remote Procedure Calls Work

As the previous section demonstrated, procedure calls require lots of mucking about with addresses: the address of the procedure being called, the address to go back to when the procedure is done, the addresses of values passed to and from the procedure. That's all fine and good when the calling program and the called procedure are on the same machine, because they'll share the same set of addresses. But how do you do this across machines with entirely separate address spaces and, perhaps, even entirely different ways of representing addresses?

The solution is fairly straightforward, but it adds some extra plumbing to any system that uses RPCs. All machines in systems based on RPC use whatever communication software they have (for example, sockets) to send RPC requests and replies back and forth among themselves. The procedure arguments in these messages must be packaged in a format appropriate for transmission from one machine to another. Packaging arguments in this way is called *marshaling*. Marshaling is done by a a piece of RPC middleware that's called a *client stub*. Once marshaled arguments are received by the server machine that will execute the RPC, they must be *unmarshaled*. Unmarshaling is done by a server stub. Besides unmarshaling, the server stub makes the call to the server procedure that will handle the RPC. After the server executes the remote procedure, its stub marshals the results and sends them back to the client. There, the client stub unmarshals the results and returns to the client program, just as if a local procedure call had been performed.

The flow of an RPC from client to server is shown in Fig. 5.1. Note that, to the client program and the server program, an RPC looks like a normal nonremote procedure call,[4] even though the flow of execution goes from one machine to the other through several extra layers of software. RPC systems provide tools for generating client and server stubs—*stub compilers*—so programmers won't have to develop all the plumbing

---

[4]This is not entirely true. The programmer of the client program needs to include code that will send calls to remote procedures through the RPC layers rather than through the normal procedure call mechanism. However, at the point where the procedure call is made, the code for a procedure called remotely may well be indistinguishable from that for the same procedure, if it were made locally.

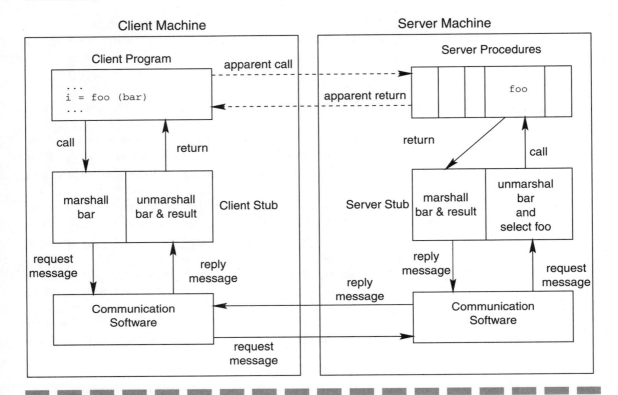

Figure 5.1
Framework for RPC.

code in Fig. 5.1 on their own. The input to a stub compiler is the *signature* of a procedure to be executed remotely, i.e., the procedure's name, its return value (if any), the types of its arguments, the order in which those arguments appear, and whether or not they can be changed by the procedure. The outputs of the stub compiler are the code for a procedure's client and server stubs.

Anything that RPCs do can also be done by implementing your own set of messages to pass between client and server. Getting everything in place for RPC can be a nontrivial job, what with all the stubs and communication modules and other glue that must be deployed. And the extra layers of RPC software add a performance penalty. So why does anyone bother with RPC at all?

In a system that has only a few operations and simple data types, it's probably better to build an application-specific message protocol by hand. However, the effort to handcraft such a protocol can quickly get out of hand if there are lots of operations or complicated data structures. In this situation, RPC acts like an automated protocol generator. The result of an RPC is, after all, just a set of messages that are exchanged between two machines, i.e., a message-based protocol. The difference is that once a programmer has gone to the trouble of getting all the RPC mechanisms in place, creating a large set of operations and adding new operations is a lot less work than handcrafting the equivalent message protocols. Also, the system software that supports RPC usually takes care of other network busywork on behalf of the programmer, such as opening and managing connections to other machines or receiving and sending messages.

RPC can also simplify the development and testing of a distributed system. During the first phases of a project, programmers can use non-remote procedure calls everywhere and do all their testing on a single machine. Once they've got that version working, they can convert the necessary portions to RPC. If things stop working after that, they know it's something in the RPC, not the basic logic, that's at fault.

5.4 Distributed Objects

A distributed object system lets software in one host or process invoke methods on objects that are in another host or process, just as though both were in the same host or process (this is called *location transparency*). Distributed objects use an RPC-like technology to call methods on remote objects. The problems of working with objects across languages and platforms are solved by using a special language to define the interfaces of distributed objects. All you need for cross-platform and cross-language portability then is the translation—a *mapping*—from the interface language to the representations used by other platforms and languages. CORBA's interface language is called *IDL (Interface Definition Language)*. IDL gives cross-platform and cross-language portability for any language that has an IDL mapping defined for it. Java RMI, Sun's Java-based, distributed-object framework, uses the Java language as its interface language. Java is platform independent but it is (obviously) not language independent, so Java RMI can be used only in all-Java systems. This has advantages and disadvantages, as we'll see in Sec. 5.6, which covers RMI in more detail.

RPC mechanisms and interface languages, by themselves, are not enough to build a distributed object system. Other things must also be handled:

- How does an application find the objects it needs when they can be located anywhere and could even move from machine to machine?
- How does a server know it's safe to garbage collect an object it thinks may no longer be in use when the clients are on other machines?

Remote objects and their clients are tracked by an *object manager*. Every remote method call request goes through the object manager, which keeps track of all remote objects in a distributed object system. The object manager makes sure requests go to the right servers and their results get back to the right clients. Other than using an object manager, the mechanisms of a remote method call are similar to those of RPC, including the use of stub compilers that handle marshaling and unmarshaling on the client and server side. All this is shown in Fig. 5.2.[5]

---

[5] Note that RPC client stubs are called *object stubs*, and server stubs are called *object skeletons*. To confuse things even more, the client stub is sometimes called a *proxy*. There doesn't seem to be a good reason for the change of terminology. The RPC terms are probably better, because people always seem to forget whether it's stubs on the client side and skeletons on the server side or the other way around.

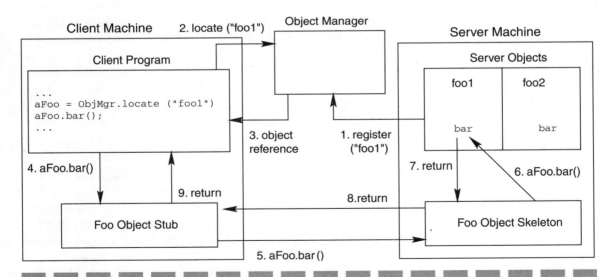

Figure 5.2
Framework for distributed objects.

5.5 CORBA

CORBA, the *Common Object Request Broker Architecture,* is the *Object Management Group's (OMG)* distributed object standard. The OMG, formed in 1989 to standardize distributed object technology in a multiplatform, multilanguage, and multiprotocol environment, has about 700 member companies and organizations defining and extending CORBA.

CORBA's object manager is called the *Object Request Broker (ORB).* ORBs provide object registration and location services. They may also provide other services, including the following:

- **Concurrency control:** Restricts object access to one client at a time
- **Licensing:** Measures use of an object for billing
- **Life-cycle management:** Provides services for creating, copying, moving, or deleting objects
- **Security:** Restricts object access to authorized clients
- **Persistence:** Gives clients the ability to store an object into a database for later access by the same or other clients

An ORB must be present on both the client and server sides of a CORBA remote method invocation. The client ORB takes client requests, marshals them, routes them to remote object servers, and passes the results back to the clients. The server ORB registers remote object implementations, unmarshals remote method call requests from client ORBs, invokes the requested methods on the server objects, marshals the results, and sends them back to the client ORB. (See Fig. 5.3.)

ORBs encode the messages they exchange, using either the *General Inter-ORB Protocol (GIOP),* CORBA's basic communication protocol, or the *Internet Inter-ORB Protocol (IIOP),* a mapping of GIOP to TCP/IP. GIOP defines seven message types. Three of these (Request, LocateRequest, CancelRequest) are sent by clients. Three (Reply, LocateReply, CloseConnection) are sent by servers. Error messages may be sent by clients and servers. GIOP also defines CORBA's *Common Data Representation (CDR),* a specification of the data types CORBA recognizes and their encoding when sent from one machine to another. GIOP can be used with any transport protocol that provides a reliable, connection-oriented byte stream. Examples of such protocols include TCP/IP, IPX, ATM, and SS7.[6]

---

[6]This is not to say that mappings of GIOP onto all these have been defined. Just that they could be.

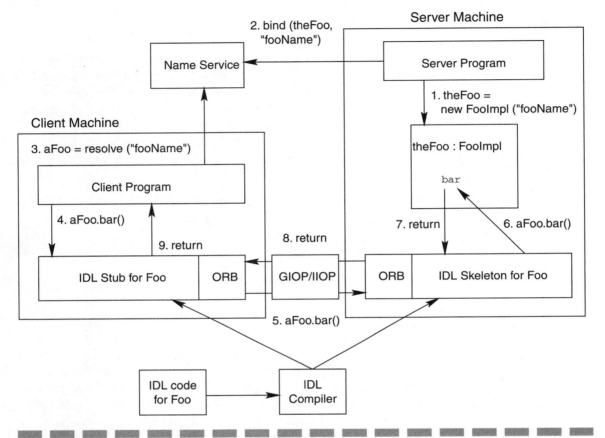

Figure 5.3
Distributed objects with CORBA.

GIOP defines the format of CORBA message headers but not that of CORBA message bodies. Thus, ORBs that exchange GIOP messages must agree among themselves on the format of message bodies. That makes multivendor and multisystem interoperability difficult in theory and unlikely in practice. CORBA release 2.0 introduced IIOP as a solution to this problem. Besides mapping GIOP to TCP/IP, IIOP specifies a detailed layout of CORBA messages, including their bodies.

To keep itself from being tied to a single language, CORBA uses an *Interface Definition Language (IDL)*. CORBA IDL (or just "IDL" when we know that it's CORBA's IDL under consideration) describes a client's view of a remote object in much the same way as languages like C++

and Java, or at least the parts of those languages that are used to define object interfaces.[7] Assuming there's a defined mapping between IDL and the languages used by a system's clients and servers, there's no problem mixing languages throughout a CORBA-enabled application.[8] IDL compilers use these mappings to generate stubs and skeletons in the desired implementation languages.

Another important part of CORBA is its *Naming Service*. The Naming Service matches remote objects with the unique names by which they are known to clients. When a server creates a remote object, it registers the object and its name with the Naming Service. When clients use a remote object, they supply its name to the Naming Service, which returns the information needed to find it. CORBA specifies only how the Naming Service behaves. The implementation details are left to vendors of CORBA software. CORBA Naming Services could be built on top of LDAP, ISO X.500, and Sun's NIS, to name just a few possibilities.

The optional CORBA *Query Service* and *Properties Service* support object searches that go beyond what the Naming Service offers. Remote objects register properties other than just their their name with the Property Service, for example, `time_last_modified`. Clients use the Query Service to find objects with specific properties, such as one with a particular name whose `time_last_modified` falls within a desired range.

To see how CORBA looks in practice, let's look at part of a CORBA implementation of a simple bank system, starting with the IDL for a bank object that has only one operation:

```
module Bank System {
   interface Bank {
      float getBalance (in string accountId);
   };
};
```

It's important to keep in mind that IDL does not define objects as such; it just defines *interfaces*—a programmer's view—for objects that will be implemented in languages like C++ or Java. Below is a Java implementation of the `Bank` object defined in the bank system IDL. In this code, `_BankImplBase` is the name of the server skeleton class the IDL compiler generated from `Bank`. `BankImpl` implements the methods

---

[7]Recall that, in C++ and Java, an object's interface is all the methods it's declared as `public`.

[8]Note that a language need not be object oriented in order to have an IDL mapping. For example, IDL has mappings for both C and Cobol.

declared in `Bank`'s IDL plus a constructor that remote object servers will use to create instances of `BankImpl`:

```java
import org.omg.CORBA.*;
import java.util.*;
import BankSystem.*;


public class BankImpl extends _BankImplBase {
    private HashSet balances;
    public float getBalance (String accountId) {
        return (Float)(balances.get(accountId)).floatValue();
    }
    public BankImpl (void) {
        balances = new HashSet();
    }
}
```

The next bit of code creates an instance of `BankImpl` and registers it with the server ORB so clients can find it. Before it can register the object, it has to find the ORB registration service ("`NameService`"). Once that's done, the program registers the new object with the name by which clients will know it ("`FirstBank`").[9]

```java
import org.omg.CORBA.*;
import org.omg.CosNaming.*;
import BankSystem.*;
public class BankServer {
    public static void main (String[] args) {
        // Initialize CORBA.
        ORB orb = ORB.init();
        BOA boa = orb.BOA\_init();
        // Create the Bank object.
        BankImpl theBank = new BankImpl ("FirstBank");
        boa.obj_is_ready (theBank);
        // Contact the naming service.
        org.omg.CORBA.Object object =
            orb.resolve_initial_references ("NameService");
        NamingContext context = NamingContextHelper.narrow (object);
        // Register the Bank object with the naming service.
        NameComponent[] name = newNameComponent[1];
        name[0] = new NameComponent ("FirstBank", "");
        context.bind (name, theBank);
        // Remote Bank object ready for use.
        boa.impl_is_ready();
    }
}
```

---

[9]Note to Java-literate readers: Kids, don't try this code at home. It won't compile. Some of the methods it calls throw exceptions that must be caught for this code to work. I've omitted exception-handling code in this and all other Java examples in this book, because it just gets in the way of explaining the main ideas.

Distributed Computing

Finally, here's a CORBA client that uses the `BankImpl` object just created to implement the `Bank` interface. Note how similar the client's code for finding the ORB registration service and the desired object is to the server's code for registering the object:

```
import org.omg.CORBA.*;
import org.omg.CosNaming.*;
import BankSystem.*;
public class BankClient {
   public static void main (String[] args) {
      // Initialize CORBA.
      ORB orb = ORB.init();
      // Contact the naming service.
      NamingContext context =
         NamingContextHelper.narrow
            (orb.resolve_initial_references ("NameService"));
      // Find the remote object by giving its name to
      // the naming service.
      NameComponent[] name = new NameComponent[1];
      name[0] = new NameComponent ("FirstBank", "");
      Bank theBank = BankHelper.narrow (root.resolve (name));
      // Invoke a method on the remote object.
      System.out.println ("Balance is "
                  + theBank.getBalance ("someAccount"));
   }
}
```

For this client code to work, the client stub for `Bank` has to be installed on the client machine. With only a few objects, that's not too much of an obstacle. But what if there are many different CORBA-enabled objects residing throughout a network? Do you really have to install all possible client stubs on every possible client machine? CORBA's *Dynamic Invocation Interface (DII)* offers a way to get around this problem. The price to be paid is that the client code is more complicated.

With DII, a client program can query CORBA services for an object's function. A functional query returns a reference to a remote object that provides the desired function, just as a query by name returns a reference to a remote object with the desired name. However, the previous client code invoked the `getBalance` method on that remote object directly, because it assumed there was a precompiled stub for `Bank`. A program can't assume that for DII. In fact, it must assume there is no such precompiled stub available. A DII-enabled client asks the remote object for information about its interface. The object returns high-level interface information. From that, the client gets a complete description of the interface. And from that, it extracts information about the methods supported in that interface:

```
org.omg.CORBA.InterfaceDef bankIF = theBank._get_interface ();
org.omg.CORBA.FullInterfaceDescription fullBankIF = bankIF.describe_interface();
```

A cautious client finds out if the remote object supports the methods it wants by searching the interface description for the names of those methods and seeing what arguments go with those methods. Following is an example of the sort of code needed just to traverse through all that information, much less check it. To implement those tests, you would have to add code that compares what's found with what's wanted. But even leaving that out, this should be enough to give some idea how much DII demands of the programmer:

```
for (int i = 0; i < fullBankIF.operations.length;i++) {
   Identifier methodName = fullBankIF.operations[i].name;
   TypeCode resultType = fullBankIF.operations[i].result;
   int numArgs = fullBankIF.operations[i].parameters.length;
   for (int j = 0; j < numArgs; j++) {
      TypeCode argType = fullBankIF.operations[i].parameters[j].type;
      Identifier argName = fullBankIF.operations[i].parameters[j].name;
   }
}
```

Assuming the `Bank` interface checks out OK, the client invokes the first of its methods, which happens to be `getBalance`:

```
ORB orb = ORB.init();
NVList resultList = orb.create_list(0);
Any result = orb.create_any();
result.type = fullBankIF.operations[0].result;
resultList.add_value ("", result, 0);
NVList argList = orb.create_list(0);
int numArgsint numArgs = fullBankIF.operations[0].parameters.length;
Any arg = orb.create_any();
arg.type = fullBankIF.operations[0].parameters[0].type;
String accountID = "1234567890";
arg.insert_string (accountID);
argList.add_value (fullBankIF.operations[0].parameters[0].name,
                   arg,
                   fullBankIF.operations[0].parameters[0].
                      mode.value()+1);
Request request = theBank._create_request(null,
                                 fullBankIF.operations[0].name,
                                 argList,
                                 result.item(0));
request.invoke();
float balance = request.result().value().extract_float();
```

That's an awful lot of code for a simple method invocation. Comparing it with the non-DII client code, it's obvious DII is not for the faint of heart. Especially when you consider that this code is considerably stripped down, completely omitting error checking and other such tedious but necessary housekeeping code. DII is a powerful mechanism of considerable generality. But like many such tools, there's a high price to be paid by those who would use it.

Another issue that's been glossed over so far is how remote objects come into being. The example showed a server program that created an object, registered it, and waited for requests. That's fine in a system with only a few objects. But what about a system with thousands or even millions of objects spread across many machines? The startup time would be intolerably long if you had to wait until all those objects were created. CORBA's solution is to let you activate objects on demand, using either the *Basic Object Adapter (BOA)* or the newer *Portable Object Adapter (POA)*.[10]

The basic idea behind BOA and POA is that requests for objects go through an *object adapter* that figures out whether or not that object is available. An object may exist but be unavailable because someone else is using it. An object may be unavailable because it doesn't exist at all. An object may be unavailable because it exists but has been deactivated (perhaps written to a file in order to save memory). In all these cases, the object adapter grants access to the object, bringing it into existence, activating it, or just letting the client go right in as necessary. For this to work, the remote object server code must be written to cooperate with the BOA or POA mechanisms, working through them to create, destroy, activate, and deactivate objects.

CORBA is a massive system and I've covered only a bit of it. Readers who want to dig deeper should see the OMG's CORBA Web site (`http://www.corba.org`).

5.6 Java RMI

Java Remote Method Invocation (RMI) is a framework for distributing Java objects. It makes extensive use of Java *object serialization,* a means for translating a Java object into a stream of characters that can be saved to a file or sent to another machine. Because Java object serialization works only with Java objects, RMI cannot be used in systems that mix other languages in with Java. However, for systems that use only Java, RMI is much easier to program and deploy than other distributed object frameworks.

---

[10]The reason there are two of these is that the BOA specification left a lot of things undefined. To fill these gaps, vendors built their own extensions. That's one reason why CORBA products from different vendors don't work well with each other. To address the compatibility problems with BOA, the CORBA designers started over again and developed POA.

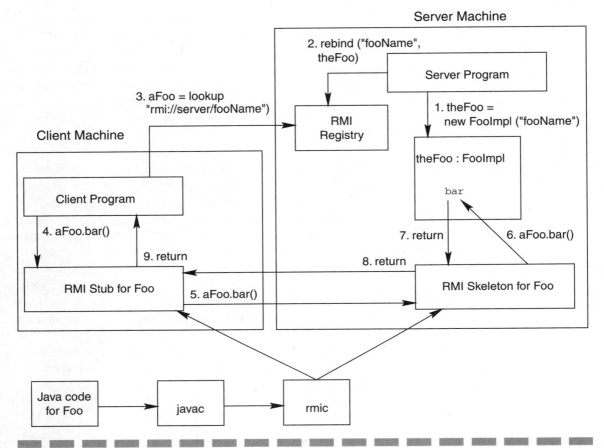

Figure 5.4
Distributed objects with Java RMI.

RMI's object manager is called the *RMI registry* (see Fig. 5.4). An RMI registry process must be running on every RMI object server. Object servers create remote objects and assign them unique names. The objects report those names to their server's RMI registry. When a client wants to use an object, it supplies the object's name to the RMI registry of the host it believes has that object. The RMI registry determines whether the server has an object with that name. If it does, the RMI registry returns a reference to the object.

For Java programmers, a big advantage of RMI is that they don't have to learn a special IDL to define remote objects. Java is RMI's IDL. The simple bank object defined previously in CORBA IDL would be defined for Java RMI as follows:

```
package BankSystem;
import java.rmi.*;
import java.util.*;
interface Bank extends java.rmi.Remote {
    float getBalance (String accountId) throws RemoteException;
}
```

Note that every remote method must throw `RemoteException`. That and extending `java.rmi.Remote` is about all there is to making the `Bank` class remotable with RMI. Once you've got that, you generate the RMI stubs and skeletons, by feeding `Bank` first into the normal Java compiler (*javac* in Sun's Java distribution) and then into the RMI compiler (*rmic*).

Now that you've defined a remote interface and generated its stubs and skeletons, the next thing you need to do is write a class that implements the interface. The main thing worth noting is that the object registers its own name in the RMI registry by calling `Naming.rebind`:

```
import java.rmi.*;
import java.rmi.server.UnicastRemoteObject;
import java.util.*;
import BankSystem.*;



public class BankImpl
extends UnicastRemoteObject
implements Bank {
   private HashSet balances;
   public float getBalance (String accountId)
      throws RemoteException {
      return (Float)(balances.get(accountId)).floatValue();
   }
   public BankImpl (String name)
   throws RemoteException {
   try {
      balances = newHashSet();
      // Register this object's name with the RMI registry.
      Naming.rebind (name, this)
      } catch (Exception e) {
         System.out.println (e);
      }
   }
}
```

Now you create an instance of the remote object and make it available to clients. There's no explicit registration of the new object in this code since it takes care of that on its own by calling `Naming.rebind` when it's created:

```
import java.rmi.*;
import java.rmi.server.UnicastRemoteObject;
import BankSystem.*;


public class BankServer {
    public static void main (String[] args) throws Exception {
        // Enable Java's security mechanisms for RMI.
        if (System.getSecurityManager() == null)
           System.setSecurityManager (new RMISecurityManager());
        // Create an object that will be remotely available from
        // this machine.
        BankImpl theBank = newBankImpl ("FirstBank");
    }
}
```

To finish, let's look at a client that uses the remote object just created. The client queries the RMI registry through the `lookup` method of the `Naming` interface, the same interface the remote object used to register itself. The client supplies a URL (`rmi://www.rmibank.com/First-Bank`) to identify the object it wants. The last part of this URL is the name under which the remote object registered itself. The first part is the name of the machine on which that object and its RMI registry are believed to be running. `rmi://` means the URL should be handled by the RMI registry running on that machine. By default, RMI uses HTTP to communicate between client and server. This is good because HTTP is the standard vehicle for Web traffic and most firewalls are configured to allow HTTP requests to pass through:[11]

```
import java.rmi.*;
import java.rmi.registry.*;
import BankSystem.*;


public class BankClient {
   public static void main (String[] args) throws Exception {
      try {
      // Enable Java's security mechanisms for RMI.
      if (System.getSecurityManager() == null)
         System.setSecurityManager (new RMISecurityManager());
      // Find the remote object from its RMI registry.
      Bank theBank =
          (Bank)Naming.lookup ("rmi://www.rmibank.com/FirstBank");
      // Use the remote object.
      System.out.println
         ("Balance is "+ theBank.getBalance ("someAccount"));
```

---

[11]Some jiggering with the firewall may be necessary to let RMI HTTP requests through, but at least they won't be blocked just as a matter of course.

```
      }
      catch (SystemException e) {
         System.err.println (e);
      }
   }
}
```

As with CORBA, this is all straightforward when the client machine already has the stubs it needs. But what if a client wants to use an object whose stub isn't already installed on the client's machine? RMI's solution is much simpler than CORBA's DII. The real problem is getting code from one place to another. Java code is already designed to be portable. Just as applet code can be downloaded from a Web server to execute in a Web browser, so RMI client stubs can be downloaded from a remote RMI registry to a client, when they are needed.

Section 3.4's discussion of Java applets presented the risks of downloading code from remote locations. The same risks come into play when RMI stubs are downloaded. That's why the client and server RMI code has the (until now) unexplained calls to `System.getSecurityManager` and `System.setSecurityManager`. Java won't let a program use remote code, unless it's first insulated itself with a security manager. Thus, you have to set up security first whenever you use RMI.

You might wonder why the server has to use a security manager, since it's the one supplying the stubs. Consider this scenario: A client calls a remote method that takes another object as an argument.[12] That object could be remote to the server. The server may not have a client stub for that object. To invoke any methods on that object, it has to download the client stub for that object. And to use that stub it has to have a security manager in place. That's why even object servers need to protect themselves from downloaded objects.

With the introduction of Java 2, RMI added support for *dynamic object activation*. With this technique, a server can create or activate a remote object only when it is needed. The mechanisms for dynamic object activation are fairly complicated, but here are some highlights.

- Rather than extending `UnicastRemoteObject`, the remote object's implementation extends `Activatable`.

- When remote objects are created, they have to do some extra work to interact with the activation system.

---

[12] A brief technical note: With one exception, arguments are passed by value to remote methods (i.e., a copy is sent to the remote method). If the argument is an object local to the client, the copy is made using object serialization. If the argument is an object remote to the client, it is passed as a reference to the remote object, not as a copy.

- There no longer needs to be a server class with the sole function of creating the remote object and registering it. Instead, there needs to be a program on the server that installs the remote object in the activation system.
- Besides the RMI registry, the server must also run a program that manages object activation (*rm*id in Sun's Java distribution).

More information on RMI may be found at Sun's Java Web site (http://java.sun.com) and in [70], [15], and [2].

5.7 DCOM

DCOM, the *Distributed Component Object Model*, is Microsoft's framework for distributed objects. It's built into every copy of Windows NT, Windows 98, and Windows 2000, and can easily be added to Windows 95 by downloading free software from Microsoft. DCOM can be hard to figure out for the uninitiated. Part of the problem is that DCOM has a rather complex ancestry and is sometimes confused with other Microsoft technologies from which it is descended.

The *Common Object Model (COM)* is Microsoft's framework for object-to-object communication on a single machine.[13] COM objects can be part of a single application, or they can be part of many applications all running at the same time. COM is the foundation for *Object Linking and Embedding (OLE)* and ActiveX. OLE lets applications share common functions (for example, a spellchecker). It also lets documents from one application appear within another application's documents (say, an Excel spreadsheet in the middle of a Word document). *ActiveX* is a technology for downloading COM objects, for example in a Web browser. It is Microsoft's alternative to Java applets.

DCOM extends COM to support communication among objects that run on separate machines. DCOM provides its users with *interfaces*, named sets of methods that are offered by *classes*. A class may implement one or more interfaces. A DCOM object is a run-time entity that provides the services of a particular class, and thus the services of one or more interfaces.

---

[13]In theory, one could use COM across a network, but it's not easy, since you have to anticipate and handle all possible RPC error conditions on your own.

The biggest hurdle a non-Windows programmer faces in trying to puzzle out DCOM is that, while it has things that look like objects, they really aren't the same as objects in languages like C++ and Java. One of the most important features of an object in these languages is its identity. Simply put, this means that two objects that are instances of the same class can be told apart, even if they contain all the same values. (Think of it as being like twins: they both look the same, but of course they aren't the same person.)

Programmers can't rely on this property in the DCOM world. Rather than using true objects, they use a DCOM object that implements a particular set of interfaces. The DCOM object may have internal data, but it's just temporary and remains valid only as long as the object is being used. There's no way to use a DCOM object in one session and reconnect with the same object later. You can get another object that does the same thing, but you can't get the same object, since it no longer exists. This is a problem in environments with unreliable connections, since it makes it that much harder to recover work that is interrupted by a network fault. DCOM's solution is to assign something it calls *monikers* to objects that need to have their own identity.

DCOM's equivalent of the CORBA IDL is *Microsoft IDL (MIDL)*. Below is the MIDL code for the Bank interface that should be familiar by now from the CORBA and Java RMI examples. By convention, the interface name begins with "I" and appears in the code as IBank. The uuids scattered throughout the code are 128-bit *universally unique IDs*.[14] DCOM uses uuids to identify both interfaces and classes. There is no such thing as an object ID in DCOM. The coclass declaration states that a class named Bank will implement the IBank interface:

```
[
    uuid (7371a240-2e51-11d0-b4c1-444553540000),
    version (1.0)
]
library BankSystem {
   importlib ("stdole32.tlb");
   [
      uuid (BC4C0AB0-5A45-11d2-99C5-00A02414C655),
      dual
   ]
```

---

[14]For the curious, a uuid is made up of the unique network identifier of the machine on which it was originally generated (as obtained from an Ethernet device on that machine) combined with a timestamp. Another name Microsoft uses for uuid is *globally unique ID (GUID)*. These IDs can be generated with tools supplied by Microsoft in its development environments.

```
interface IBank : IDispatch {
   HRESULT getBalance
      ([in] BSTR p1, [out, retval] float * rtn);
}
[
   uuid (BC4C0AB0-5A45-11d2-99C5-00A02414C655),
]
coclass Bank
{
   interface IBank;
};
};
```

Next comes the code for `Bank`, a class that implements the `IBank` interface. Note that the `CLSID` of the `Bank` class is the same as the `uuid` that precedes `coclass Bank` in the MIDL code above. This is important, because that's how a client using this class gets to a DCOM instance of it:

```
import com.ms.com.*
import BankSystem.*;
public class Bank implements IBank {
   private static final String CLSID =
      "BC4C0AB0-5A45-11d2-99C5-00A02414C655";
   private HashSet balances;


   public float getBalance (String accountId)
      return (Float)(balances.get(accountId)).floatValue();
   }
   public Bank() {
      balances = new HashSet();
   }
}
```

And here is a client that uses a remote instance of `Bank`. It's similar to the Java RMI client code in Sec. 5.6:

```
import BankSystem.*


public class BankClient {
   public static void main (String[] args) throws Exception {
      try {
         IBank theBank = (IBank) new Bank();
         System.out.println
            ("Balanceis" + theBank.getBalance ("someAccount"));
      }
      catch (com.ms.com.ComFailException e) {
         System.err.println (e.getHResult());
         System.err.println (e.getMessage());
      }
   }
}
```

Distributed Computing

There is nothing corresponding to `BankServer` in DCOM. Instead, after compiling the `Bank` class, one installs information about it in the Windows Registry. (If it's a Java class, Microsoft's JavaReg tool takes care of that.) Requests to find and instantiate DCOM objects can then be handled by a DCOM element known as the *Service Control Manager (SCM)*, which corresponds roughly to CORBA's ORB.

5.8 SOAP

5.8.1 Why SOAP?

SOAP, the *Simple Object Access Protocol*, is different from the distributed object technologies covered so far. Where CORBA, RMI, and DCOM provide complete environments for deploying distributed objects, SOAP provides only a message format for sending remote method calls from one machine to another. There is no SOAP API and there are no SOAP object managers. If there's so little to SOAP, then why is it of any interest, when compared with its more fully featured siblings?

The first thing to keep in mind is that distributed object systems use a client-server model. Object servers offer remote objects, which clients use. This is all very tidy until one realizes that object servers can themselves be clients of other object servers. Thus, you could have a situation in which client A requests the services of object 1 on server B, and object 1 in turn requests the services of object 2 on server C. In this case, B acts as both server (of object 1) and client (of object 2). This is called *server-server communication*.

CORBA, RMI, and DCOM have proven themselves in closed server-server situations. In these, a single organization controls all the servers in a *server farm*. There's been far less success doing this on the Internet, with millions of clients out there requesting services. There are several reasons for this:

- Conventional wisdom is that it's best to use a single CORBA or DCOM vendor for the distributed object software in clients and servers. Although there are multivendor solutions, reviews have been mixed so far. RMI requires you to use a single language (Java) in clients and servers. There is little or no way to guarantee that all potential clients will have or want to install the needed software for CORBA, RMI, or DCOM.

- CORBA, RMI, and DCOM rely on an environment that is highly administered, especially with regard to security. It's not easy to get two random computers to talk to each other with these technologies, without a lot of tweaking on both ends. Imagine trying to do that with millions of computers out on the Internet.
- CORBA, RMI, and DCOM have authentication and firewall issues that make them hard to work with on the Internet. Most administrators block CORBA and similar traffic to and from the Internet. And the Internet is where all the potential clients are sitting with their PCs and Macs.

All of these factors can be controlled within a server farm operated by a single organization. The designers of SOAP wanted to break out of the server farm and find a way to offer distributed objects on the Internet.

A fundamental principle of SOAP was to "first invent no new technology"; in other words, use technology already likely to be present on client machines. The two things needed were to represent remote method calls as messages and get those messages from one machine to another. SOAP's designers chose XML as the means to represent remote method calls. While clients may not have XML now, it can be expected they will have it by the time SOAP is widely deployed, since it's a W3C standard that's being built into all the major Web browsing products. As for passing SOAP messages around, HTTP was chosen because it's the foundation on which World Wide Web message exchange is built. Clients better have HTTP or they're not going to be on the Web at all. Just as important, HTTP traffic can get through firewalls with little trouble.[15]

It's important to emphasize once again that SOAP is nothing more than a tool for passing remote method call messages and their results between two machines. What SOAP does is put these messages in a format both can understand and provide them a route that's not likely to be blocked at either end. SOAP says nothing about how objects are identified, how a server invokes a method on the correct object, how clients find out about remote objects, how remote objects come into being, or how objects are destroyed when they're no longer needed. In other words, SOAP says nothing about many topics that concerned us when we were looking at CORBA, RMI, and DCOM.

---

[15]SOAP has since expanded the range of message exchange protocols it supports beyond HTTP. However, HTTP remains for now (and probably will remain for the foreseeable future) the primary message transfer protocol for SOAP.

This is not to say those topics are not relevant in a SOAP-based system. They are. It's just that SOAP doesn't say anything about how they should be handled. From SOAP's perspective, they're implementation details. In the following sections, you'll see a few hints of how it might be done. However, none of it is cast in stone. Some of the solutions are merely conventions the SOAP community is developing as it gains experience. Since SOAP is such a new technology, it's likely other conventions will continue to be developed and some old ones abandoned. Keep that in mind when reading about SOAP. With a rapidly developing technology like SOAP, new information will usually be the most reliable. Even a month or two can make a big difference.

5.8.2 SOAP and XML[16]

SOAP's creators originally started to design a new XML-based language for describing SOAP messages. This language, known as the *Component Description Language (CDL)*, was similar in many ways to CORBA's IDL. As it turned out, XML schemas were addressing the same things as CDL, but in a more general way. CDL was abandoned and XML schemas adopted as the means for defining SOAP messages.[17]

Just like any remote method call message, a SOAP message must indicate the method to be called, the object on which it is to be called, and any arguments and results that are to be exchanged. Before looking at how schemas can be used to define SOAP messages, we'll look at the messages themselves.

How might the `getBalance` example used for CORBA, RMI, and DCOM be handled with SOAP? There's no such thing as a SOAP stub compiler to generate SOAP messages, so we'll have to work it out on our own. Since SOAP is based on XML, this isn't too difficult.

---

[16]It's assumed that the reader is familiar with XML in general and XML schemas in particular, as described in Section 4.3.2. We cover only the bare essentials of SOAP here. More details can be found in a reference work, such as [87]. Because SOAP is so new and is still being defined, one would do well to keep abreast of developments at the W3C Web site (http://www.w3c.org).

[17]SOAP does not require the use of XML schemas. However, schemas make things a lot easier. They will almost certainly become the sole vehicle for SOAP in the future, so we won't discuss any other ways to define SOAP messages here.

A SOAP message consists of several XML elements. The outermost of these is the `<SOAP:Envelope>` element, in which everything else is packaged:

```
<SOAP:Envelope
  xmlns:SOAP = "http://schemas.xmlsoap.org/soap/envelope/"
  SOAP:encodingStyle = "http://schemas.xmlsoap.org/soap/encoding/">
   body-element optional-header-element
</SOAP:Envelope>
```

The `xmlns:SOAP` attribute defines the XML namespace that will be used for all SOAP tags in the message. Any tag beginning with `SOAP:` should be interpreted as a SOAP tag, according to the rules set forth by SOAP's designers.[18] The URL associated with the SOAP namespace, `http://schemas.xmlsoap.org/soap/envelope/`, is a unique identifier, one that SOAP's designers control. This URL differentiates SOAP tags from other tags that might have the same names (for example, `SOAP:Envelope` versus `USPS:Envelope`). If there were several versions of the SOAP specification, this URL would indicate which one was being used in this message. It would be the job of the sending and receiving machines to use that information to figure out how the message is to be handled.

The `SOAP:encodingStyle` attribute defines the format in which arguments for the remote method will be encoded. The receiving machine uses this information to convert arguments into a format it can use. The value shown in the example tells the sending and receiving machines to use SOAP's default encoding.

The SOAP envelope contains an optional header element and a mandatory body element. The body element contains one of three things:

- Remote method call information from the client to the server (name of method, arguments, and so forth)
- A remote method call response from the server to the client
- Fault information from the server to the client when a remote method call has failed

---

[18]Another string that's commonly used to define the SOAP namespace is `SOAP-ENV`. One would then see the following attribute in the `<SOAP:Envelope>` tag:

   `xmlns:SOAP-ENV="http://schemas.xmlsoap.org/soap/envelope/"`

Remember that the value of the string used to label a namespace doesn't matter. It's just used to differentiate tags.

Once you add the body to the envelope, a SOAP *method call message* for the `getBalance` method of a remote `Bank` object looks like this:

```
<SOAP:Envelope
  xmlns:SOAP="http://schemas.xmlsoap.org/soap/envelope/"
  SOAP:encodingStyle="http://schemas.xmlsoap.org/soap/encoding/">
  <SOAP:Body>
    <bankapp:getBalance
        xmlns:bankapp="http://www.mycompany.com/bankapp">
      <bankapp:accountId>
        ABC123DEF
      <bankapp:accountId>
    </bankapp:getBalance>
  </SOAP:Body>
</SOAP:Envelope>
```

`<SOAP:Body>` tags bracket the body element. The body contains an element tagged with `<bankapp:getBalance>` that names the method to be called. The attributes of this tag set up a namespace—bankapp—that is associated with the URL `http://www.mycompany.com/bankapp`. As usual, it doesn't matter if the URL points to an actual Web document or not. It just has to be globally unique. The `<bankapp:getBalance>` element contains one element for each argument passed to it. Each argument has its own tag. Here there is only one argument with the tag `<bankapp:accountId>`. The content of this element is the value of that argument, encoded as indicated by the envelope's `SOAP:encodingStyle` attribute.

So far, so good. But you may be wondering about which object the `getBalance` method gets called. We'll defer that discussion for a while, since it's part of SOAP's HTTP transmission, which is covered later in this section. For now, you'll just have to trust that it does get handled.

`<SOAP:...>` tags are defined by the SOAP specification and its XML schema. Any tags that appear inside `<SOAP:Body>` must be defined specially for each SOAP-based application. As noted earlier, XML schemas are the best way to do that. Here is an XML schema that defines the tags needed for the `getBalance` method call:

```
<xsd:schema xmlns:xsd="http://www.w3.org/2000/08/XMLSchema">
  <xsd:complexTypename="GetBalanceType">
    <xsd:sequence>
      <xsd:elementname="accountId" type="xsd:string"/>
    </xsd:sequence>
  </xsd:complexType>
  <xsd:element name="getBalance" type="GetBalanceType">
</xsd:schema>
```

A similar schema must be defined for every remote method in a SOAP-based application. The only thing worth noting about this schema is that the type of the `accountId` argument is specified as `xsd:string`. This tells any client software sending this message that the account ID value should be encoded as a string according the rules given by the `encodingStyle` attribute of the message's envelope. Similarly, it tells any server software that processes this message how to decode that argument.

Now we'll look at the *response message* a server might send after executing the `getBalance` method call. Note that the response message tag (`getBalanceResponse`) is the same as that for the method call (`get-Balance`), except that `Response` has been added at the end. There is no rule in SOAP to enforce this, but it is an accepted convention. The name of the return value need only be something the client and the server have agreed on beforehand. Other than this, the method call and response messages are similar:

```
<SOAP:Envelope
xmlns:SOAP="http://schemas.xmlsoap.org/soap/envelope/"
SOAP:encodingStyle="http://schemas.xmlsoap.org/soap/encoding/">
  <SOAP:Body>
    <bankapp:getBalanceResponse
        xmlns:bankapp="http://www.mycompany.com/bankapp">
      <bankapp:balance>
         2593.72
      <bankapp:balance>
    </bankapp:getBalanceResponse>
  </SOAP:Body>
</SOAP:Envelope>
```

Below is the schema for the response message. The type of the return value is declared as `xsd:decimal`. As with the method call message, this tells the server how to encode the value and it tells the client how to decode what it gets:

```
<xsd:schema xmlns:xsd="http://www.w3.org/2000/08/XMLSchema">
  <xsd:complexType name="GetBalanceTypeResponse">
    <xsd:sequence>
      <xsd:element name="balance" type="xsd:decimal"/>
    </xsd:sequence>
  </xsd:complexType>
  <xsd:element name="getBalanceResponse" type="GetBalanceResponseType">
</xsd:schema>
```

Besides the body, a SOAP envelope may also include a `<SOAP:Header>` element. The header provides information that is not part of the body proper. For example, a header could be used to request an execution trace for a method. As with tags in the body, tags in the header are application specific and must be defined in the application's schema:

```
<SOAP:Envelope
  xmlns:SOAP="http://schemas.xmlsoap.org/soap/envelope/"
  SOAP:encodingStyle="http://schemas.xmlsoap.org/soap/encoding/">
  <SOAP:Header>
    <bankapp:traceRequest
        xmlns:bankapp="http://www.mycompany.com/bankapp">
      full
    </bankapp:traceRequest>
  </SOAP:Header>
  <SOAP:Body>
    <bankapp:getBalance
        xmlns:bankapp="http://www.mycompany.com/bankapp">
      <bankapp:accountId>
        ABC123DEF
      <bankapp:accountId>
    </bankapp:getBalance>
  </SOAP:Body>
</SOAP:Envelope>
```

The *fault message* brings this brief tour of SOAP messages to an end. A fault message is indicated by a `<SOAP:Fault>` element within the `<SOAP:Body>` element. SOAP defines four subelements of `<SOAP:Fault>`:

`faultcode`. Mandatory. Identifies the type of fault. SOAP defines four basic fault codes. Other application-specific codes may be created by appending strings that further refine the meaning of these codes. For example, `Server.BankApp.AccountIdNotFound` refines the built-in `Server` code to say that a problem occurred on the server in its `BankApp` service and that the specific problem was that a supplied account ID could not be found. It's up to the designers of a SOAP application to define their own faultcodes and make sure all parts of the system understand and use them consistently. The following are SOAP's built-in faultcodes:

`VersionMismatch`. Message received with invalid namespace for `<SOAP:Envelope>` element.

`MustUnderstand`. Subelement of `<SOAP:Header>` element not understood or not obeyed, even though its `SOAP:mustUnderstand` attribute was "1."

`Client`. Message was incorrectly formed or did not contain appropriate information.

`Server`. Message appeared to be correctly formed, but could not be processed for other (probably application-specific) reasons.

`faultstring`. Mandatory. A human-readable description of the fault.

faultactor. Optional if generated by the system that was the ultimate destination of the message that caused the fault; mandatory if generated by any other system en route to the ultimate destination. Contains the URI of the system that generated the fault.

detail. Mandatory if the fault was caused by a failure to process the contents of `<SOAP:Body>` (which is usually the case). Contains application-defined information that further describes the fault.

Here's an example of a fault message:

```
<SOAP:Body>
   <SOAP:Fault>
      <faultcode>SOAP:Server.BankApp.AccountIdNotFound<\faultcode>
      </faultstring>
         Could not find any records for account ABC123DEF
      </faultstring>
      <detail>
       <bankapp:faultDetails
            xmlns:bankapp="http://www.mycompany.com/bankapp">
         <class>com.mycompany.BankApp</class>
         <method>getBalance</method>
       </bankapp:faultDetails>
      </detail>
   </SOAP:Fault>
</SOAP:Body>
```

5.8.3 SOAP and HTTP

For SOAP to be useful, it needs a robust transport protocol to get method requests and responses from client to server and back. This protocol has to be something already installed on the clients, it has to scale to support millions of clients, and it has to be able to get through firewalls. Just such a protocol already exists: *HTTP, the HyperText Transfer Protocol,* that's used to request and return Web pages. Every Web browser supports HTTP. The Web demonstrates that HTTP scales well. And firewalls generally let HTTP requests through with little or no trouble.

HTTP behaves much like a remote procedure call. A client makes an HTTP request for some information (RPC equivalent: invokes a remote procedure). The server tries to satisfy the request (RPC equivalent: executes the remote procedure). Finally, the server responds by sending back the requested information or an error message (RPC equivalent: returns the results of the procedure).

HTTP is a *stateless* request-response protocol. This means that, once a client gets a response from a server in answer to a request, there is nothing in HTTP that can be used to correlate later requests and responses with the earlier one. This is not to say there's no way to correlate a series of HTTP exchanges between a client and server. The server and the client can keep track of messages themselves to get around this limitation. But there's nothing built into HTTP to help them do that.

This is important to keep in mind when you build Web applications. For example, suppose you go to a Web page from which you log into an online brokerage account. Your login identity is sent to the server. The server authenticates you and returns a Web page with information about your account. So far, so good. But now you click on that page to sell one of your stocks. That request is sent to the server. At the HTTP level, it's just a random request coming in from the Internet. How is the server to know it came from the same person who logged in to your account earlier?

There are several ways to handle this situation. All require storing extra information on the client and/or server machines and adding information to the payload of HTTP messages.

HTTP REQUEST AND RESPONSE MESSAGES The first line of an HTTP request has three parts:

- An HTTP command, called a *method,* that tells the server what to do. I will focus on two of these: GET and POST.
- A *Universal Resource Identifier (URI)* that tells the server to what the method is to be applied. URIs identify resources in the Internet. *Universal Resource Locators (URLs),* are just one sort of URI. A URL refers to a specific resource at a particular location on the Internet. URLs are most familiar as the addresses one enters into a Web browser when navigating to a Web page.
- A *protocol identifier* that tells the server what version of HTTP the client is using.

Here's an example of an HTTP request that uses the GET method to request the document /catalog.html using version 1.0 of HTTP:

```
GET / catalog.html HTTP/1.0
```

Following the request line, there may be one or more optional header lines that tell the server more about the client. Each header line is of the form:

keyword: value

Header lines are commonly used to let servers know what kind of browser is sending the request and the kinds of data it can accept in return. For example:

```
User-Agent: Mozilla/4.61 [en] (X11; U; Linux 2.2.12-20i686)
Accept: image./gif, image/jpeg, text/*, */*
```

The header section ends with a blank line, followed by any more data the server needs to process the request (the *request entity*) and a final blank line that ends the request.

After processing the request, the server sends back a response. The first line of the response contains the version of HTTP used by the server, a status code, and a brief text explanation of the status code. For example:

```
HTTP/1.0 200 OK
```

indicates a successful response. After that come a sequence of response headers. These give the client information about the content of the response:

```
Date: Wed, 08 Nov 2000 20:00:55 GMT
Server: GWS/1.10
Content-Type: text/html
Content-Length: 1880
```

The response headers end with a blank line. The rest of the response is the data returned to the request. This may be the result of a successful request, such as some HTML to be displayed, or it may be some text with further explanation of why things went wrong.

GET AND POST   GET and POST are the most commonly used HTTP messages. GET's main purpose is to retrieve information. It can send small bits of data (say, a few values taken from a form), but only in limited quantities.[19] When a lot of information has to be included in a request, perhaps a form with many fields, the POST method should be used. Here's a simple GET request for some server's /index.html Web page:

```
GET /index.html HTTP/1.0
Accept: */*
```

---

[19]The actual amount depends on how the client and server machines are configured. A good rule of thumb is that GET requests should be no longer than 240 characters.

Distributed Computing

Suppose you have a Web form containing a text input field and a button labelled "Search." Here is the GET request that might result when a user enters `linux` in the form's text field and presses the search button. The extra information from the form is appended to the request URI in a *query string* (the stuff after the question mark):

```
GET /search.cgi?query=linux&button=search HTTP/1.0
Accept: */*
```

This request goes to `/search.cgi` on the server and provides it with the two values `query=linux` and `button=search`. It's reasonable to assume that `/search.cgi` is a program that looks for the string `linux` in some information source. The result of this request might be a Web page built by `/search.cgi`.

The POST method is used when there's more data than can be handled by GET using query strings. The request entity section of a GET message is always empty. The request entity section of a POST message is always present and contains the same query string information that would be appended to the request URI in a GET message:

```
POST /search.cgi HTTP/1.0
Accept: */*

query=linux&button=search
```

Note the blank line between the header and the entity sections. Note also that short query strings can be sent with either GET or POST, but long query strings must be sent with POST.

HTTP MESSAGES FOR SOAP The content of a SOAP request tends to be fairly long, at least when compared with the amount of data that can be sent with GET. For that reason, SOAP uses POST. Here's an example:

```
POST /Bank?bankId=95834098 HTTP/1.1
Host: www.bank.com
Content-Type: text/xml[20]
Content-Length: nnnn
SOAPAction: http://www.mycompany.com/bankapp\#getBalance
<SOAP:Envelope
  xmlns:SOAP="http://schemas.xmlsoap.org/soap/envelope/"
```

---

[20]This is sometimes seen as `Content-Type: text/xml-SOAP`. That's a relic of the early days of SOAP.

```
        SOAP:encodingStyle="http://schemas.xmlsoap.org/soap/encoding/">
   <SOAP:Body>
     <bankapp:getBalance
         xmlns:bankapp="http://www.mycompany.com/bankapp">
       <bankapp:accountId>
           ABC123DEF
       <bankapp:accountId>
     </bankapp:getBalance>
   </SOAP:Body>
</SOAP:Envelope>
```

Recall that in our earlier discussion of SOAP method call messages, the matter of how a client told a server what object was to be used for a method call was left hanging. We can at last tie up that loose end. Just as POST can be used to invoke a CGI program, it can be used to tell software on the server what object is to be the target of a SOAP method call:

```
POST /Bank?bankId=95834098 HTTP/1.1
```

Like a CGI program, the server should translate `/Bank` to refer to a program that can handle the method call. That program will find or create the proper object, using the information after the question mark, `bankId=95834098`, which is presumably the ID of an object that represents a specific bank. Once the required object is at hand, the software translates the rest of the message into the desired method call on that object, makes that call, and returns the results.

The mandatory `SOAPAction` field in the POST message's header is new. It's used to inform recipients about the purpose of the message (for example, to make a method call to the method named in the `SOAP-Action` field). Putting this information in the header lets the recipient to do some basic filtering of the message without having to parse all its SOAP content. If this information were put in the SOAP portion of the message, all sorts of software (for example, firewalls) would have to be enhanced to understand SOAP's syntax. By putting it in the header, any software that understands HTTP can get at it.

Assuming the firewall, or any other software sitting in judgment, approves of the indicated `SOAPAction`, the message is passed on to the software responsible for handling the message. That software, which should know how to parse SOAP messages, will compare the `SOAP-Action` value:

```
SOAPAction: http://www.mycompany.com/bankapp#getBalance
```

with the name of the method being invoked and the URL of its namespace:

```
<bankapp:getBalance xmlns:bankapp="http://www.mycompany.com/bankapp">
```

If they don't match, it should send back a SOAP fault response. Doing verification in two stages like this allows firewalls to do a quick check of what messages to let through without having to understand SOAP. At the same time, it keeps messages that are traveling with forged papers from doing any mischief.

For the sake of completeness, I should mention that SOAP messages may also be sent with HTTP's M-POST method. M-POST allows the recipients of HTTP messages to enforce the use of certain HTTP headers. With SOAP, a common use of this feature would be firewalls that want to ensure SOAP messages have a `SOAPAction:` field for them to check in the header.

M-POST is part of the HTTP Extension Framework and is not recognized by every server. In SOAP v1.0, clients had to use the POST method first. They could try M-POST if that failed with the HTTP status *510 Not Extended.* In SOAP v1.1, it's left to developers to decide if they want to try POST and then M-POST, or just use M-POST.

5.9 Comparing Distributed Object Technologies

Like so many decisions in the computing world, choosing distributed object technologies often leads to the start of a religious war. The problem is that all the systems we've looked at have advantages and disadvantages, often complementary, so there is no perfect solution. The following highlights some of their merits and faults.

- All things being equal, CORBA is the preferred solution for a multilanguage environment. This is often so when legacy systems (for example, based on COBOL) are being integrated into a newer architecture.

- Interworking CORBA products from different vendors is a risky proposition. Though things have improved, unbiased observers still recommend using a single CORBA vendor whenever possible.[21]

---

[21] Lack of interworking is one reason there isn't a significant third-party market for CORBA components, as there is for JavaBeans and DCOM. You would need separate versions of your components for each CORBA vendor's product. Even assuming improvements in the interworking situation that would eliminate the need for different versions, you would still need to test components against a variety of CORBA products.

- RMI is the easiest to use of all these technologies (for example, compare Java's object downloading mechanisms with CORBA's DII), and it has the best multiplatform support. However, RMI is suitable only for systems that are built entirely or predominantly with Java. If that's so, RMI is the preferred solution.
- DCOM is a good choice for Windows-only environments, since it's free (once you buy Windows) and because there's a huge supply of third-party DCOM components. The large pool of programmers who already know Windows and its programming environments, coupled with the support for DCOM in those environments, also make DCOM attractive from a development perspective.
- Despite attempts to make DCOM into a multiplatform system, it's still, practically speaking, a Windows-only solution. Ports to other platforms have been attempted, with mixed results.
- Bridges can be used in environments where distributed object systems must be mixed: DCOM to CORBA, CORBA to RMI via IIOP, Java to DCOM using Microsoft's own JVM. The problem is that each of these (CORBA, RMI, DCOM) is similar to, but not identical with, the other, so the bridges don't make a perfect match. They usually do just fine on basic functions, but aren't as good when you try to integrate more advanced capabilities.
- SOAP has many attractive features, but it's really a protocol for communicating within a distributed object system, not a distributed object system, as such. Because SOAP says nothing about many basic services of a distributed object system, an application that uses SOAP will have to ensure on its own that these things are taken care of and that all parts of the application have the same understanding of what's been done to handle them. Also, SOAP is a young technology. Vendor support is limited and the protocol may change as experience is gained.

5.10 For More Information

Just about any book on UNIX programming will cover sockets in some fashion. Matthew and Stones have a good brief chapter on sockets in their *Beginning Linux Programming* [65]. For in-depth discussion of RPC and other topics, one of the standard texts on distributed computing is

Colouris, Dollimore, and Kindberg's *Distributed Systems: Concepts and Design* [10]. There are nearly as many books on CORBA as there are on Java. One that combines both topics in a single thick volume is Orfali and Harkey's *Client/Server Programming with Java and CORBA* [70]. DCOM is similarly well-represented in the literature. Two books on that subject that have been favorably received are Thai's *Learning DCOM* [90] and Grimes' *Professional DCOM Programming* [20]. SOAP is so new that not much has been published about it yet. An early entrant is Scribner and Stiver's *Understanding SOAP* [87]. The best source of the most recent information is the W3C's Web site (http://www.w3c.org).

CHAPTER 6

Directories

If you don't find it in the index, look very carefully through the entire catalogue.
(Sears, Roebuck, and Co. Consumer's Guide, 1897)

6.1 Basic Concepts

Directories are simple. You know the name for something. You want to find out something about whatever it is that has that name. You tell the directory what it is you want to know and give it the name. The directory looks up the entry for that name and returns the information (or an error if it doesn't have that information for that name). The phone company white pages are an obvious example of a directory with which most people are familiar. It maps names to phone numbers and addresses. On the Internet, the *Domain Name System (DNS)* maps names such as `somemachine.somecompany.com` to an *IP address*—a sequence of four numbers separated by periods, such as 192.168.0.39—that's used to locate that machine.

The growth of computer networks in the past twenty years has been accompanied by a growth in the need for directories to keep track of all sorts of resources—machines, files, services, people with e-mail addresses, and so forth—that have been attached to them. Given the size of these networks and the pace at which things appear or disappear on them, paper directories obviously won't do. Because DNS was designed to solve one problem (finding machines on the Internet), it's neither easy nor practical to extend it to cover every new directory application. This has led to the development of full-fledged *directory services*.

A directory service consists of the directory: a repository for names and the information associated with them, a protocol for exchanging information in the directory, and a way to describe the information in the directory by means of a *schema* or *data model*. Nearly all directory services being developed today are based on either the ITU's *X.500* standard or the IETF's *Lightweight Directory Access Protocol (LDAP)*, a relative of X.500. Directory services resemble databases in many ways. Like a database, a directory service lets you store and retrieve information. However, directories have some characteristics that set them apart from general-purpose databases:

- Directories are designed to be read more often than they are written, as much as 1,000 to 10,000 times more in some instances. This is not to say that directories are inefficient at writes. Nevertheless, when designing a directory, read performance takes precedence. Every aspect of a directory, including the types of data stored in it, should reflect this. A directory shouldn't be used to store data that changes frequently (though the precise definition of "frequently" might vary from one directory or application to another). For example, it probably makes sense to store addresses or birth dates in a directory. It probably makes no sense to store the running output of a medical sensor of some sort (for example, temperature, pulse) in a directory.

- Just about any sort of information can be kept in a database. There's nothing to prevent one from using a directory as a general-purpose database (and in fact, the backend of a directory—the piece that actually stores data—is often a general-purpose database). However, directories are best used for applications in which, given one piece of information about someone or something, you want to find out a few other things about it. Examples: given a name, return the address and phone number; given a phone number, return the name; given an Internet address, return the kind of machine it is and the name of the person responsible for it.

- It's easy to extend the types of data that are kept in a directory. For example, entries in a directory that once contained only name, address, and phone number might be extended to include e-mail address and wireless number. This extension could be made without affecting entries already in the directory that don't use the new fields.

- A directory may distribute and/or replicate its data across many systems to improve performance and availability.

- Databases are generally owned and operated by a single organization. It should be possible to treat a collection of directories owned by many different parties as if they were a single large directory.[1] Thus, interoperability standards are even more important for directories than they are for databases.

It's not always obvious whether a directory or a general-purpose database is the right tool for a particular job. These are some general principles for making that decision:

- A database is preferred when the items being stored have complex relationships with each other, relationships that will be the target of queries by database users who want to generate reports or analyze data (for example, list every employee in Austin whose manager works in Chicago).

---

[1]This leads to the dream of a single, worldwide, integrated directory, the Holy Grail of directory designers. A worthy goal, even if it eventually proves to be out of their reach.

- A directory is preferred when the items being stored are, for the most part, independent of each other (for example, a directory of machines on a network).
- A directory is preferred when the kinds of information stored in an entry may have to be extended or otherwise changed someday.
- All things being equal, a directory is better at handling data that is distributed across more than one server.

Sometimes an application straddles the line between database and directory, and neither is obviously more appropriate. There are several ways to handle this:

- *Build the application on top of a database.* The disadvantage is that applications that already know how to use directories (for example, e-mail) may not know how to use a database.
- *Build the application on top of a directory.* This is probably the best solution if the data must be distributed. However, if some or all of the data already resides in an existing database, and the data can't be moved out of the database, a pure directory solution won't work.
- *Duplicate the data in both a database and a directory.* This is a straightforward solution, but it will take extra effort to build the two systems and keep them synchronized.
- *Build a hybrid, a directory service with a database underneath.* If the data doesn't have to be distributed, this may be worthwhile. However, a database may not offer the read performance and schema extension capabilities of a pure directory solution.
- *Build gateway functions between the directory and the database.* With this technique, the directory asks the database for certain pieces of data, and vice versa. Examples would include a database that looks up security credentials in a directory, and a directory that gets personnel data from a human resources database.

6.2 Domain Name System (DNS)

The *Domain Name System (DNS)* is the directory system used to store and retrieve information about hosts that are attached to the Internet. Though DNS is not a general-purpose directory service, like X.500 or LDAP, I'll be covering it in some detail, because several current and emerging standards use DNS to find call participants on the Internet. Among these are ENUM and SIP, which are covered later in this book.

In the early days of the Internet, there weren't all that many attached hosts (on the order of 10s of them in the earliest days), and all the information about them could be kept in a single file called *HOSTS.TXT*. The most up-to-date version of *HOSTS.TXT* was kept at a central location and periodically retrieved by other hosts. Not surprisingly, this became impractical as the Internet grew. The problem turned out to be not the size of *HOSTS.TXT* itself (though that would have become a problem eventually), but rather the time it took to keep up with a constant stream of modifications as hosts were added, removed, or otherwise changed.

DNS was introduced as a replacement for *HOSTS.TXT* in the mid-1980s. DNS is a distributed database. Because it's distributed, local system administrators can update their own master copy of information about the hosts for which they are responsible. DNS then makes this information available to the entire Internet.

6.2.1 The DNS Namespace

To ensure all hosts have unique names, DNS uses a hierarchical *namespace* organized as a tree, much like the UNIX file system. Figure 6.1 shows an example of a portion of this tree. The branches of this tree connect *nodes*. Every node has a *label*. The node at the bottom of the tree (labeled "" in Fig. 6.1, because its label is empty) is called the *root* of the tree.

Branching out from the root node are *child nodes*, labeled com, edu, gov, org, and net. The root node is the *parent* of those children. In turn, the edu node is the parent of child nodes berkeley and uiuc, and berkeley is the parent node of two nodes labeled physics and math. Nodes like physics and math that don't have any children are called *leaf nodes*.

No two children of the same parent can have the same name. Thus, there can be only one physics child below berkeley, and one physics below uiuc, but there's no problem having two nodes labeled physics under different parents. The full name of a node is formed by joining together all the names of its ancestors starting at the root of the tree, separated by periods, with the names closest to the root coming last. For example, the full name of the uiuc node is uiuc.edu, and the full name of the University of Illinois physics node is physics.uiuc.edu.

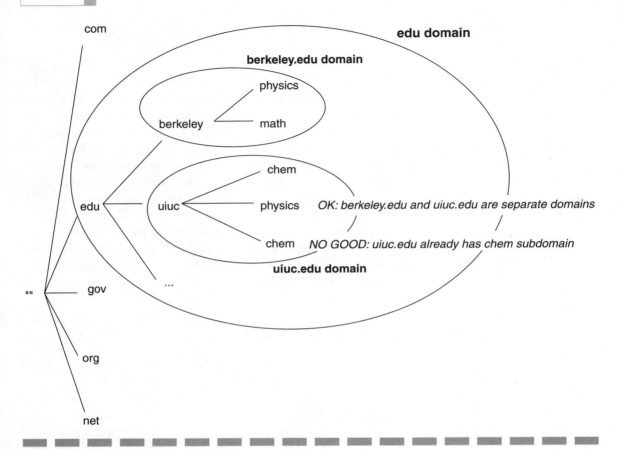

Figure 6.1
DNS's hierarchical namespace.

Every node represents a DNS *domain* whose *domain name* is the same as the full name of that node. Each child node represents a *subdomain* of its parent's domain. Leaf nodes typically contain information about specific hosts within the node's domain. At the higher levels of the tree, nodes contain information about an entire domain and its subdomains—usually references to child nodes, but they may contain information about specific hosts as well. For example, BigCo.com might represent the domain for the BigCo Corporation and it might also represent the host that is BigCo's main Web server.

Each domain can be administered by a different organization. Each of those organizations can divide its domain into subdomains and then delegate responsibility for managing those to still other organizations. For example, an organization called InterNIC manages the `edu` domain. InterNIC likely gives the University of California at Berkeley responsibility for the `berkeley.edu` subdomain. The University of California at Berkeley may further subdivide its domains. Figure 6.1 shows a possible division of domains and subdomains along these lines.

A node is not the same as the domain for which it is responsible. For example, the `berkeley.edu` domain in Fig. 6.1 contains three nodes, even though only one node sits at the top of the domain. Because domains are hierarchical, a domain near the bottom of the namespace tree is contained in all the domains above it. For example, `math.berkeley.edu` is in both the `berkeley.edu` and the `edu` domains.

Until recently, there were eight primary domain names (plus a large group of domains assigned to individual countries) just under the root of the DNS namespace tree. Though there are now more than eight primary names, the original set still represents the most commonly used of these:

com Commercial organizations, for example, companies such as IBM (`ibm.com`) and SBC (`sbc.com`)

edu Educational organizations, for example, universities and colleges, such as UC-Berkeley (`berkeley.edu`)

gov U.S. government organizations, for example, NASA (`nasa.gov`) and the U.S. Patent and Trademark Office (`uspto.gov`)

mil Military organizations, for example, the U.S. Navy (`navy.mil`)

net Networking organizations, for example, NFSNET (`nfs.net`)

org Noncommercial organizations, for example, the IETF (`ietf.org`)

int International organizations, for example, the ITU (`itu.int`)

arpa A special domain that's used for reverse lookups (going from an address to a name)

Because the Internet started in the United States, names at the top of the DNS namespace have a strong bias towards U.S. organizations. To remedy this situation, two-letter abbreviations for country names, as defined in the ISO 3166 standard, are reserved as top-level domain names. For example, the United Kingdom uses the top-level domain name `uk`, while the tiny nation of Tuvalu uses `tv`.[2]

---

[2]This domain name is a lucky break that Tuvalu has exploited by offering names in the `tv` domain to those who want to take advantage of a perceived synergy between the two defining media of our day: television and the Internet.

6.2.2 Nameservers

DNS stores information about a domain's namespace on a *nameserver*. Each nameserver holds information about one or more *zones*. Each zone is a subset of the entire domain namespace over which a nameserver is said to have *authority*. Domains and zones are easily confused. A domain is an entire subtree of the global namespace tree. Information about that domain is usually spread across one or more nameservers, each responsible for one or more zones within that domain. A zone corresponds to the information held on a nameserver. If a zone is at the bottom of the namespace tree, that information will include references to hosts within the zone. If a zone is higher up in the hierarchy, the associated nameserver may have information about hosts in the zone, but it's more likely to have records that point to other nameservers to which it has delegated authority for one or more of its subdomains.

In Fig. 6.2, the `edu` domain covers all servers and records in the entire `edu` namespace. The `edu` zone, on the other hand, is a single server that contains only references to the servers for two other zones: `berkeley.edu` and `uiuc.edu`. Similarly, the `berkeley.edu` zone server contains references to two other zone servers: `physics.berkeley.edu` and `math.berkeley.edu`. But the `berkeley.edu` domain encompasses all those zones.

There are two kinds of DNS nameservers. A *primary master* nameserver for a given zone stores a master copy of that zone's data. A *secondary master* (or *slave*) nameserver for a zone gets its data from another server, called its *master nameserver* (often, but not necessarily, the primary master for that zone). When a slave starts up, it first does a *zone transfer* by obtaining its zone data from its master. Once the transfer is finished, the data in a slave is just as authoritative as that in its master. Slaves improve the performance and reliability of DNS by providing a measure of redundancy.

Besides offering information about their own zones, nameservers can also search the DNS namespace for information outside the scope of their authority. DNS clients, called *resolvers*, ask nameservers for information about domain names. If the nameserver that's asked has the information in its own database, it sends that back. Otherwise, it sends out queries that traverse nameservers in the namespace tree until it finds one with the desired information.

For example, let's say you want to access a host named `micawber.math.berkeley.edu`. Unless you're attached to the UC-Berkeley network yourself, the nameserver you use probably doesn't

Directories

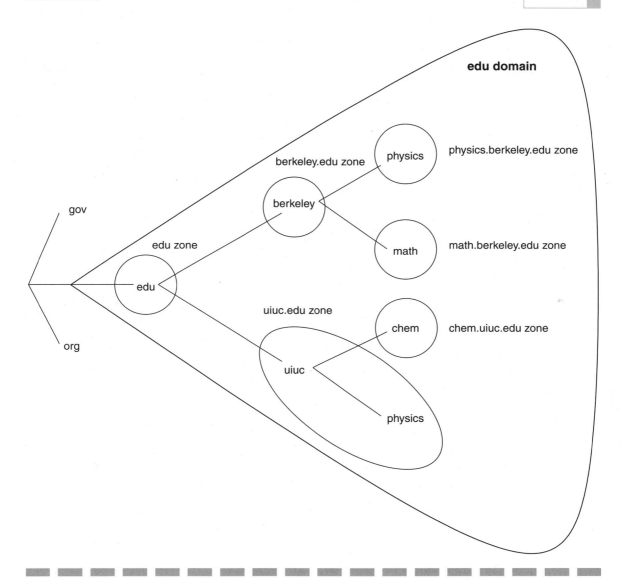

Figure 6.2
DNS domains and zones.

have a record of that machine. When your nameserver gets the request and finds it doesn't have that data, it forwards the request to a nameserver that knows about the highest level zone in the name `edu`. That nameserver sends back a referral to a nameserver that oversees the `berkeley.edu` zone, thus getting you one step nearer your goal. Your nameserver sends the request to that server, which sends back a referral to a nameserver for the `math.berkeley.edu` zone. Your nameserver sends the request to the `berkeley.edu` server, which presumably does know about `micawber.math.berkeley.edu`. The `math.berkeley.edu` server returns the information, which your nameserver returns to you.[3]

You now have a system that can get you an IP address when you know a machine name. But what if you want to go the other way, i.e, find a machine's name if all you know is its IP address? To support these *reverse lookups*, DNS reserves a special part of its namespace, the `arpa` domain. Within the `arpa` domain, subdomains are established to catalog different kinds of addresses. For example, the `in-addr.arpa` subdomain keeps track of IP addresses. A host whose IP address is 192.168.0.99 is represented by a record associated with the name `99.0.168.192.in-addr.arpa`. The numbers in an IP address are reversed in its corresponding `in-addr.arpa` name. In IP addresses, the items between periods go from high-level to low-level in the IP address hierarchy, as you go from left to right. With DNS domain names, you move from high-level to low-level items by starting at the right and moving left. The `arpa` domain is particularly interesting for convergent networks, because standards are being established to keep track of phone numbers in the newly established `e164.arpa` subdomain (see Sec. 8.5).

6.2.3 Resource Records

Entries in the DNS database are called *resource records*. Resource records may be of various types, depending on the kind of data they hold. To

---

[3] One point worth mentioning is that, for this to work, a nameserver must know where to find nameservers for the topmost zones of the Internet, such as `edu`, `com`, and `gov`. Not surprisingly, these nameservers are heavily used. It should also be pointed out that a local nameserver doesn't have to go all the way to the topmost zones for queries it can't handle on its own. In our example, your local nameserver might already know where to go for information about the `berkeley.edu` zone. One reason it might already have that information is that it had queried for a machine in the `berkeley.edu` zone recently. Rather than throw away the information as soon as it's used, a nameserver can store the information for some period of time, in case it's needed again, a technique that's called *caching*.

see how this works, let's look at some of the records that might appear on a nameserver that's running the *BIND (Berkeley Internet Name Domain)* software at the University of Southern North Dakota. The first entry BIND needs is an *SOA (start of authority)* resource record, that specifies the zone over which the nameserver has authority:

```
usnd.edu. IN SOA primary.usnd.edu. root.primary.usnd.edu. (
    1; 10800 3600;604800; 86400 )
```

The first field of the SOA record above, `usnd.edu.` (the period at the end of the name is not a typo), gives the zone name. The next field, `IN`, which stands for "Internet," identifies the record's class of data. `IN` is the only thing you're likely ever to see. Following that is the record type, `SOA`. The name that follows, `primary.usnd.edu.`, is the name of the primary master server for this data. Then comes `root.primary.usnd.edu.`, the e-mail address (with @ changed to .) of the person who maintains the data on this server. The rest of the record contains various timeout and other values used by the nameserver that we needn't go into.

After the SOA record, the next thing BIND needs are *NS (nameserver)* records. There should be one of these for each nameserver in the zone:

```
usnd.edu. IN NS nserver1.usnd.edu.
usnd.edu. IN NS nserver2.usnd.edu.
```

Once again, `IN` stands for "Internet," and it's the only thing you're likely to see here.

Next come records that actually refer to machines. First, some *A (address)* records that map names in the `usnd.edu` zone to IP addresses:

```
copperfield.usnd.edu. IN A 192.168.0.33
nickleby.usnd.edu.    IN A 192.168.0.34
```

CNAME (canonical name) records map an *alias* machine name to the machine name with which its IP address is stored:

```
david.usnd.edu.    IN CNAME copperfield.usnd.edu.
nicholas.usnd.edu. IN CNAME nickleby.usnd.edu.
```

PTR (pointer) records support reverse lookups:

```
33.0.168.192.in-addr.arpa. IN PTR copperfield.usnd.edu.
34.0.168.192.in-addr.arpa. IN PTR nickleby.usnd.edu.
```

DNS is used for other things besides matching host names and IP addresses. For example, DNS keeps track of e-mail servers with *MX (mail exchange)* records. Given a domain name, these identify one or more *mail exchangers* that serve it. A mail exchanger routes mail within its domain, either delivering it within that domain or forwarding it somewhere closer to its eventual destination. It's not uncommon to have several mail exchangers handling a single domain so that, if one fails, another can pick up the slack:

```
usnd.edu.  IN MX  0  mail1.usnd.edu.
usnd.edu.  IN MX 10  mail2.usnd.edu.
usnd.edu.  IN MX 10  mail3.usnd.edu.
usnd.edu.  IN MX 20  mail4.usnd.edu.
```

The first field of an MX record gives a domain name, and the last field names a mail exchanger for that domain. You've seen the IN field before, and the meaning of MX should be obvious. The numbers in the fourth field are preference values that help to select among multiple mail exchangers for the same domain. The more highly preferred a mail exchanger is within a domain, the lower its value here will be. The best possible preference value is zero. Mail exchangers can share the same preference value. All mail exchangers with the same preference will be tried before moving to one that's less preferred (has a higher preference value in its MX record). For example, `mail4.usnd.edu` will not be tried until both `mail2.usnd.edu` and `mail3.usnd.edu` have failed.

DNS has many other features of which I've only discussed a few. While it can't handle queries like "give me all the persons with a last name of Jones who work for BigCo in Cincinnati," it does what it does very well and isn't going away. Nevertheless, some applications need more than DNS can provide. That's where X.500 and LDAP come in.

6.3 X.500

X.500 is a series of standards that grew out of work begun in the mid-1980s by the CCITT (later, the ITU) to standardize directory services for phone, e-mail, and OSI networks. X.500 has a reputation for being large and complex. This is true, but at the same time it provides a framework by which separate directories can be integrated to act as one massive distributed directory. Nevertheless, while X.500 was designed to support distributed directories, and while this is the source of much of its complexity, it has no trouble handling centralized directories as well.

Directories

Each X.500 directory server, known as a *Directory System Agent (DSA)*, holds a portion of the data in this distributed directory. Taken together, all this data is referred to as the *Directory Information Base (DIB)*. The DIB may be split across DSAs (distributed), or every DSA may have a copy of the same data (replicated). When an X.500 client sends a request for information to a DSA, the DSA checks to see if it has the requested data. If it does, it sends back the results. If it doesn't, it either *chains* the request to another server or *refers* the client to the other server (see Fig. 6.3). If it

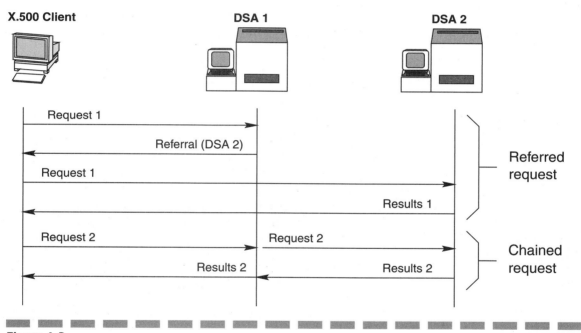

Figure 6.3
Referred and chained requests in an X.500 directory.

chains the request, the chained-to DSA handles the request and returns results to the first DSA, which returns them to the client as though it had handled the request itself. If it sends back a referral, the client uses that information to contact the other DSA on its own.

The first version of X.500 was approved in 1988, with updates in 1993 and 1997. Much of the 1988 standard describes two protocols needed to distribute data across DSAs:

- The *Directory Access Protocol (DAP)* specifies how clients get at information in the DIB.
- The *Directory Service Protocol (DSP)* specifies how a DSA that can't handle a request should pass it on to other DSAs.

In 1993, the ITU refined DAP and DSP and added some new directory administration and replication features:

- *Access controls* use public and private key encryption plus digital signatures to keep unauthorized users from seeing or modifying directory data.
- *Collective attributes* allow a single attribute, say a phone number, to be stored and modified in one place, while appearing as a part of many entries.
- The *Directory Information Shadowing Protocol (DISP)* allows information in a master DSA to be replicated (shadowed) in other DSAs. This improves directory performance by putting data closer to clients. It improves reliability by ensuring that if one copy is lost, another is likely to be available.

The 1997 standards further extended and refined the standards, but introduced no features anywhere near as important as those in the 1988 and 1993 versions.

In addition to protocols, X.500 also specifies an *information model* for organizing directories. As stated previously, the contents of an X.500 directory are called its *Directory Information Base (DIB)*. A DIB contains a set of *entries*, one for each entity—person, business, computer, and so forth—represented in the directory. Each entry is a collection of *attributes* that represent facts about that entity. Each attribute has a *type* and one or more *values*. The type determines the set of allowed values for that attribute. X.500 includes a predefined set of standard attributes and types. For example, X.500 defines `surname` as an attribute, the values of which must be strings: `surname=Doe` or `surname=Smith`.

Every entry in an X.500 directory belongs to an *object class*, the name of which is given in the entry's `objectClass` attribute. An object class specifies the attributes that must be present in entries with that `objectClass` attribute and those that are optional. For example, entries of the `person` object class must have `surname` and `commonName` attributes. Other attributes, such as `telephoneNumber`, are optional for those entries.[4]

---

[4] X.500 object classes are similar to the classes of object-oriented programming, as described in Section 2.2. Just like classes in programming languages, X.500 object classes can be extended through inheritance to produce class hierarchies of related classes.

The entries in a directory are organized as a tree, the *Directory Information Tree (DIT)*. If you imagine X.500 as its creators did, as a single universal directory, there would be a global DIT that encompassed every X.500 directory in existence.[5] Every tree must have a root. So where would you put the root of this global DIT and who would be in charge of it? Rather than designate some sort of global superuser who has control over every directory in the world, X.500's designers specified an imaginary *root entry*, below which every other entry in the world sits. The root entry has no attributes. It's only purpose is to make the X.500 model complete.[6]

Figure 6.4 shows what a global X.500 DIT might look like. Every country is assumed to be in charge of its own piece of the global directory. Within each country, directories are further split into government and corporate categories, and so on, each with its own administrative authorities. Each entry on the tree has a unique *distinguished name (DN)*. For example, BigCo, a hypothetical U.S. company has the following DN:

```
country=US, orgClass=com, org=BigCo
```

BigCo's Boston location has the same DN with the location information appended:

```
country=US, orgClass=com, org=BigCo, loc=Boston
```

Adding a name to that DN gives the DN of Jim Smith who works for BigCo in Boston:

```
country=US, orgClass=com, org=BigCo, loc=Boston,name=Jim Smith
```

Jim Smith who works for the British Ministry of Wax has a different DN:

```
country=UK, orgClass=gov, org=Ministry of Wax, name=Jim Smith
```

The bits of a DN that are separated by commas are called its *components*. Each component is an attribute of the entry to which the DN refers. The DNs for both Jim Smiths above use the component attributes `country`, `orgClass`, `org`, and `name`.

The rightmost component in a DN is its *relative distinguished name (RDN)*. The DN for BigCo has the RDN `org=BigCo`. Its Boston location has RDN `loc=Boston`. Both Jim Smiths have RDN `name=Jim Smith`.

---

[5]In theory. In practice, we're nowhere close to it.
[6]Yes, this is a bit fussy. But standards are supposed to be fussy.

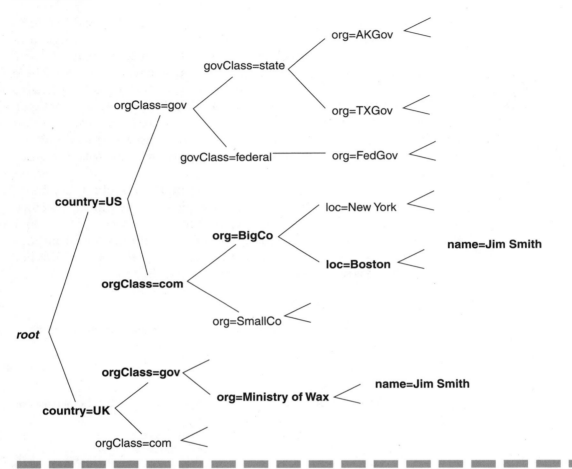

Figure 6.4
An example of an X.500 Directory Information Tree (DIT).

Note that two different DNs can have the same RDN even though every DN must be unique. Two DNs are are said to be *peers* in the DIT if they have the same immediate parent entry. Peer DNs must have unique RDNs. Nonpeer DNs can share the same RDNs. The immediate parents of the two entries with the RDN `name=Jim Smith` are:

```
country=US, orgClass=com, org=BigCo, loc=Boston
```

and

```
country=UK, orgClass=gov, org=Ministry of Wax
```

These are different, so there is no problem with their having the same RDN. If two or more persons named Jim Smith worked for BigCo in Boston or for the British Ministry of Wax, some way would have to be found to distinguish between them. One way would be to use a multi-valued RDN (say, name plus ID number). Another would be to use a property that is guaranteed to be unique (say, ID number) for the RDN instead of name.

X.500 provides three types of operation to clients that want to access the contents of a directory:

- SEARCH/READ. Retrieve attributes from an entry using some search criteria.
- MODIFY. Add, delete, or change attributes; add or delete entire entries; or change the RDN key of an entry.
- AUTHENTICATE. Identify a client to the directory and set up or tear down a session with it. Once a client is authenticated, each operation and its result may be signed, using public key authentication technology.

There's much more to X.500 than what's covered here, but this should be enough to give a sense of what X.500 does and how it works, and to introduce LDAP, the subject of the next section.

6.4 LDAP

LDAP (Lightweight Directory Access Protocol) started out as a simplified version of X.500 DAP. Early users of X.500 found that DAP didn't fit well on desktop machines of the time (circa 1990). But most potential users of X.500 had desktop machines. Two separate groups set out to design directory access protocols that were simpler than DAP. Their efforts resulted in the *Directory Assistance Service (DAS)* protocol [23] and the *Directory Interface to X.500 Implemented Efficiently (DIXIE)* protocol [24]. Of the two, DIXIE was more successful. However, both were limited by the fact that they depended on Quipu, a specific implementation of X.500.

DIXIE and DAS showed that simpler directory access protocols could be used with X.500 and thus stimulated work on LDAP. The first LDAP specification [26] was published in July 1993. This was followed by the specification for LDAPv2 [28], the first widely used version of LDAP.

The first implementation of LDAP, which came to be known as *U-M LDAP*, was developed at the University of Michigan and made freely available on the Internet in the early 1990s. Besides client software that ran on a wide variety of machines, U-M LDAP included a C language API for writing LDAP client applications. This API was documented by the IETF [29] and has since become a de facto standard.

Because LDAP started out as a way to access X.500 directories, it uses a lot of the same ideas: directory trees, entries as collections of attributes, using DNs and RDNs as keys, and treating entries as instances of object classes. LDAP can be used to access non-X.500 directories, but the servers in front of those directories must make them look like an X.500 directory to the LDAP client.

LDAP simplifies DAP in several ways:

- LDAP implements only the most important of DAP's operations, eliminating redundant and seldom-used features. Because they don't have to support as many functions, LDAP clients can be a lot smaller. Furthermore, the functions they do support are often simpler than their DAP counterparts.

- LDAP transmits most data as text strings. DAP transmits everything using ASN.1's BER (see Section 2.4.2). While LDAP wraps data in a simplified version of BER, it stops far short of the degree to which DAP uses it. A full implementation of BER, as DAP requires, takes up a significant chunk of software. LDAP's simplified BER is a lot smaller.

- LDAP runs on top of TCP/IP. DAP originally ran on the much larger and more complicated OSI protocol stack. Versions of DAP that run on TCP/IP are now available, but LDAP has by now established itself in this territory.

A session between an LDAP client and server starts with a BIND REQUEST sent by the client. This request includes the identity of the client and may include other authentication data (for example, an encrypted password). After the server processes the the request, it either sets up a session with the client or rejects the request. Either way, it returns a BIND RESPONSE to tell the client what it's done. Assuming a session has been established, the client can issue operation requests to the server. When the client is finished, it ends the session with an UNBIND REQUEST.

LDAP clients can request the following operations of a server:

- SEARCH. Look for object(s) that match(es) supplied criteria.
- MODIFY. Change an entry.

Directories

- ADD. Add a new entry.
- DELETE. Remove an entry.
- MODIFY RDN. Change the last component of an entry's DN.
- COMPARE. Find out if a condition is true of an entry.
- ABANDON. Tell the server to stop working on an outstanding request.

Before 1995, LDAP clients had to talk to an X.500 server, as shown in Fig. 6.5. X.500 servers were large, complicated pieces of software that need-

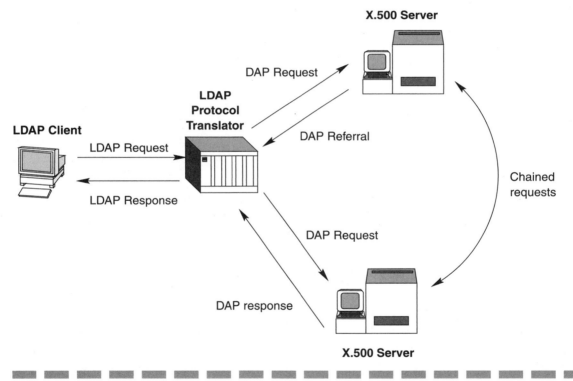

Figure 6.5
LDAP as a gateway to an X.500 directory.

ed a lot of care and feeding. In 1995, the U-M LDAP implementation added *SLAPD (standalone LDAP daemon)*, the first native LDAP server. SLAPD is a simple directory server that can easily be installed and run on inexpensive machines. Furthermore, SLAPD is a more efficient LDAP server, since it eliminates the conversion from LDAP to DAP (see Fig. 6.6). With the introduction of SLAPD, users who lacked access to an X.500 server could take advantage of directory services. This was an important step for all directories, because the barriers to entry for developers and users were virtually eliminated. It's likely that, without SLAPD, LDAP would not be nearly as popular as it is today.

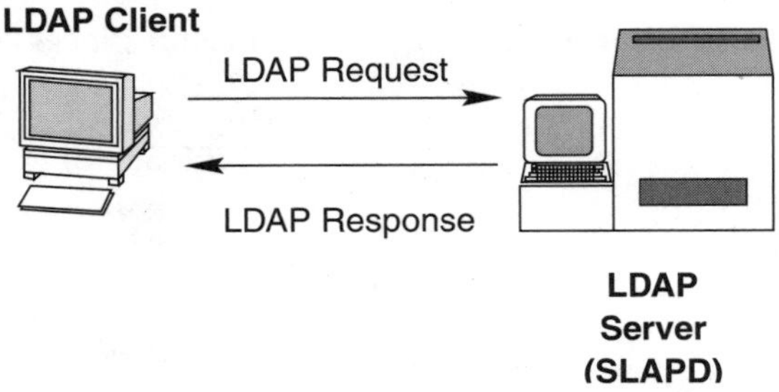

Figure 6.6
LDAP with standalone SLAPD directory server.

In late 1997, a new version of LDAP, LDAPv3, was approved as a proposed Internet Standard [37]. LDAPv3 adds several features to LDAPv2:

- LDAPv3 messages use the UTF-8 character set, which makes it possible to transmit text for any world language.

- LDAP servers can return a referral to another server in response to an LDAP request. Just as in X.500, the client can either ignore the referral or follow it.

- LDAPv3 adds security features by using the *Simple Authentication and Security Layer (SASL)* [36], a method for adding authentication support to connection-based protocols, and *Transport Layer Security (TLS)* [40], an IETF extension of Netscape's *Secure Sockets Layer (SSL)*.

- LDAPv3 provides a standard mechanism for adding new features to existing LDAP operations and for adding new operations. This mechanism makes it unlikely there will have to be an LDAPv4 standard in the foreseeable future.

- LDAPv3 servers use a special directory entry—the *root DSE (Directory-Server Specific Entry)*—in which they publish the versions of LDAP they support and other useful information, including schemas they support. These let clients and other LDAP servers discover the capabilities of an LDAPv3 server to make better use of it.

6.5 Future Trends

6.5.1 X.500 and LDAP: Partners or Rivals?

X.500 was to have been the foundation for a single worldwide directory service. While admirable, this goal led X.500's designers down some paths that slowed its acceptance. The sheer size of the problem was such that X.500 ended up as a large and sometimes confusing set of standards. Early implementations suffered from bugs and poor performance. As if this weren't enough, X.500 products from different vendors had problems working with each other. X.500's reliance on the OSI protocol stack was another factor that held it back. Once upon a time, some thought the OSI stack might replace TCP/IP. They were wrong. That left X.500 sitting atop a stack of orphan protocols. X.500 can now run on top of TCP/IP, so this is no longer a meaningful objection, but its initial dependence on OSI protocols was considered a detriment. Finally, DAP's heavy computing requirements left open a window of opportunity for the simpler LDAP to overtake it as the dominant protocol for client directory access.

LDAP's success and X.500's problems, especially in the early days, have led many to conclude that LDAP should replace X.500 entirely. In their opinion, X.500 is much too complicated and resource hungry. However, it's important to remember that LDAP is Lightweight *DAP*, not Lightweight *X.500*. Its creators wanted to make it easier to deploy clients that could access X.500 directories. Where X.500 handles three different areas of directory protocols—user to server (DAP), server to server (DSP), and data replication (DISP)—LDAP handles only user to server. While the referral features in LDAPv3 give it a way to deal with requests a server can't handle, there are advantages to the chaining approach supported by X.500's DSP:

- It's a lot easier for a client to send a single request and get back a definitive answer, regardless of where that answer comes from. Clients that have to handle referrals will be more complicated.

- Servers can be programmed to detect loops in a chain in which one server passes a request to another, which passes it to yet another, and so on until it gets back to the original server, without having obtained an answer. Centralizing this function also makes for less code in the clients.

- When the answer to a request combines results from more than one server, the original server can merge these and eliminated duplicates. It can also support client time limits on a request, sending back results received "so far," when the time limit has been reached. Again, centralizing these functions makes for simpler clients.

LDAP has a simple replication mechanism that's provided by a process, SLURPD, acting in concert with SLAPD. As a directory is modified, the SLAPD process serving it generates a change log. The SLURPD process reads this log periodically and then replays all the changes against another server, by acting as an LDAP client generating the corresponding requests. This mechanism works, but there are things it can't do:

- Every change in the original directory must be applied to the replicated directories. There's no way to replicate just the most important—and likely to be requested—information, while leaving less commonly used items on their original server.
- There's no way to determine which server has the master data and which has a replicated copy. Concurrent updates to the master and a copy are especially hard to manage.
- If the goal is to eliminate referrals, all information on all servers must be replicated throughout the system.

The bottom line in all cases is that, if LDAP is to replace X.500 entirely, it will have to be extended to support many of the same features as X.500. In the process, it's likely to become just as complex as X.500 and to duplicate many of the solutions X.500 has already arrived at. It's certainly true that X.500 is complex. But it's solving complex problems. Simpler solutions may be possible on the server, but simple solutions probably are not. This is not necessarily a bad thing, when the goal is to centralize complexity with an eye toward simplifying clients. LDAP's original designers seem to have realized this by concentrating on its user-to-server protocol, where the simplifications were obvious and useful. In so doing, they made directory technology much more accessible than it would have been otherwise.

6.5.2 Directory Enabled Network (DEN)

The so-called *Directory Enabled Network (DEN)* is a directory application that's been the focus of much attention in the past few years. DEN was first proposed in 1997 by John Strassner, a Cisco fellow, and embraced by Microsoft soon after. The motivation for DEN is the need for networks and network services to keep track of many types of data: subscriber information, service parameters, network configurations, and so on. While individual networks may have adequate means of cataloging this data, there's little or no way for networks to share it among themselves. Even if two networks store the same types of data, they may use different names or formats for it. The goal of DEN is to establish a common layout for all this data and its relationships. Once this is done, vendors, customers, and services providers could all exchange that information among their systems. It's a simple idea. But it's turned out to be extremely complicated to carry out in practice. Nevertheless, the advantages are clear and and the idea hasn't gone away.

6.6 For More Information

Reference [1] provides excellent in-depth coverage of DNS and its setup with BIND. The LDAP RFCs may be found at the IETF's Web site (http://www.ietf.org). A good introduction to LDAP and directory technology may be found in Howes, Smith, and Good's *Understanding and Deploying LDAP Directory Services* [22]. The LDAP API is covered at length in Howes and Smith's *LDAP: Programming Directory-Enabled Application with Lightweight Directory Access Protocol* [21]. The X.500 series of recommendations, starting with the core document that describes the X.500 architecture [59], may be obtained from the ITU (http://www.itu.int). Reference [6] is a more digestible version of these. Data Connection Ltd. has an excellent white paper on its Web site [12], with many useful insights that influenced Section 6.5.1's comparison of LDAP and X.500.

CHAPTER 7

Telephony for Programmers

TELESCOPE, n. A device having a relation to the eye similar to that of the telephone to the ear, enabling distant objects to plague us with a multitude of needless details. Luckily it is unprovided with a bell summoning us to the sacrifice.
—Ambrose Bierce

I'm trying to use the phone!

—Peewee Herman
("Peewee's Big Adventure")

7.1 The Voice Network

Up to this point, I've focused on technology that comes from the data/Internet side of the convergent network landscape. We now turn to the voice side of things, starting with the traditional (nonconvergent) voice network. The traditional voice network is the first point of contact for convergence. Before we look at convergence, we'll first need to go over some background material on the voice network.

The voice network is commonly referred to as the *PSTN (Public Switched Telephone Network)*. *POTS (Plain Old Telephone Service)* is the basic service offered by the PSTN. POTS covers all the basic features and services of the phone network, including such things as:

- Dial tone
- The ability to make calls from rotary and pushbutton telephones
- Local, long distance, and international calls
- Operator services
- 911 emergency services

Strictly speaking, the PSTN is only that part of the voice network operated by public phone companies. In common usage, this distinction is ignored, and the PSTN is understood to include private voice networks (for example, those owned by corporations, universities, or government organizations) that are attached to the public network. A couple of new names—*GSTN (Global/General Switched Telephone Network)* and *SCN (Switched Circuit Network)*—have been introduced in the past few years. Though more accurate, so far, they have yet to supplant PSTN. (I use the older term, despite its inaccuracy, simply because it is so familiar).

The PSTN has four major elements:

- *CPE (Customer Premises Equipment)* is anything that customers can connect to the PSTN: phones, modems, answering machines, and PBXs.
- *Transmission facilities* are the links that connect elements of the PSTN to each other: wires, microwaves, fiber optics, and so forth. Most CPE connects to the PSTN on a transmission facility, called a *line* or *loop*.
- *Switching systems* set up connections between elements of the PSTN. Most switching systems are owned by phone companies, but

Telephony for Programmers

some—*Private Branch Exchanges (PBX)*—are owned by organizations such as companies, universities, and government bodies.[1] Switching systems connect to each other on transmission facilities called *trunks*. (See Fig. 7.1.)

---

[1] A PBX supports basic phone service within the organization that owns and operates it. It also provides special features, like abbreviated dialing (instead of dialing all seven digits to call a phone attached to the PBX, one need dial only the last few). Phone companies offer a package of central office-based services, called *Centrex*, that is similar to what a PBX supports.

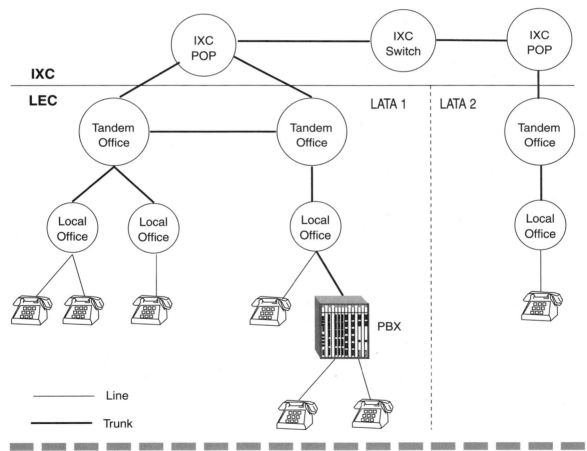

Figure 7.1
Switching systems in the PSTN.

- *Operations, Administration, and Management (OA&M) systems* keep all the pieces of the PSTN up and running.

A phone company switching facility is called a *central office (CO)*. A central office with lines connecting it to CPE is called a *local exchange*, or *end office (EO)*, or *Class 5 office* (a holdover from the days before divestiture, when the Bell System organized its switching systems in a hierarchy with five levels).[2] A central office that connects only to other central offices is called a *tandem office*, or *Class 4 office* (another holdover from the old Bell System terminology).

Offices that support local phone service are grouped together into *Local Access and Transport Areas (LATAs)*. Their operators are called *Local Exchange Carriers (LECs)*. *Incumbent LECs (ILECs)* are LECs that were offering local service in a given area before the introduction of competition into the phone market. *Competitive LECs (CLECs)* are LECs that have entered a local phone service market to compete with an ILEC.

Phone traffic between LATAs is carried by *Interexchange Carriers (IXCs or IECs)*. An IXC connects to a LATA network at a switching office, called a *Point of Presence (POP)*.

7.2 Switching: Analog and Digital

Switching is the process of routing calls to the correct network destination and then connecting their endpoints to each other. A switch is a network element that performs switching. Switches connect the endpoints of a call through a *switching matrix* or *switch fabric*. The points at which a connection passes into and out of a switching matrix are called *ports*.

The first switches were human-operated *cordboards*. *Step-by-step switches*, the first mechanized switching systems, were introduced in the late 1800s, followed in the 1930s by the faster *crossbar switches*.[3] All these are examples of *analog switches*. An analog switch takes the analog electrical signal for a call on an input port of its switching matrix and physically connects that port to an output port. As long as the call is in progress,

---

[2]"Office," "exchange," and "switch" are, for the most part, used interchangeably.

[3]Step-by-step switches, or *steppers*, were so called because each dialed digit in a phone number triggered the movement of a set of mechanical switches that took the call one step closer to its output port. Crossbar switches had smaller, faster, and quieter components than the stepper.

there is a dedicated physical connection between the ports. This is called *space-division switching*.

Step-by-step and crossbar systems are also examples of *electromechanical switches*. Electromechanical switches have an analog switching matrix under hard-wired control. To reprogram an electromechanical switch, one must go in and rewire it. *Stored program control (SPC)* switches, which first appeared in the 1960s, replaced the hardwired control of electromechanical switches with computer controls. Their switching fabrics remained analog, but reprogramming them took only a software change, not a a complete rewiring.

From the mid-1970s through the 1980s, switching started to take advantage of the ability to transmit voice digitally, with the introduction of *digital switches*. A digital switch takes a digitized voice signal on an input port of its switching matrix and, through time and/or space switching techniques, connects it, still digitized, to an an output port.

These switches and their switching matrices are just computers that have been specialized for telephony. Thus, digital switches have been able to take advantage of the same price/performance improvements seen elsewhere in the computing world (though they are still by no means inexpensive pieces of equipment). Furthermore, maintenance of a digital switch is not nearly as labor-intensive as it is on an analog switch. All analog switches deteriorate as dirt and wear accumulate on their electrical contacts and moving parts, so lots of time must be spent simply keeping them clean.

Analog voice is converted to digital format using *pulse code modulation (PCM)*. Figure 7.2 shows a simple example of how this works. First, the

Figure 7.2
A simple example of pulse code modulation (PCM).

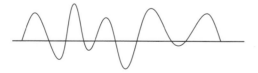

Original analog signal ...

... sampled and quantified ...

8 11 12 10 8 4 2 3 4 12 14 8 ...

... yields a stream of numbers ...

1000101111001010100 ...

... that can be transmitted digitally.

original analog signal is sampled at a rate of 8000 times per second. Each sample is assigned a value that depends on its amplitude. In the example shown, I've quantified these samples in a range from 0 to 15. In reality, samples are quantified in a range from 0 to 255, which makes 8 bits per sample, or 64 Kbps (this rate is called *DS0*, the fundamental unit of voice transmission in the PSTN). The stream of numbers that results from quantification is converted to binary digits and transmitted as a stream of bits.

More than one PCM stream can be multiplexed onto a single digital link using *time division multiplexing (TDM)*, as shown in Fig. 7.3. A hierar-

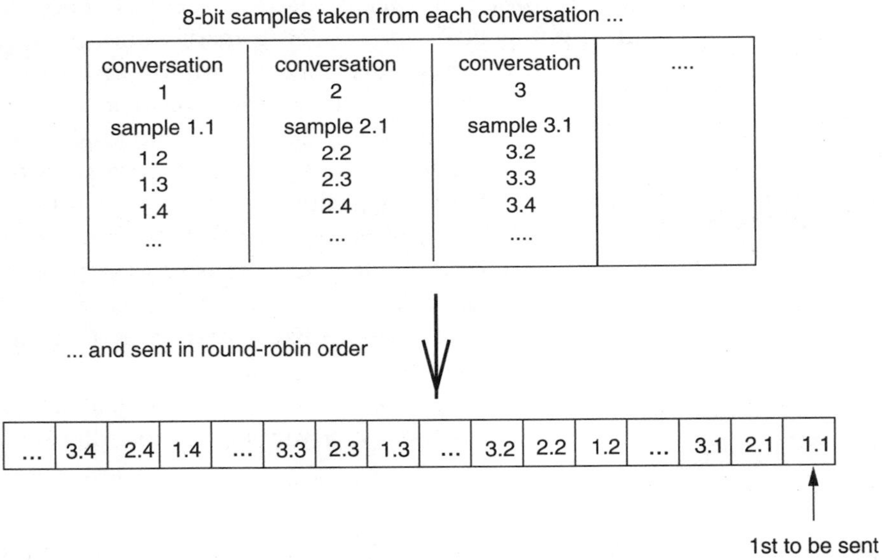

Figure 7.3
Time division multiplexing (TDM) of multiple PCM streams onto a single link.

chy of TDM carriers has been standardized in the U.S. for multiplexing DS0 *voice channels* on a single digital link:

T1. 24 DS0 voice channels with an aggregate 1.544 Mbps data rate. (The comparable rate outside the U.S. is called *E1,* which carries 32 voice channels at an aggregate rate of 2.048 Mbps).

T2. 4 T1 channels with an aggregate 6.312 Mbps data rate.

T3. 6 T2 channels with an aggregate 44.736 Mbps data rate.

T4. 6 T3 channels with an aggregate 254.176 Mbps data rate.

7.3 Signaling and Call Processing

Call processing is all the steps that must be performed in the PSTN to set up, maintain, and tear down a phone call. *Signaling* is the exchange of messages that support call processing between network elements. The processing of a simple two-party call proceeds as shown in Fig. 7.4 (it's assumed that both parties are served by the same end office):

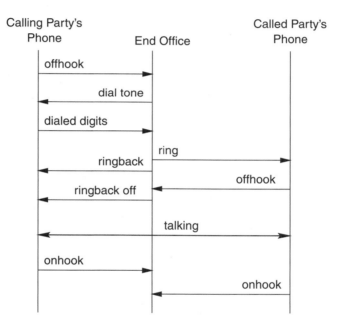

Figure 7.4
Circuit-associated signaling in a simple intraswitch phone call.

1. The calling party lifts the receiver of the phone (sends an *offhook* signal to the end office). This tells the end office the calling party wants to make a phone call. This step is called *origination*.
2. The end office sees that the calling party's line is offhook and marks it "busy."
3. The end office plays dial tone back to the calling party. This means the switch is ready for the calling party to start dialing digits.
4. The calling party dials the first digit of the called (pronounced with two syllables—"call-ed"—by those in the know) party's phone number. Once the switch sees that digit, it turns off dial tone.

5. The calling party continues dialing digits. The switch analyzes the digits to decide where the call should go *(digit translation)*. If the dialed digits can't be translated, the switch sends the call to a special tone or announcement that tells the calling party something went wrong.

6. Successful digit translation produces a *route*. The end office uses this to find and reserve a path to the called party through its switching matrix. If no such path can be found, the end office plays a fast busy tone back to the calling party. Otherwise, the end office checks the status of the called party. If it's busy, the end office plays a busy signal back to the calling party. If the called party is idle, the end office plays *audible ringing* or *ringback* tone to the calling party.[4] At the same time, the end office *alerts* the called party (rings that phone).

7. If the called party answers, the end office sees an offhook signal on that line. The end office recognizes it and turns off ringing on that line, as well as the ringback being played to the calling party.

8. The end office connects the calling and called party's lines along the path reserved for them through the switching matrix during routing. Once that's done, the two parties can begin talking. This is called "establishing a speech path" or "voice path," or simply "cut through."

9. When either party finishes talking, that party hangs up, sending an *onhook* signal to the end office.

10. The end office frees the resources it was using for the call. Those may now be used for new calls.[5]

The following signals were used in this simple call:

- *Offhook.* Sent by the calling party to originate and by the called party to answer the call
- *Dial tone.* Sent by the end office to let the calling party know it's ready to receive digits

---

[4] If the call routes through other switches, ringback tone is played back from the switch that serves the called party. There is no correlation between the ringing of the called party's phone and the ringback heard by the calling party. That's why it's possible to call somebody and have them answer before you even hear ringback.

[5] This is an extremely simplified version of what really goes on. Tearing down a call is one of the more complicated parts of call processing, since there's no telling who will hang up first, and much of what has to be done depends on that, especially when a call is being billed.

- *Dialed digits.* Sent by the calling party to identify the called party
- *Ringback.* Sent by the end office to let the calling party know the call has been placed and is waiting for an answer
- *Ringing.* Sent by the end office to tell the called party there is an incoming phone call
- *Onhook.* Sent by either party to end the party's participation in a phone call
- *Others.* Various busy signals and announcements sent by the end office when a call can't be placed

The processing of calls that pass through more than one switch is basically the same except that, as a call is passed from one switch to another, they exchange information and the speech path flows along trunks rather than lines.

7.4 Phone Numbers

A phone number is what the PSTN uses for routing to the called party in a phone call. Common acronyms for phone numbers include *TN (Telephone Number)* and *DN (Directory Number).* Often, but by no means always, a phone number corresponds to a particular phone. A single phone number may refer to a collection of phones. A single phone may have more than one phone number. To complicate matters, phone numbers may also be used to indicate a service (for example, the so-called *star codes,* such as *72 for invoking call forwarding) or to select the carrier for a phone call (for example, dialing 10-10-xxx before a long distance number, where the digits "xxx" identify a specific carrier).

The ITU-T standards for international phone numbers are set out in Recommendation E.164. Phone numbers compatible with this standard are called *E.164 numbers.* E.164 divides the world into nine geographic *World Zones,* numbered 1 through 9. Each country within a World Zone has a unique 1-, 2-, or 3-digit *country code.* The first digit of a country code is the same as the World Zone in which that country is located.

The layout of everything after the country code, referred to as either the *national* or *significant number,* is the responsibility of some administrative body within the country. The standards they set are called a *national numbering plan.* A country code plus a national number form an *international phone number.* International phone numbers can be, at most, 15 digits long.

In North America, the national numbering plan is called the *North American Numbering Plan (NANP).* The NANP is administered by the *NANP Administration (NANPA),* a private group overseen by the FCC. NANP phone numbers are 10 digits long: a 3-digit area code (also called the *Numbering Plan Area (NPA)),* a 3-digit office code, and a 4-digit station number. Most NPAs correspond to a geographic area (hence, "area code"). The exceptions are NPAs, such as 800, 888, 411, and 911. NPAs like 411 and 911 are a special case, since they stand by themselves. The 3-digit office code may or may not refer to an actual central office. These digits are sometimes called an *NXX,* since the first digit can be anything from 2 through 9 (*N digit*), and the next two can be anything from 0 through 9 (*X digit*). The format used to be NNX, so you'll occasionally see that term also. The last four digits are the station number. Pretty much anything goes here, though it's common practice to reserve station numbers that start with the digit 9 for payphones.

7.5 ISDN

The *Integrated Services Digital Network (ISDN)* is a technology that supports voice and data on existing home and business access lines. With ISDN, the bandwidth of these lines is split into 3, 24, or 31 digital channels. A 64 Kbps ISDN *B channel* carries either data or voice. A 16 or 64 Kbps ISDN *D channel* carries out-of-band signaling for ISDN services that use the B channels. ISDN B and D channels are combined in two common service offerings:

- *Basic Rate Interface (BRI)* offers two B channels and one D channel, and is often called 2B+D. BRI is targeted for homes and small businesses. Each of the two B channels can carry a voice conversation or a 64 Kbps data connection. Since the two B channels are independent of each other, it's possible to have two voice conversations, two data connections, or a data and a voice connection at the same time. It's also possible to aggregate the two B channels for a single 128 Kbps data connection.
- *Primary Rate Interface (PRI)* offers 23 B channels and one D channel (23B+D) in the U.S. and Japan, and 30 B channels and one D channel (30B+D) in Europe. PRI was designed largely for businesses with PBXs.

ISDN has had limited success in the U.S. The rise of the Internet gave ISDN a boost by increasing demand for high-speed data access. ISDN's data rate of 64 to 128 Kbps was far above the speed of most analog modems at the time. Unfortunately for ISDN, modem speeds soon increased to the point where they came within shouting distance of the rate of a single B channel. And not too long after that, true broadband service arrived with the introduction of cable modems and the various flavors of *DSL (Digital Subscriber Line)*. These offered speeds many times faster than ISDN at comparable prices. ISDN's been much more successful in Europe. Furthermore, PRI access is the basis for many network services, and that will continue to be the case in the U.S. and elsewhere. It should also be noted that, even if ISDN hasn't been an unqualified success, by demonstrating that traditional copper loops could support simultaneous voice and high-speed data, it stimulated the research that led ultimately to DSL. If for no other reason, that would be enough to cement ISDN's importance in the history of communications.

However, DSL is not ISDN's sole legacy. The ISDN D channel protocols for out-of-band signaling will be with us for quite some time, even if they aren't always used with ISDN. Rather than using electrical signals as POTS does (offhook, onhook, ringing, and so forth), ISDN exchanges messages between an ISDN phone and an ISDN-equipped switch. These messages are defined in the ITU-T's Q.931 [58] protocol. As it turns out, Q.931 is also the basis for some important IP telephony protocols.

Figure 7.5 shows a basic phone call that uses Q.931 signaling as described below:

1. The calling party goes offhook. The phone sends a Q.931 SETUP message that tells the switch this will be a voice call on the B1 channel for the indicated calling number. The SETUP message also includes a *call reference number* that the switch and the phone use to keep track of activity on the originating side of the call.

2. The switch replies with SETUP ACKNOWLEDGE. This message echoes the call reference number and channel ID from the SETUP and tells the phone to play dial tone.

3. The caller starts dialing digits. These are sent to the switch as INFO messages. After the switch gets the first INFO message containing a dialed digit, it sends back an INFO message of its own that tells the phone to turn off dial tone.

4. INFO messages go from the phone to the switch until there are no more digits to be sent. Some ISDN phones collect all the digits

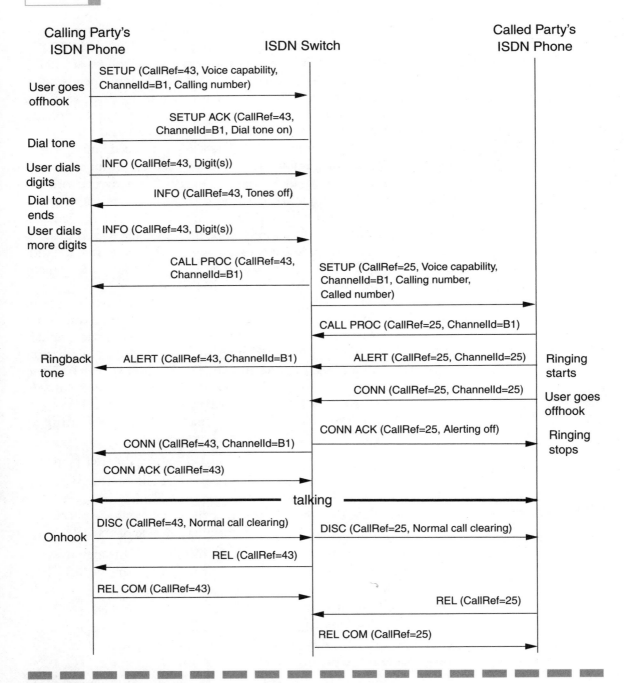

Figure 7.5
Basic phone call with ISDN Q.931 signaling.

and send them in a single INFO message, a technique known as *en bloc sending*.

5. The switch sends a CALL PROCEEDING message to the phone once it has enough digits to route the call.

6. The switch sends a SETUP message to the called party's phone. Like the originating SETUP message, this includes a call reference number. It need not be the same as the originating side's call reference number since it's just a way for the phone and the switch to keep track of things between themselves. Besides the call reference number, this message includes a B channel assignment, an indication that this is a voice call, and the calling and called phone numbers.

7. The called party's phone responds with a CALL PROCEEDING message. Shortly thereafter, it sends an ALERTING message to tell the switch it is ringing.

8. The switch sends an ALERTING message of its own to the calling party's phone. This message tells the phone to start playing ringback.

9. The called party answers the call and that phone sends a CONNECT message to the switch. The switch responds with a CONNECT ACK that tells the phone to stop ringing.

10. On the other side of the call, the switch sends the calling party's phone a CONNECT message. The phone stops playing ringback and responds with a CONNECT ACK message.

11. The switch establishes a talking path between the calling and called parties, and the call is set up.

12. Eventually, one of the parties goes onhook and ends the call. Assume it's the calling party. That phone sends a DISCONNECT message to the switch. The switch sends a DISCONNECT message of its own to the called party's phone.

13. The switch returns a RELEASE message to the calling party's phone, which responds with a RELEASE COMPLETE, and that side of the call is done. A similar exchange ends the called party's side of the call.

This example describes a pure ISDN call, in which the calling and called party's phones take care of many things a switch handles in the non-ISDN world: dial tone, ringing, and ringback, for example. It's common for the switch to handle these things, just as it does in POTS, while still exchanging the same Q.931 messages with the phones.

7.6 SS7

7.6.1 Circuit-Associated and Common-Channel Signaling

For most of its history, the PSTN used *circuit-associated signaling,* carrying signals on the same links as voice. Circuit-associated signals may be further classified as *in-band* (carried in the same frequency range as voice) or *out-of-band* (carried in a different frequency range than voice). While circuit-associated signaling has served the phone network well, it does have some drawbacks:

- Circuit-associated signals are tones of various frequencies and lengths. They're inherently slow, since they have to be played long enough for the equipment on a switch to see and recognize them.
- While dial tone, ringback, and the like are meaningful to humans, other circuit-associated signals, especially those used to communicate between switches, are just so much noise that contribute nothing to a customer's experience.
- While circuit-associated interswitch signals are meaningless to most people, some enterprising individuals figured out how to imitate them to trick the phone network into placing free phone calls and other such skullduggery. Since the signals were carried on the same links as voice, a phone line was all it took to gain access to the signaling infrastructure.

In the mid-1960s, *common channel signaling* was introduced by the former Bell System as a replacement for circuit-associated interswitch signaling. Instead of traveling on the same links as voice, common channel signals travel on special signaling links that are entirely separate from voice links. A single such signaling link can be shared by many voice channels, hence the name *common channel.*[6] Common channel signaling has several advantages over circuit-associated signaling:

- There is less opportunity for fraud, since signals travel on entirely separate links from voice. These are links to which customers typically have no access.

---

[6]Common channel signaling is often called out-of-band signaling. This is not strictly accurate, since "band" refers to frequency at which the signal is transmitted, not to the physical link.

- Rather than trying to make voice links serve a dual purpose—voice transmission and signaling—signaling links can be optimized for signaling. Since they no longer need to support voice traffic, signaling links can be designed to carry data, which is what signals really are, when you think about it. Once you've replaced signaling tones with data messages, you increase the speed of the links and the number of different messages sent on them.

The first common channel signaling system—*Common Channel Interoffice Signaling (CCIS)*, also known as *Signaling System 6 (SS6)*—was introduced in the 1960s. SS6 transmitted data at a rate of 2400 bps (later upgraded to 4800 bps) in fixed-length messages. In the early 1980s, the newer *Signaling System 7 (SS7)* was introduced, and SS6 was phased out.

SS7 is similar in many ways to SS6. One difference is that it transmits data at higher speeds, 56 Kbps and higher. Another difference, and this is arguably even more important, is that SS7 supports variable-length messages. This considerably increases the number of possible messages and makes it much easier to expand the SS7 message set.

7.6.2 SS7 Network Architecture

The basic elements of the SS7 network are as follows:

Signaling Point (SP) Any PSTN element that can communicate with another using SS7. Every signaling point has a unique SS7 network address called a *point code*. SS7 messages contain point code fields that identify their source and destination signaling points, the *signaling end points (SEPs)*. Each point code consists of three 8-bit numbers:[7]
- The *network number* identifies the carrier toward which the message is being sent. Large carriers operate their own SS7 networks, each with its own network number. Since there can be no more than 256 unique network numbers, a company's SS7 network must be of a certain size before it is assigned a network number. Smaller networks are assigned one or more cluster numbers within network numbers 1, 2, 3, and 4. Very small networks are assigned point codes within network number 5, with their cluster number determined

---

[7]What's described here is the ANSI flavor of SS7 used in North America. Other versions, each with its own slight variations, include Chinese, Japanese, and ITU (Europe and elsewhere). Traffic that passes from a network using one version of SS7 to a network using another must be converted as it crosses the boundary between the two.

Figure 7.6
SS7 network architecture (simplified).

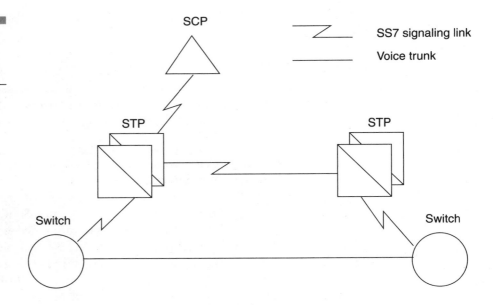

by the state in which they are located. Network number 0 is not used at all. Network number 255 is reserved for future use.
- The *cluster number* identifies a group of up to 256 signaling points within an SS7 network. A network can have up to 256 clusters.
- The *member number* identifies a unique signaling point within a cluster.

Linksets A collection of one to sixteen *signaling links*. Linksets connect adjacent signaling points. Signaling links transmit data at a minimum rate of 56 Kbps.

Switches Network elements that originate phone calls, terminate phone calls, and act as midpoints, or *tandems*, for phone calls. From SS7's perspective, there are several different kinds of switches:

Common Channel Signaling Switching Office (CCSSO) A switch equipped to use ISUP (see Section 7.6.3), the SS7 protocol that handles call setup, management, and teardown.

Service Switching Point (SSP) A switch that can stop a call while it's in progress, launch a query to an SCP, and continue processing the call using the results it gets back. SSPs use the SS7 TCAP protocol (see Sec. 7.6.3) for database queries and responses. SSP and CCSSO functions are complementary and often coexist on the same switch.

Operator Services System (OSS) A switch with the equipment needed to provide operator assistance to callers. OSSs make TCAP queries to SCPs to check things like credit card validity.

Signaling Transfer Point (STP) A packet switch that routes SS7 messages to and from other signaling points. STPs also sit between SS7 networks, acting as an SS7 firewall of sorts that filters the messages they exchange.

Service Control Point (SCP) A signaling point that provides information to be used during call processing. One of the first SCP-based applications of SS7 was 800 number database lookup. Before 800 numbers were invented, area codes always indicated where a call should be routed. But an 800 number can be anywhere, and its 800 area code gives no clue where that might be. A database in an SCP correlates 800 numbers with their actual phone numbers. When you call an 800 number, the 800 area code tells your end office it needs to query the 800 number database by exchanging SS7 messages with the SCP that holds that information.

SCPs and STPs are usually deployed in mated pairs to provide redundancy and protect against loss of phone service, should there be an isolated SCP or STP failure. Links between signaling points are also deployed in pairs to increase reliability and support load sharing of SS7 traffic. If a link fails, its signaling traffic is diverted to one of its mates. The SS7 protocol has numerous error correction and retransmission features that support failure recovery on behalf of signaling points and links.

7.6.3 SS7 Protocols

SS7 is a layered protocol. However, it was designed before the OSI 7-layer protocol model was developed, so an exact mapping of SS7 onto that model is not possible. Nevertheless, it's useful to keep the OSI model in mind for comparison with the SS7 protocol stack in Fig. 7.7.

MESSAGE TRANSFER PART (MTP) The *Message Transfer Part (MTP)* sits at the bottom of the SS7 protocol stack. MTP corresponds to the physical, data link, and network layers of the OSI model. As the fundamental SS7 protocol, it must reside on every SS7 signaling point. MTP provides a connectionless message delivery service. MTP messages

Figure 7.7
Core protocols of the SS7 stack.

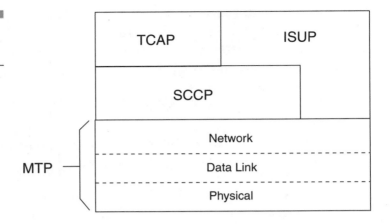

contain an *Origination Point Code (OPC)* and *Destination Point Code (DPC)* to identify the SP that sent the message and the SP for which it is destined.

SIGNALING CONNECTION CONTROL PART (SCCP) The *Signaling Connection Control Part (SCCP)* sits on top of MTP. It's used mostly for routing TCAP messages.[8] Where MTP is strictly connectionless, SCCP provides a connection-oriented service as well. These, however, are little used.

One of SCCP's most important features is an enhanced version of SS7 routing, called *global title translation (GTT)*. A *global title* is a sequence of digits—an 800 number, a calling card number, and so forth—that SCCP translates into a destination point code and *subsystem number*. A subsystem number identifies an application or database at the destination signaling point. With global title translation, an originating SP can send a message even when it doesn't know the point code or subsystem number for which the message is bound. All it has to supply is a global title—dialed digits or other information—and SCCP will figure out the destination.

An interesting example of global title translation in practice is 800 number dialing. Recall that 800 numbers don't represent actual physical addresses in the phone network. Before calls can be routed to them, they

---

[8]The SS7 standards show that SCCP may transport ISUP, but this has not yet been implemented and, at this time, SCCP is used only with TCAP. It's not clear when, if ever, ISUP on SCCP will be implemented.

must first be translated to actual phone numbers. To do this, you might have every switch keep track of these translations on its own. It should be obvious why this hasn't been done.

A better solution would be to keep this information in one place and give all switches access to it. This was plausible before there was any long distance competition and one company (AT&T) handled all 800 numbers. As other long distance carriers entered the market, they also wanted to offer 800 number service. Even with many long distance carriers, you could still let a single database keep track of all 800 numbers. While that would make things easier for switches, since all they need to know is where to find that database, it would be hard to keep updates pouring in from all those carriers accurate and consistent. Errors, delays, and confusion would be the likely result.

To avoid those problems, each carrier that wanted to offer 800 number service was assigned a block of 800 numbers it could manage as it saw fit. That meant each carrier took care of its own 800 number database. Now switches had to first check in what range a dialed 800 number fell before deciding whose database to query. They also had to keep track of where to find these databases.

This worked, but there was still a problem. Because specific carriers were assigned specific 800 numbers, anybody with an 800 number who wanted to change carriers would also have to change that 800 number as well. Thus, *800 number portability* was introduced to prevent this. Rather than make switches keep track of where to go for 800 number translations based on the range within which a number falls, separate databases were set up for this. When a switch gets an 800 number call, it first queries one of these databases to find out who handles the 800 number in question. Once it has this information, it sends a query to the 800 number's handler. The handler returns the actual routing number.

With global title translation, switches needn't know where to find any of these servers. For the first query, the one to find an 800 number's handler, a switch just sends out a message whose SCCP "to" field says it needs help resolving an address. That address is set to the 800 number to be resolved. Global title translation in the SS7 network routes the query to the right database, which returns the address (point code) of the relevant 800 number handler. The switch uses that address to make its second query, the one to find the actual address that corresponds to the 800 number.

ISDN USER PART (ISUP) The *ISDN User Part (ISUP)* supports SS7 call signaling.[9] ISUP defines the messages and procedures needed to set up, manage, and release trunks for calls involving two or more switches. Despite its name, ISUP handles both ISDN and non-ISDN calls.

In a simple ISDN call between two switches, ISUP signaling proceeds as shown in Figure 7.8.[10]

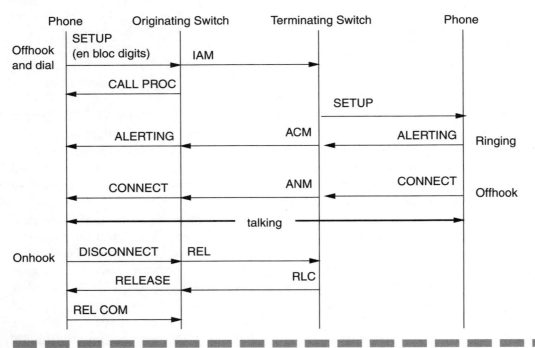

Figure 7.8
SS7 ISUP signaling for a simple interswitch ISDN phone call.

1. After the calling party finishes dialing, their switch does routing and translation. Since ISUP isn't used for intraswitch calls, we can assume the call routes to another switch. The originating switch reserves an outgoing voice circuit for the call and sends an *Initial*

---

[9] A precursor to ISUP was the *Telephone User Part (TUP)*. TUP handles only analog circuits.
[10] It would look much the same with POTS phones. We've chosen to show ISDN signaling to demonstrate SS7/ISDN interworking and the similarity between those two protocols.

Address Message (IAM) to the terminating switch through the SS7 network. This message contains the originating point code, the destination point code, the ID of the voice circuit reserved by the originating switch, the dialed digits and, optionally, the calling party's phone number and name.

2. The terminating switch checks the dialed number to see if it's busy. If it isn't, the terminating switch returns an *Address Complete Message (ACM)* to the originating switch. This lets the originating switch know that the other end of the voice circuit has also been reserved. The terminating switch signals the called party's phone to start ringing and plays ringback to the calling party.

3. When the called party answers the phone, the terminating switch stops ringing and sends an *Answer Message (ANM)* to the originating switch. The two parties are connected and may talk.

4. If the calling party hangs up first, the originating switch releases its end of the voice circuit and sends a *Release Message (REL)* with the ID of that circuit to the terminating switch. If the called party hangs up first, or if the the called party's line was busy upon receipt of IAM, the terminating switch releases its end of the voice circuit and sends a REL message to the originating switch. No matter who sends it, the REL message includes a reason for the release (for example, normal disconnect of a call in progress or calling party busy).

5. Once either end gets a REL message, it sends back a *Release Complete Message (RLC)* message. The recipient of the RLC message idles the circuit, the ID of which was in the REL message it sent.

TRANSACTION CAPABILITIES APPLICATION PART (TCAP)
The *Transaction Capabilities Application Part (TCAP)* provides a means for one network element to request that operations be performed at another network element. Every SSP and SCP must support TCAP. SS7 network elements use TCAP to invoke services or request information from other SS7 network elements:

- SSPs request information from SCP databases and applications that supply routing data, such as for 800 numbers, or customer information, such as calling card number, PIN, or calling name. In wireless networks, TCAP carries *Mobile Application Part (MAP)* messages that support user authentication, handset identification, and roaming.

- One SSP may query another about the status of a line (for example, is it busy or idle?).
- One SS7 network entity (not necessarily a switch) may invoke a feature or application in another.

Network elements request operations and return results by carrying out what's called a *dialogue* or a *transaction*.[11] Many operations may be active within a single dialogue, each at different stages of processing. Many operations may be active and at different stages of processing within a dialogue. A typical dialogue consists of a request message and its corresponding response message. TCAP also allows dialogues with multiple requests and responses. TCAP messages transmit operations and results in *components,* of which there are two basic kinds:

INVOKE Request to perform an operation. Other fields in the component specify the operation and its arguments. ANSI TCAP has two types of INVOKE: regular INVOKE and INVOKE_NL (for "invoke not last"). When a request has only one component, it's marked INVOKE_L. When it has more than one component, all but the last are marked INVOKE_NL. The last one in the sequence is marked INVOKE_L. ITU TCAP handles both situations with INVOKE.

RESULT Response to a successful INVOKE. Other fields in the component contain the result values. In both ANSI and ITU TCAP, a RESULT can be either intermediate (RESULT_L, for "result last") or final (RESULT_NL, for "result not last"). A RESULT_NL component will be followed by more RESULT components. A RESULT_L component is the last or only component in the sequence.

Besides these, TCAP also supports an ERROR component, used when a request cannot be carried out, and a REJECT component, used when a network element refuses to carry out an operation on behalf of the requester.

TCAP messages are at most 256 bytes long, of which half or more can be overhead. Because not everything needed to request an operation or return its results may fit in 256 bytes, a single request or result may have to be spread across more than one component. TCAP software stores components it gets requests or results from until it gets a *dialogue primitive*. Dialogue primitives mark key points in an ongoing dialogue, and their number varies, depending on whether you are using ANSI or ITU TCAP. However, they all do the same things:

---

[11] Applications that use TCAP call it a *dialog*. TCAP software talking to other TCAP software calls it a *transaction*.

UNIDIRECTIONAL (UNI) Sends information from one network element to another with no reply expected.

BEGIN Starts a TCAP dialogue. The receiver may end the transaction. ANSI TCAP splits this into Query With Permission and Query Without Permission primitives. Query With Permission starts a simple request and response dialogue, where no further messages will be exchanged after results have been sent back. Query Without Permission starts a request in which the receiver may not end the dialogue on its own (presumably, the sender expects to send more messages for that dialogue).

CONTINUE Continues a TCAP dialogue begun earlier. ANSI TCAP splits this into Conversation With Permission and Conversation Without Permission. The receiver of Conversation With Permission may end the dialogue. The receiver of Conversation Without Permission may not.

END Ends a TCAP dialogue. ANSI TCAP calls this Response.

ABORT Ends a TCAP dialogue before it has had a chance to complete.

Once TCAP software gets a dialogue primitive from an application, it transmits all the components it's saved up for that application as a single TCAP request or response. TCAP software on the receiving end unpacks these and sends them up to the receiving application as shown in the request and response dialogue in Fig. 7.9. (I use ITU names for the primitives in that diagram, because I find them easier to understand than their ANSI equivalents.) On the receiving end, TCAP software passes the dialogue primitive up first, then the components in the order they were passed down from the sending application. Figure 7.10 shows a unidirectional dialogue in which the receiver does not send a response. Figure 7.11 shows a dialogue that uses CONTINUE to convey results across more than one TCAP transmission.

There can be many TCAP dialogues going on at the same time. Every dialogue has a unique transaction ID. Every message in a dialogue carries the same transaction ID.[12] Within a dialogue, components have *component* or *invoke IDs* that associats them with the INVOKE component that requested them. Component IDs are assigned by the requester.

TCAP messages don't have to follow a particular path through the SS7 network (as compared with ISUP messages that have to pass through all

---

[12]Actually, there is an an *origination transaction ID* created by the request originator and a *responding* or *destination transaction ID* created by the responder. But it's normal for the responder to use the same ID as the one created by the originator.

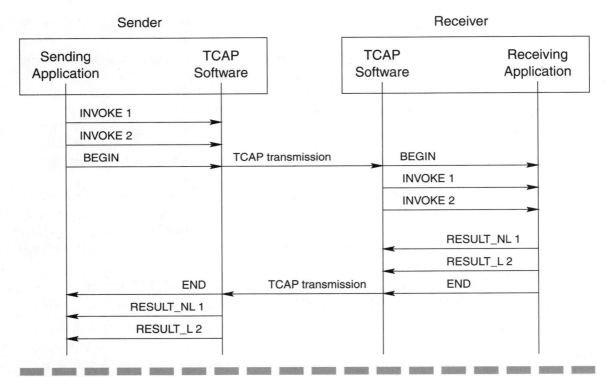

Figure 7.9
A simple TCAP request and response dialogue. TCAP transmissions may contain more than one component.

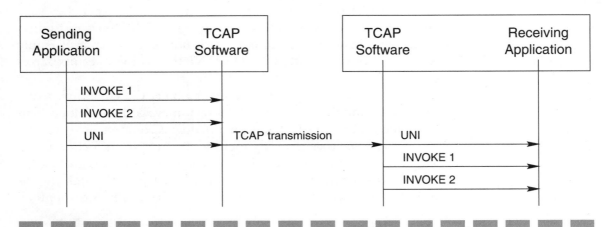

Figure 7.10
A UNIDIRECTIONAL dialogue doesn't have a response.

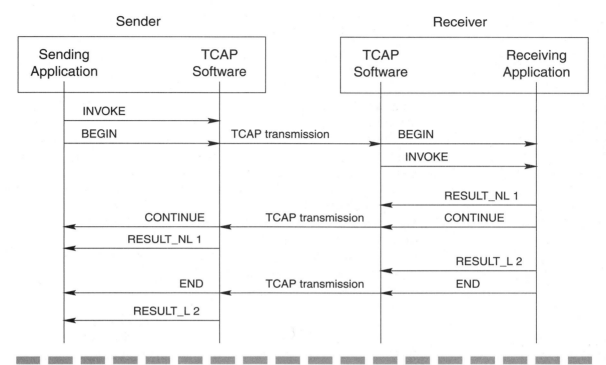

Figure 7.11
A TCAP dialogue may extend across many transmissions.

the switches in a call's voice path). TCAP uses SCCP as its transport protocol because the destination of a TCAP message may be a signaling point with a single point code that supports more than one database or application. SCCP's global title translation and subsystem addressing allow TCAP messages to go to different recipients on a single point code.

7.7 Intelligent Network (IN)

7.7.1 The Road from IN2 to IN0

That's not a typo. The Intelligent Network did in fact go from version 2 to version 0. To see how that came about, let's look at a bit of history. The introduction of stored-program control and digital switches in the 1970s and 1980s turned the development of voice services into a programming task like any other. However, until recently, only those few companies that actually built switches could write that software. There were many reasons for this. Probably the most important was that the sheer complexity and size—on the order of many millions of lines of code—made it impractical for anyone other than the switch vendors themselves to modify and add to that software. Thus, whenever service providers wanted to offer a new service, they had to go to the switch vendors, convince them to add it to their development plans, and then wait the years it took to be developed and tested. Finally, they could buy the new software and offer the service to their customers. If they used more than one kind of switch, they had to repeat this process for each one.

Not surprisingly, service providers weren't entirely happy with this situation. *Intelligent Network (IN)* is a collection of technologies first developed in the late 1980s and early 1990s, the purpose of which is to allow service providers to build and offer phone services themselves. IN's roots go back to the early 1980s, when the first SCP-based network database for 800 numbers was introduced. Ownership of this database wasn't an issue when there was just one Bell System. After divestiture in 1984, the newly created *Regional Bell Operating Companies (RBOCs)* saw it would be in their best interests eventually to deploy their own 800-number databases. Bellcore, the Bell Labs clone created to provide R&D for the RBOCs after divestiture, set about developing requirements for these localized 800-number databases. This was the first system to which the name "Intelligent Network" was applied.

The desire of the newly independent RBOCs to do their own service development led Bellcore to explore ways in which SCP-based IN could be expanded to support more than just routing and translations. In the mid-1980s, Bellcore came up with something they called *IN2* (now referring to the earlier 800-number service as *IN1*). IN2 envisioned an entire suite of advanced capabilities, including the ability to intervene in the middle of call processing and to manipulate multiparty conference calls at will. IN2 was an ambitious piece of work. Too ambitious as it turned out, as switch vendors could see no way IN2 could be implemented.

In the late 1980s, IN entered a new phase with the launching of the *Multivendor Interaction Forum*. Through a series of meetings among representatives of Bellcore, the RBOCs, and the switch vendors, a new specification for IN was developed, *Advanced Intelligent Network (AIN) Release 1.0*. Despite their participation, the switch vendors still felt that, like IN2, AIN 1.0 was too much to build all at once. Bellcore's response was to shrink AIN 1.0 to a series of phased releases known as AIN 0.0, AIN 0.1, and so on.[13] IN was eventually deployed in the North American network through these phased releases.

At about the same time the Multivendor Interaction Forum was distilling IN2 down into AIN 1.0, the ITU-T started a project to develop Intelligent Network standards of its own. The ITU-T was strongly influenced by AIN activities in North America. The ITU-T IN standards are the basis for IN throughout the world. However, North American AIN and ITU-T IN, while similar, have some significant differences (for details, see [16]).

The first ITU-T IN standard defined what's called *Capability Set 1 (CS-1)*. The SS7 protocol that supports CS-1 is called *Intelligent Network Application Part (INAP)*. Just like AIN 1.0, CS-1 defines a comprehensive set of capabilities that are being introduced in stages. The ITU gave ETSI the task of developing a base set of capabilities to satisfy immediate needs, which they named *Core INAP*.

7.7.2 The Architecture of IN

Given its roots in 800-number databases, it should come as no surprise that IN depends heavily on SS7. Where traditional phone services run entirely on switches, IN divides service logic between an SCP and an IN-equipped switch. The switch, referred to in the IN architecture as a *Service Switching Point (SSP)*, is still in charge of basic call processing. Under IN, the SSP executes a call processing state machine called the *Basic Call State Model (BCSM)*. The BCSM comes in originating *(O-BCSM)* and terminating *(T-BCSM)* flavors.

The primary states in the BCSM are called its *points in call (PICs)*. PICs appear at significant stages of call processing. Associated with certain PICs are *detection points (DPs)* that are passed through as a PIC is left during call processing because an event has occurred. For example, in Fig. 7.12, which shows a portion of the AIN 0.1 originating BCSM, a call starts in the Null PIC. When the calling party goes offhook, their BCSM moves from the Null to the Authorizing Origination Attempt PIC by way of the Origination Attempt DP.

DPs may be either *armed* or *unarmed*. When a call reaches an armed DP, the SSP suspends its BCSM and sends a TCAP message requesting

---

[13]It's not clear whether the AIN number sequence was to have reached 1.0 from 0.9 as a step function, or whether it would have approached it asymptotically.

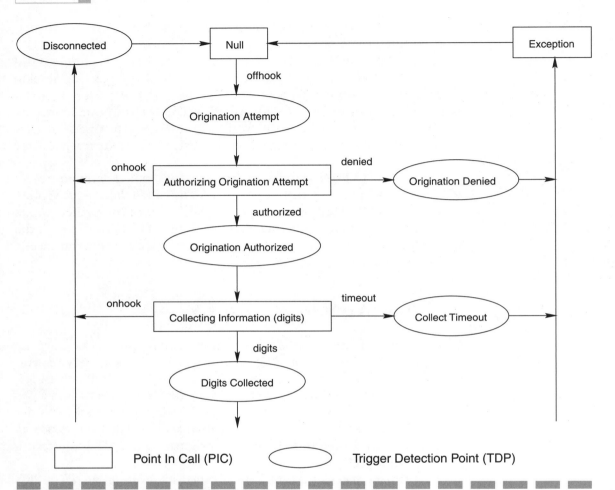

Figure 7.12
A portion of the AIN 0.1 originating BCSM.

further instructions to an SCP. To satisfy this request, the SCP executes *service logic* and sends back a reply. The SSP picks up where it left off processing the BCSM, using the information in the SCP's reply.

A DP may be armed *statically* by the service provider's provisioning systems, or it may be armed *dynamically* by the SCP. A statically armed DP is called a *trigger detection point (TDP)*. The event that causes a call to enter a TDP is called a *trigger*. (Just to confuse things, TDPs are also sometimes called triggers). A dynamically armed DP is called an *event detection point (EDP)*.

A *subscribed* or *line-based trigger* is armed for any calls that originate or terminate on a specific customer's line. A *group-based trigger* is armed for any calls that originate or terminate on a group of lines. A *switch-based trigger* is armed for any calls that originate or terminate on an entire switch.

In Figure 7.12, as we've already seen, originating calls start in the Null PIC with the subscriber onhook. When the SSP sees a subscriber go offhook, their BCSM advances to the Origination Attempt TDP. If this TDP is armed, the SSP asks an SCP for instructions before proceeding to the Authorizing Origination Attempt PIC. In this PIC, the call may be either authorized or denied. If it's denied, the call goes through an exception handler via the Origination Denied TDP and the BCSM returns to the Null PIC. If it's authorized, the call advances to the Collecting Information PIC via the Origination Authorized TDP. Once enough digits have been collected, the BCSM goes through the Digits Collected TDP, and from there to the rest of the call. Digit collection may timeout, in which case the call goes through an exception handler via the Collect Timeout TDP and the BCSM returns to the Null PIC. All the PICs after call origination may also be left if the caller goes back onhook, passing through the Disconnected TDP and then back to the Null PIC.

Besides the SCP and SSP, the IN architecture specifies other network elements:

SDP Service Data Point. Provides database capabilities. Usually incorporated into an SCP, but it may be separate.

IP Intelligent Peripheral. Supports what's sometimes called the *specialized resource function (SRF)*. These are special capabilities needed by IN services: digit receivers, announcements, conference bridges, fax transmission, voice recognition, and so on.

AD Adjunct. Executes service logic like an SCP, but connects to SSPs by some means other than SS7 (for example, TCP/IP).

SN Service Node. Executes service logic like an SCP and provides specialized resources like an IP, but connects to the SSP via ISDN rather than SS7.

IN services are developed within a *Service Creation Environment (SCE)*, which provides tools that are tailored for writing IN service logic. A *Service Management System (SMS)* provides the interface between IN and a service providers' *Operations Support Systems (OSSes)*. An SMS can be used to monitor service performance and to update IN service logic or data. (See Fig. 7.13.)

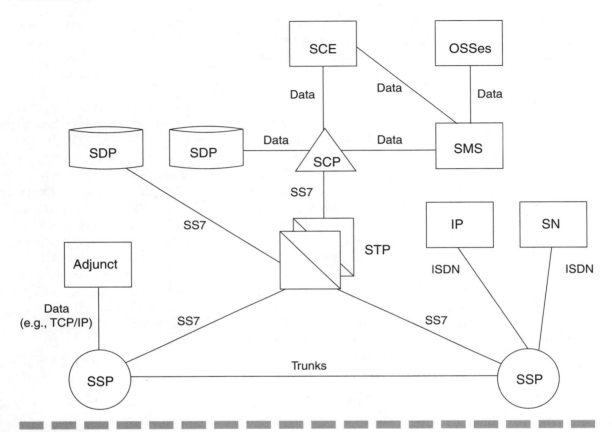

Figure 7.13
IN network elements.

7.7.3 An Example of an IN Service

Voice Activated Dialing (VAD) is a good example of how an IN service can work (see Fig. 7.14):

1. The calling party subscribes to VAD. The SMS statically arms the Offhook Immediate trigger in the subscriber's SSP and sets up other information in the SCP.
2. The subscriber initiates a call by going offhook. The SSP hits the Offhook Immediate trigger detection point while executing the O-BCSM and finds it armed for this customer.

Telephony for Programmers

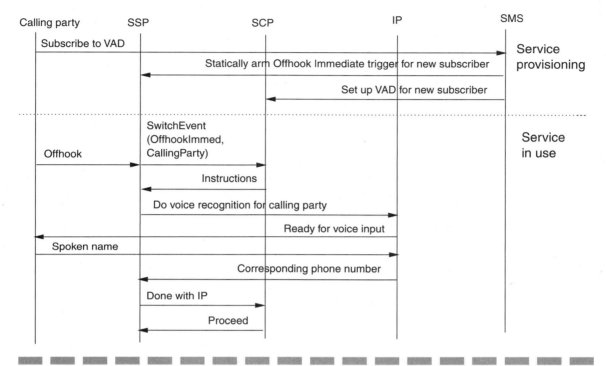

Figure 7.14
(Greatly) simplified message flow for IN-based Voice Activated Dialing (VAD) service.

3. The SSP sends a message to the SCP telling it to run VAD on behalf of the subscriber.
4. The SCP executes the VAD service logic and returns instructions to the SSP.
5. The SSP continues processing the call according to the instruction it got back from the SCP. These tell it to signal the IP to do voice recognition on a voice path set up between it and the subscriber.
6. The IP signals to the subscriber that it is ready for voice input. This signal could be a beep, a voice prompt, or some other indication.
7. The subscriber speaks the name of the person they are calling.
8. The IP interprets the name and returns a result to the SSP, either an error or the phone number that corresponds to the spoken name.

9. The SSP tells the SCP it has successfully finished its interaction with the IP.
10. The SCP tells the SSP it can proceed with call processing.
11. The SSP continues executing the O-BCSM, using the phone number it got from the IP for routing and translation.

7.8 For More Information

There are, surprisingly, few books the sole topic of which is the technology underlying the PSTN. Reference [9] provides a comprehensive overview of telecommunications ranging from the PSTN to the Internet, with particular emphasis on telephony.

Russell's *Signaling System #7* [85] is the standard one-volume reference work on SS7. Faynberg et al.'s *The Intelligent Network Standards: Their Application to Services* [16] is an excellent introduction to Intelligent Network history and technology. While it concentrates on the ITU-T's version of IN, it is nonetheless invaluable as a source of background material. Furthermore, because the Internet is inherently global, standards for convergent networks have to account for both local and international practices. Furthermore, the ITU-T's IN vocabulary is the basis for many discussions of convergent network technology. ITU-T standards may be obtained directly from the ITU (http://www.itu.int).

CHAPTER 8

IP Telephony

This 'telephone' has too many shortcomings to be seriously considered as a means of communication. The device is inherently of no value to us.

(Western Union Internal Memo, 1876)

8.1 What Is IP Telephony?

IP telephony is a loosely defined term. Basically, it's a set of technologies that support the equivalent (or a reasonable facsimile) of voice network phone services on data networks, private and public. The potential for cheap phone service was one of its first attractions. Companies liked finding new uses for under-utilized private data networks. Individuals liked making "free" long-distance calls via the Internet. Recently, people have turned from the purely monetary advantages of IP telephony (some of which are disappearing with telecom deregulation) and are paying attention to its potential to enable new services that blend IP and traditional telephony.

Included within—and sometimes equated with—IP telephony are several different categories of technology and service:

- *Voice over IP (VoIP).* The technologies, such as H.323, that support voice telephony on IP networks.

- *Voice over the Internet.* The application of VoIP technology to the public Internet. Because Internet performance cannot be predicted and varies widely from place to place and from time to time, the quality of a VoIP phone call on the Internet cannot be guaranteed.

- *Private VoIP networks.* The application of VoIP technology to privately managed IP networks. Such networks try to emulate the services of the PSTN or a PBX, using IP technology in place of PSTN infrastructure.[1]

While IP telephony is often seen primarily as a technology for voice calls, it actually embraces a full gamut of multiparty, multimedia communication. A significant advantage of IP telephony is that much of it has been designed to work with a wide variety of media. If nothing else, this alone offers service designers plenty of new territory to explore.

IP telephony has several features besides multimedia that make it attractive as an adjunct to, or substitute for, traditional PSTN telephony:

- Everything in the PSTN is based on 64 Kbps DS0 voice channels. With IP telephony, bandwidth can be adjusted to suit individual needs and circumstances.

---

[1]Some refer to this application of VoIP as "IP telephony" to distinguish it from Voice over the Internet service.

IP Telephony

- Phones in the PSTN have an admirably simple user interface. But while they're good at providing simple access to simple services, they aren't at all suitable for more advanced services, even ones that have been around for a while, such as call forwarding. IP telephony, with its roots in the computer world, offers the potential for an entirely new generation of communication terminals that take advantage of modern user interfaces. When properly designed, these should make services easier to use and, at the same time, make it feasible to offer even more sophisticated services.

- Services and data are highly localized in the PSTN. Features you have on your work phone are probably not available on your home phone. Even if the same features are available in two locations, any data they use has to be entered separately at each location (for example, speed calling lists). Because the Internet is inherently a global network, it's much easier to provide service and data portability with IP telephony.

- Intelligence in the PSTN is highly centralized. Beyond the switches and SS7 elements of the core network, equipment attached to the PSTN is given little responsibility for service logic. With IP networks in general, and IP telephony networks in particular, there are few restrictions on where intelligence goes. It may be totally centralized on servers; it may be pushed out entirely to the end points; or (most likely) it may use a hybrid approach with central servers and endpoints collaborating to offer services.

- Protocols for IP telephony services, such as SIP, are generally text based and frequently simpler than the binary protocols used in the PSTN, such as SS7 AIN. This makes it easier to develop protocol stacks and service creation tools.

Figure 8.1 shows many of the protocols to be covered and their relationship to each other, with respect to layers and functions. At the bottom of the stack are physical media: Sonet, ATM, Ethernet, dialup, and so forth. Just above these are protocols, such as PPP, that adapt IP for use on a specific medium. The next two layers up the stack—IP and TCP/UDP—provide the fundamental IP environment. Above those are the protocols that are the subject of this and the next few chapters.

Many IP telephony protocols do only one or a few things. Frequently, there is more than one protocol to carry out a given function. The advantage of this modular approach is that protocols can be selected to fit the problem at hand. Furthermore, if this modular approach is to work at all, protocols have to be designed ahead of time to be loosely

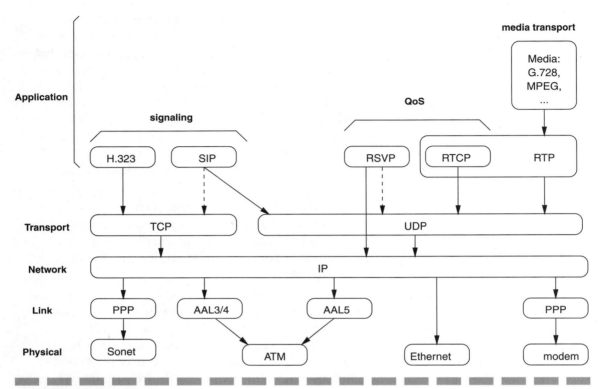

Figure 8.1
IP telephony protocols.

coupled with each other. If they are built this way from the start, it makes it that much easier to swap out old protocols and introduce new ones without affecting anything else.

In the next section, I'll briefly go over quality of service issues in packet networks and how they are being addressed by media transport technologies like RTP, media encoding techniques like G.728, and protocols aimed directly at quality of service, such as RSVP and RTCP. Then I'll turn to H.323, one of the first protocols for call signaling in a packet environment. SIP, the other major signaling protocol for IP telephony, is of such importance that I devote a couple of chapters specifically to it.

The protocols shown in Figure 8.1 provide enough infrastructure to do some interesting things but, to be truly useful, IP telephony needs to be integrated with the existing voice network. That's the subject of the final sections of this chapter, which cover IP telephony gateways—the crucial network elements that bridge the gap between PSTN and IP voice networks—and the routing and translation problems that must be solved to carry calls from one network to the other.

8.2 QoS for IP Telephony

8.2.1 Why Is QoS Such a Big Deal?

Some of the biggest differences between IP and traditional telephony result from the fact that IP networks use packet switching, while the PSTN uses circuit switching. Because it uses circuit switching, the PSTN has no problem giving callers the bandwidth needed to support a phone call, once they've established a circuit to their called party. The downside is that such a circuit may not be available on the required route. If that's so, the call won't go through, even if there's spare capacity elsewhere in the network.

On packet-switched networks, with *best effort* service, voice data is almost sure to get through (sooner or later). But since it travels on a shared resource, quality may suffer. An entire class of protocols has evolved to help packet-switched networks mimic as best they can the *Quality of Service (QoS)* of a circuit-switched network like the PSTN.

When you're sending packet-switched voice, listeners won't notice missing data (up to a point), but they will quickly notice data with *jitter* (arrives irregularly) or *latency* (takes longer than usual to arrive) ("He......llo. How a.......re y..............ou. . . . LONG AWKWARD PAUSE . . . I'm fine"). The human ear is very sensitive to both conditions. Missing data makes speech sound less clear. Jitter can make it unintelligible, and it's a real possibility on the Internet. Latency destroys the smooth flow of conversation. Both occur because there's no predicting what route any given packet in a conversation will take and what it will run into along the way. Some packets may travel on a short route and others on a long route. Even if every packet takes the same route, a node along that route might get bogged down with traffic from time to time and, as a consequence, delay some packets or release them at an irregular rate.

What this means is that voice data needs to get from sender to receiver as quickly and as steadily as possible. Delay is a real problem when voice data is sent through public IP networks like the Internet. It's turned out to be a lot harder and more expensive problem to tackle than some may have expected in the early days of IP telephony. A full discussion of IP telephony's QoS problems and their solutions could easily fill an entire book by itself. Since the primary focus here is service technologies, not the infrastructure on which they sit, the discussion will be just enough to give some sense of what's involved.

One of the simplest techniques for handling IP telephony QoS is simple over-engineering, i.e., providing so much bandwidth it's unlikely any packet will have trouble making it through in a reasonable time. While this is effective, when scaled up, it can be expensive. Furthermore, it works best when done within a single enterprise, since the network can then be managed as they see fit. On the public Internet, there's no telling what you'll run into. For that, more sophisticated methods are required, which will be the subject of the rest of this section.

8.2.2 Speech Encoding

Recall from Sec. 7.2 that pulse code modulation turns a standard analog voice signal into a 64 Kbps digital stream by sampling the amplitude of the analog signal 8000 times per second, an encoding standardized by the ITU-T in its Recommendation *G.711*. Sixty-four Kbps is no problem for the PSTN, since its basic DS0 channel is set to 64 Kbps, and every call gets the full use of one of those channels. On the other hand, 64 Kbps is a lot of data to put on a public packet-switched network like the Internet. Fortunately, there are ways to reduce how much voice data you send without degrading its quality (much).

The simplest technique is to send only the differences between successive PCM samples, a method that's called *differential PCM (DPCM)*. Consecutive samples don't usually differ by much, so fewer bits are needed to represent their differences than for the values themselves. For example, it takes 16 bits to send 106 followed by 109 as two 8-bit numbers. It takes only 3 bits to send their difference: 2 bits for the absolute difference of 3, and 1 bit to say whether it's a difference up or down. *Adaptive DPCM (ADPCM)* adds some enhancements to the basic DPCM algorithm to get the data rate for voice down to 32 Kbps. The ITU-T has standardized ADPCM in its Recommendation *G.726*. Beyond this, there are sophis-ticated techniques that use mathematical models of human speech to reach even lower rates. Among those standardized by the ITU-T are *G.7232.1* (6.3 and 5.3 Kbps), *G.728* (16 Kbps), and *G.729* (8 Kbps).

With all these, there must be a *coder* on one end of a transmission that converts analog speech to digital data, and a *decoder* on the other end that converts it back to analog speech. These two functions are usually combined in a single (hardware or software) entity called a *codec*. Codecs reduce the size of the IP telephony QoS problem, by reducing the data that must be dealt with on a shared IP transmission path. Codecs are the first line of attack in getting IP telephony's QoS to acceptable levels.

8.2.3 RTP and RTCP

Another way IP telephony can improve its QoS is by using UDP, rather than TCP, to transmit voice data.[2] Unlike TCP, UDP makes no guarantee that packets will get to their destination in the correct order or that they will get there at all. Even though this might seem a drawback, it's an advantage for IP telephony. Most of the time, UDP moves data from point A to point B faster than TCP would, since it doesn't spend any time on TCP's bookkeeping: keeping track of what's arrived or not and in what order, or retransmitting lost packets.

Although UDP is preferred over TCP for IP telephony, its very efficiency introduces a new problem, because it gets rid of all the packet sequencing information TCP has. The software at the receiving end of a media transmission can deal with lost and out-of-order packets, but it still has to know what's been lost and how to put it back in the right order. The *Realtime Transport Protocol (RTP)* [30] fixes this problem. Running on top of UDP, RTP adds a header to packets that carry audio or video data. This header adds the sequence numbers that UDP omits plus some extra information about how the data has been encoded, so the receiver will know what to do with it. The receiver of an RTP stream uses the sequence numbers to figure out what packets to discard (the ones that are out of order).

RTP handles jitter by putting timestamps in the header. The receiving machine buffers packets as they arrive and then, using their timestamp information, plays them back out at a steady speed. The timestamps, plus statistics it keeps about lost and out-of-order packets, help the receiver figure out if it can play back what it's getting with acceptable quality.

Receivers return the information they've gathered about lost and out-of-order packets back to senders using the *Real-Time Control Protocol*

---

[2]Signaling information is usually sent via TCP, since it has to be error free. Which would you rather pay for: a brief garbled call, or a clear call where the switch never gets the signal that tells it to stop billing?

(RTCP) that's built into RTP. The two ends of a connection can also use RTCP to negotiate changes that might help matters when quality gets too low, for example, agreeing to switch to a codec that generates fewer bits.[3] Besides helping to maintain transmission quality, RTCP can be used to synchronize audio and video streams, identify call participants (name, phone number, e-mail address, and so forth), and exercise control over sessions (for example, allowing a user to leave a conference call without dropping everyone else from the call).

Another important feature of RTP is its support for *mixers* and *translators*. A mixer combines several input media streams into one output media stream. Translators convert a media stream from one format to another, for example, going from a high bit rate to a lower one.

8.2.4 INTSERV, DIFFSERV, RSVP, and MPLS

Before leaving the subject of QoS in packet networks, I should mention a few ongoing efforts to emulate the circuit-switched behavior of the PSTN in a packet-switched world—as much as that's even possible. Protocol designers are coming at this problem from two directions:

1. Give each *flow*—for example, a stream of packets from one party in a call to another—its own QoS. In essence, what you do is create an end-to-end private lane for that traffic. Every router along its path must help to create and monitor the individual links along the way. No media stream packets are allowed to proceed unless the entire route says it can handle them at the required level of service. This technique gives the best QoS, but it takes lots of work by lots of network elements, all cooperating with each other. That makes it harder to scale in a large network. Within the IETF, the Integrated Services (`intserv`) and Resource Reservation Setup Protocol (`rsvp`) working groups are using this method.

2. Give each packet its own QoS. Platforms at the edge of the network then look at each packet as it tries to enter and filter out those that

---

[3] So why doesn't the sender just use the most efficient available codec for the entire call? Generally, the fewer bits transmitted, the lower the quality of the signal that can be played back to the receiver. While the quality you get from a low-bandwidth codec may be acceptable, it may not be as good as possible. To maximize quality, the sender may start off with a codec that favors quality over compression, backing down to a lower bandwidth codec only if it finds the network can't keep up.

would put the QoS of higher-priority packets at risk. This may not provide QoS as good as the previous method, but it scales better because only elements at the edge are involved. The IETF Differentiated Services (`diffserv`) working group is following this path.

Since this is not meant to be a treatise on QoS, I won't try to cover all these groups and their activities. To give some sense of what they're up to, we'll take a brief look at the `rsvp` working group's *Resource Reservation Protocol (RSVP)* [35]. RSVP tries to offer something that behaves like a circuit-switched connection, by classifying packets and then fixing the paths along which they will travel. RSVP can't guarantee QoS, but it does try to give all packets of a given classification equal and predictable handling.

Unlike the PSTN, where a request for a circuit goes from the calling to the called end of a call, RSVP requests are sent by the receiver. The sender may propose a path to the receiver, but the receiver makes the actual reservation. That reservation applies only for traffic flowing from the sender to the receiver. To make a two-way phone call with RSVP, two separate paths must be reserved, one from each end of the call. Like many other IP telephony protocols, RSVP is an evolving technology. Perhaps the most important thing to remember about RSVP is that for it to work at all, every router in an RSVP-reserved path between sender and receiver must support the protocol.

Finally, the IETF Multiprotocol Label Switching (`mpls`) working group is also addressing QoS issues in packet-switched backbone networks. MPLS, the protocol they are defining, can work with both the `intserv` and `diffserv` approaches. It started as a means to route IP traffic over ATM. It is now being extended beyond packet switching to other domains—TDM, optical cross-connect, and spatial switching—under the name *GMPLS (Generalized MPLS)*.

8.3 H.323

8.3.1 H.323 Architecture

ITU-T Recommendation *H.323* [62] is a family of standards that includes *H.225.0* for call setup and *H.245* for conference control and capability exchange between H.323 terminals. In its original form, *H.323* set standards for multimedia conferencing on LANs, as suggested by its name when first approved in 1996: *Visual Telephone Systems and Equipment for Local Area Networks (LANs) that do not Provide Guaranteed Quality of Service (QoS)*. Two years later, in 1998, it had a new release and a new name—*Packet-Based Multimedia Communications Systems*—to show that it now covered the entire range of packet networks, from LANs to the Internet.

H.323's network architecture has four elements, as shown in Fig. 8.2:

1. *Terminal.* The H.323 terminal is a user's point of access to H.323 call and conference services. An H.323 terminal is similar to a telephone in the PSTN. It must support voice streams. It may also support video and data streams. The components of an H.323 terminal are shown in Fig. 8.3:

    Audio codec(s). Because it has to support voice, an H.323 terminal must have an audio codec to translate an analog voice signal to digital format and back. A terminal may have more than one codec. It must have the G.711 codec that digitizes voice for 64 Kbps DS0 voice channels in the PSTN. An H.323 terminal will almost certainly include other codecs that generate fewer bits than G.711. Data flows in and out of a codec as an RTP stream on top of UDP.

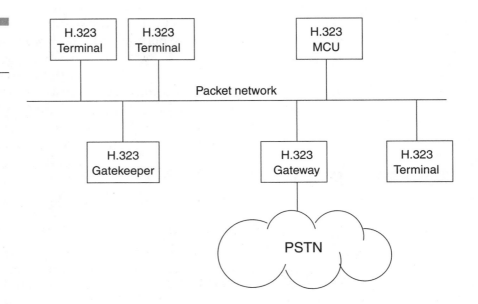

Figure 8.2
H.323 network architecture.

IP Telephony

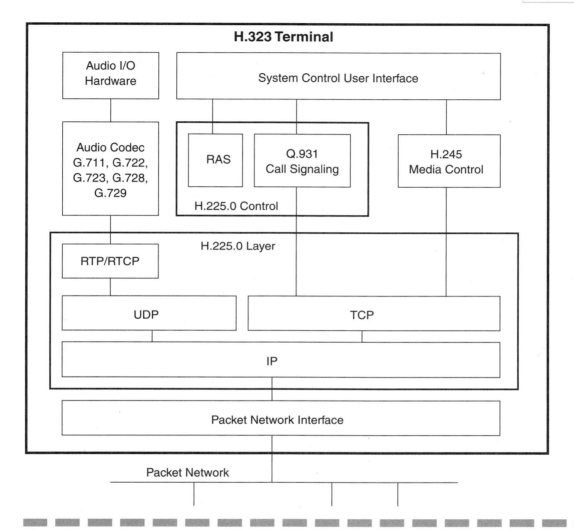

Figure 8.3
Components of an H.323 terminal.

Control. The control component handles signaling between the terminal and other endpoints in a call. This information is passed via TCP so it won't be lost en route. There are two parts to the control component, one for each of the two primary protocol functions in H.323:

H.225.0 control. Provides terminal registration and call control. It is divided into two parts:

Q.931 call signaling. H.323 terminals use a version of ISDN's Q931 call signaling protocol.

RAS. The *Registration, Admission, and Status (RAS)* protocol is used by terminals to register themselves with a gatekeeper (see below) when there is such an element in a terminal's H.323 network.

H.245 media control. Handles various end-to-end matters, such as capability negotiation (for example, making sure both ends of a call are using the same codecs) and seeing to it that a receiving terminal can keep up with a sending terminal.

H.225.0 layer. All the packaging of H.323 messages for UDP and TCP is done in the *H.225.0 layer* of the terminal (not to be confused with *H.225.0 control*).

2. *Gateway.* An H.323 gateway sits between an H.323 network and the PSTN. It allows terminals in each network to make calls to terminals in the other by translating their transport and control messages. Gateways and their protocols are discussed at more length in Secs. 8.4.1 and 8.4.2.

3. *Gatekeeper.* The gatekeeper is optional in an H.323 network. When present, terminals register themselves with gatekeepers using RAS. Gatekeepers provide various services to H.323 terminals:

   Address translation. Converts an address given as a phone number or e-mail address to the physical address of an H.323 terminal.

   Admission control. Determines who is and is not allowed access to called parties on an H.323 network.

   Call authorization. Determines who is and is not allowed to place calls from an H.323 network. Authorization can be based on factors such as calling party, called party, time of day, day of week, and the like.

   Zone management. Keeps track of a *zone*, the endpoints in an H.323 network that are overseen by a single gatekeeper.

   Call management. Provides alternative call dispositions based on predefined rules. For example, calls to a busy terminal may be redirected to another terminal.

4. *Multipoint Control Unit (MCU).* Since H.323 started out as a standard for multimedia conferences, it makes sense for there to be some element that supports conferencing features. The MCU is that element. Besides managing conferences of three or more endpoints, it also supports *mixing,* the combining of multiple media streams into a single stream, and *switching,* the selecting of a particular media stream to send to a particular endpoint. The MCU uses the H.245 protocol.

8.3.2 H.323 Call Processing

An H.323 *call* is a point-to-point communication between two *endpoints*. An endpoint can originate and terminate a call. It can send, receive, or send and receive *media streams*. Each media stream travels between the endpoints on a single *channel*. H.323 signaling messages also travel on channels. *Reliable channels* carry signaling messages on TCP.[4] *Unreliable channels* carry media streams on UDP.

Figure 8.4[5] is a simple example of an H.323 call setup using a gatekeeper. An H.323 call has five phases:

---

[4]Although H.323 version 3 introduced the ability to carry signaling messages on UDP.

[5]H.323 also defines a *Fast Connect* method of setting up a call, in which H.245 messages are eliminated by adding extra information to Q931 messages.

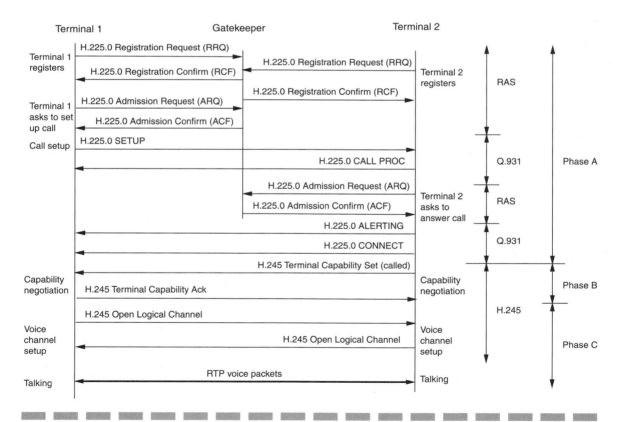

Figure 8.4
H.323 call setup with gatekeeper participating.

Phase A. Call initiation and setup. H.225.0 messages open a connection, sometimes called the *Q.931 channel*, from the calling to the called party on some well-known port. Q931 messages (as defined in H.225.0) are exchanged on this channel during call setup.

Phase B. The first communication between endpoints, during which they find out the capabilities each supports and which of them will act as the master if there are conflicts. H.245 is used during this phase.

Phase C. Establishment of audio and/or visual communication.

Phase D. Call services, such as changing bandwidth or adding new endpoints to a conference call. H.245 is used during this phase.

Phase E. Call termination. H.245 and H.225.0 are used during this phase.

8.4 IP Telephony Gateways

8.4.1 Gateway Architectures

As we've seen, IP networks and the PSTN differ from each other in significant ways. Where the PSTN uses dedicated, circuit-switched connections, IP networks use best-effort packet switching. The PSTN exchanges call signaling information using Q931 and SS7. IP networks use protocols such as H.323. On the PSTN, voice data is typically represented by an analog signal in the local loop and by G.711 digital encoding on DS0 channels in digital trunks. IP networks always use digital encoding for voice, such as G.711, G.723.1, or G.729. Finally, the PSTN uses SS7 point codes and E.164 telephone numbers to identify nodes and terminals. IP networks use a wide variety of naming schemes, including 4-byte IP addresses (e.g., 192.168.0.42), domain names (e.g., nowhere.org), e-mail addresses (e.g., nobody@nowhere.org), and URLs (e.g., http://www.nowhere.org).[6]

Because of these differences, there has to be some way to translate voice and signaling data when a call moves between the PSTN and an IP network. This is done by a network element called a *gateway*. The gener-

---

[6] Of course, all forms of Internet address resolve to 4-byte IP addresses, in the end.

IP Telephony

ally accepted architecture for an IP telephony gateway has three elements, as shown in Fig. 8.5:

Media Gateway (MG) The MG gets voice (or other multimedia) data from one network to the other by means of several functions:

- The MG converts data streams from the sending network's format to the receiving network's format, a procedure that's called *transcoding*.
- The MG takes data streams off channels of the sending network and puts them on channels of the receiving network, for example,

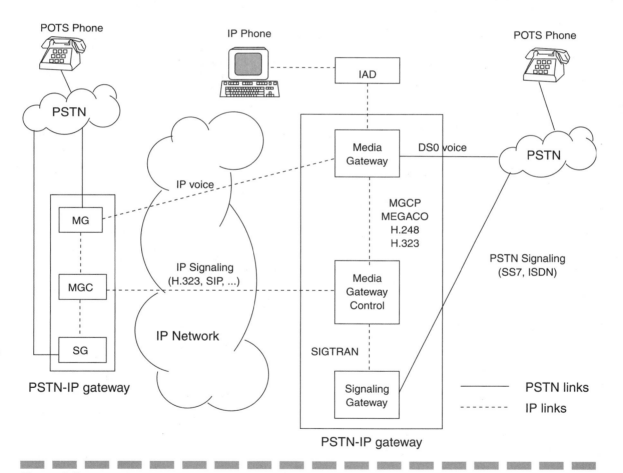

Figure 8.5
Architecture of an IP telephony gateway. (There are two clouds labeled PSTN to keep lines from crossing, not to imply there are two PSTNs.)

going from a DS0 multiplexed on a T1 or comparable link in the PSTN to an RTP stream in an IP network.
- The MG interfaces to an *Integrated Access Device (IAD)*, which acts as a voice gateway coming out of a home or business.
- The MG sets up and manages connections when instructed to do so by an MGC.
- The MG provides special resources, such as conference bridges and voice response units.
- The MG may carry out any other required activities that relate specifically to a media stream, such as echo cancellation, silence suppression, event detection, and signal generation.

Signaling Gateway (SG) The SG takes PSTN signaling information (for example, SS7 ISUP and TCAP messages) and passes it on to the *Media Gateway Controller (MGC)* in a format that element understands. Conversely, it takes signaling messages from the MGC and converts them to PSTN format before sending them out.

Media Gateway Controller (MGC) The MGC oversees one or more MGs within its *domain* by issuing them commands and deciding how to deal with the events they report. The MGC oversees call control, connection control, and resource management within a gateway. The MGC also interacts with one or more SGs to process and generate PSTN signals. If a call involves MGs or SGs not in its domain, it interacts with the MGCs responsible for them. *Call agent* is another name for a network element that supports the MGC function.

This architecture places no restrictions on the physical design of a gateway. The MG, SG, and MGC may all reside on a single machine or they may be spread across two or three separate machines. Figure 8.6 shows an IP network with two gateways. Both are *trunking gateways;* they connect central offices just as normal PSTN trunks would. The gateway on the left is fully distributed. All three of its elements—MG, SG, and MGC—are spread across separate hosts. The one on the right is partially distributed. Its MGC is in a separate host, while its SG and MG are combined in a single host. Figure 8.7 shows a different arrangement. Here, the MG is by itself, while the MGC and SG are combined. This combined MGC/SG is sometimes called a *softswitch*.[7]

---

[7]According to the International Softswitch Consortium, a softswitch is any platform that controls a communications network over IP networks.

IP Telephony

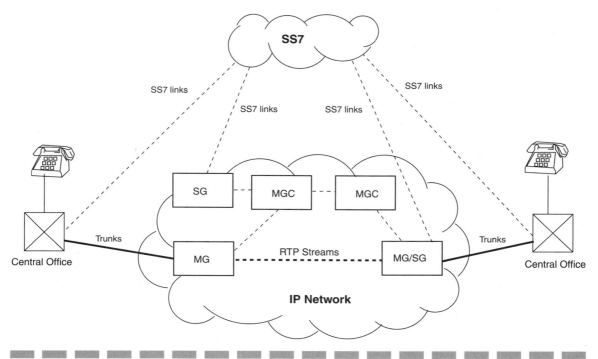

Figure 8.6
Trunking gateways with varying degrees of distribution.

8.4.2 Gateway Protocols

The data streams flowing in and out of a gateway are determined by the networks to which it is attached. We've already looked at some of these on the PSTN side (DS0, SS7, and ISDN). We've looked at some on the IP side also (H.323), and we'll look at more in the next chapter. Within the gateway, we need to look more closely at the messages passed on the MG/MGC and SG/MGC interfaces.

Two protocols have emerged at the MG/MGC interface: MGCP and MEGACO/H.248. Both support a similar set of gateway control functions, including:

Resource control An MGC must be able to allocate and deallocate resources for a call. An MG must be able to report the resource availability and status to an MGC.

206   **Chapter 8**

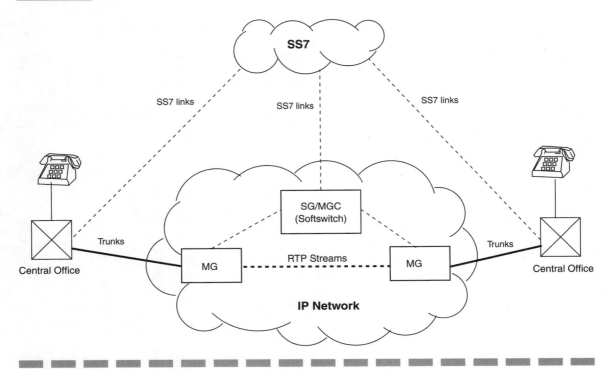

Figure 8.7
A combined SG and MGC is called a *softswitch*.

Connection management   An MGC must be able to create connections between any combination of circuit-switched and packet-switched media streams carried on any combination of transports: analog, T1, Ethernet, ATM, frame relay, and so forth.

Media processing control   An MGC must be able to specify appropriate processing and conversions for each media stream in a call, including such things as echo cancellation, silence suppression, and tone detection. An MGC must be able to control the playing of announcements and other insertions into a media stream.

Signal and event processing   An MGC must be able to tell an MG what events to monitor and which signals to apply or remove on a media stream. An MG must be able to tell an MGC about events it sees. An MGC must be able to tell an MG what to do when a particular event occurs.

Statistics   An MG must be able to report information collected during its operation (QoS, call duration, number of bits transferred, and so forth). An MGC must be able to request statistics anytime during a call.

Association management There must be a way to put a specific MG under the control of a specific MGC.

The job of protocols at the SG/MGC interface is fairly simple: carry the signaling messages of the PSTN to and from the MGC. At the SG/MGC interface, the IETF `sigtran` working group is designing the SCTP protocol.

MGCP The *Media Gateway Controller Protocol (MGCP)* was developed in the IETF Media Gateway Control (`megaco`) working group. The MGCP specification was released as an Informational Draft in October 1999 [42]. MGCP was not the first protocol proposed for the MG/MGC interface. In 1998, IP telephony service provider Level3 formed a body it called the *Technical Advisory Council (TAC)*, whose purpose was to promote its own *IP Device Control (IPDC)* protocol in the IETF and ITU. As its name suggests, IPDC's focus was device rather than call control. At about the same time, others—Cisco and Telcordia in particular—were developing the *Simple Gateway Control Protocol (SGCP)*. SGCP was similar to IPDC, but it had more call control features. In fact, Level3 had been part of the original SGCP design group. Ultimately, the various parties agreed to merge IPDC and SGCP to form MGCP.

As an Informational Draft, MGCP is not on a standards track within the IETF. Consequently, it is not being updated or revised. Nevertheless, because it was the first generally agreed-upon protocol for the MG/MGC interface, it was adopted by several vendors and industry groups for their gateway platforms—among them CableLabs and the International Softswitch Consortium—and it is now implemented in a variety of products.

MGCP models a call as one or more point-to-point or multipoint *connections*. Connections link together *endpoints*, such as trunks, which send and receive data. Endpoints can be of many kinds: packet-switched or circuit-switched; one-way (simplex), simultaneous two-way (full duplex) or alternating two-way (duplex); point-to-point or multipoint; audio, video, or text; and so forth. The characteristics of specific types of endpoints, for example, a DS0 or an RTP stream, can be put into predefined *packages*.

MGCP gives the MGC a means to tell an MG to allocate and deallocate the endpoints and resources (for example, IVR or text-to-speech units) needed for a given call:

- Resources may be specifically named by the MGC or selected from a pool by the MG.
- The MGC can get the status of resources from the MG.
- The MGC can manage create, modify, and delete media stream connections within the MG. MGCP uses SDP (see Sec. 9.4) to specify the properties of these media streams.
- The MGC can specify how a particular medium is to be processed (for example, echo cancellation, tone detection, silence suppression). The MGC can also specify announcements and other content to be inserted into a media stream or information, such as dialed digits, that is to be extracted from a media stream.
- The MGC can tell the MG what to do when certain events occur, including reporting them back to the MGC.
- The MGC can tell the MG to apply particular signals (for example, dialtone) to a media stream and when to stop applying them (for example, timeout, occurrence of event, receipt of instruction to apply a different signal).
- The MGC can tell the MG how to collect dialed digits.
- The MG can report events to the MGC.
- The MGC can request statistics and the MG can report them back.

Like many protocols defined for convergent networks, MGCP shows the influence of HTTP, the protocol used to exchange content on the World Wide Web (see Sec. 5.8.3). Like HTTP, MGCP is a stateless protocol. A command is sent, a response is sent back, and that's the end of that. Like HTTP, MGCP is text based. MGCP messages consist of a header and a body separated by a blank line. Both header and body contain text lines separated by carriage returns. MGCP messages travel on UDP.

MGCP defines eight different commands, which it calls *verbs*:

CreateConnection (CRCX)   Sent by the MGC to tell an MG to create a connection that terminates on one of the MG's endpoints

ModifyConnection (MDCX)   Sent by the MGC to tell an MG to change parameters for a previously established connection

DeleteConnection (DLCX)   Sent by either the MGC or an MG to tear down an existing connection

NotificationRequest (RQNT)   Sent by the MGC to tell an MG it should be prepared to handle a specific event on a specific endpoint

Notify (NTFY)   Sent by an MG to tell the MGC it has seen an event previously indicated in a NotificationRequest

AuditEndpoint (AUEP) Sent by the MGC to request endpoint status information from an MG

AuditConnection (AUCX) Sent by the MGC to to request connection status information from an MG

ReStartInProgress (RSIP) Sent by an MG to notify the MGC that the MG or a set of its endpoints is either being taken out of service or placed back in service

Figure 8.8 shows an example of the MGCP messages that would be exchanged between MGC and MG to set up a call between two endpoints. Note that, according to the rules of MGCP, every command must be followed by an ACK from its receiver:

1. The MGC first sends a RQNT to the MG telling it to watch for an offhook on Endpoint 1.
2. The MG sees an offhook on Endpoint 1 and notifies the MGC with a NTFY message.
3. The MGC sends a CRCX message to the MG telling it to apply dialtone, watch for and report digits, and watch for and report onhook on Endpoint 1. The message includes a *digit map* that gives the MG a pattern for the digits it should receive.[8] The MG replies with an ACK that includes an SDP description of the capabilities of Endpoint 1. This will be used to negotiate the connection to the called endpoint.
4. The MG collects digits according to the digit map it received and returns them to the MGC in a NTFY message.
5. The MGC sends an RQNT message that tells the MG to keep watching Endpoint 1 for onhook.
6. The MGC translates the digits to determine the called party (Endpoint 2) and sends a CRCX message to the MG telling it to ring that end of the call and watch for offhook there. It includes the SDP description it got from Endpoint 1. The MG replies with an ACK that includes an SDP description of the capabilities of Endpoint 2. The MGC picks a matching set of capabilities between Endpoints 1 and 2 on which to set up a connection. If

---

[8]A digit map is, for the most part, a regular expression, similar to those used for pattern matching in many editors and utility programs. The following is an example of a digit map that recognizes seven digit local phone numbers and ten digit long distance numbers (1+ and 0+):

D: [2-9]xxxxxx | [01][2-9]xxxxxxxxx

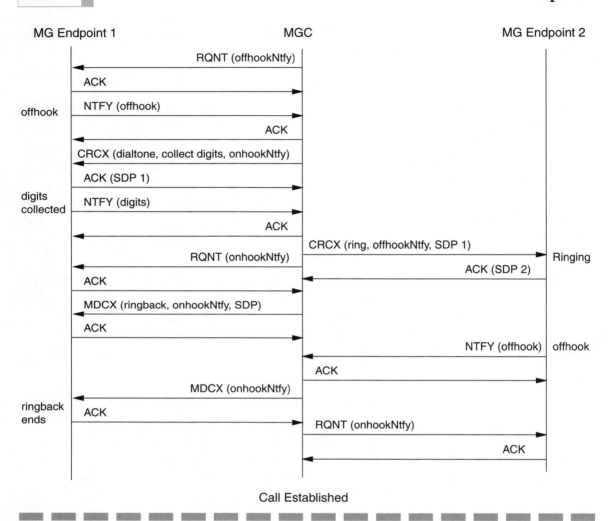

Figure 8.8
Call setup with MGCP.

there is no match, there must be further negotiation with the MGC before a connection can be established.

7. The MGC sends an MDCX message to the MG telling it to play ringback to Endpoint 1 and watch for onhook. It also supplies an SDP description of the capabilities Endpoint 1 is to use in connecting with Endpoint 2.

8. The called party answers, and the MG sees an offhook on Endpoint 2. It reports this to the MGC in a NTFY message.

9. The MGC turns off ringback on Endpoint 1 by sending the MG a MDCX message telling it to keep watching for onhook. The normal behavior is for signals to end, unless otherwise indicated, when the MGC tells the MG to do something, even if the MG is already doing it, as shown here.

10. The MGC tells the MG to watch for onhook on Endpoint 2. The call is established.

MEGACO/H.248 MGCP was never endorsed as a standard by the IETF. It did get an early start in the market, since it was the only accepted gateway protocol available when vendors, eager to capitalize on the potential growth of IP telephony, needed some common ground upon which to build. MGCP, while not perfect, was good enough for them to get started. At the same time, many argued that MGCP wasn't flexible enough to handle all the gateway functions that were needed. Thus, even as MGCP was under discussion, other gateway protocols were proposed to remedy its perceived shortcomings.

At about the same time MGCP was being developed in the IETF, Lucent introduced its own *Media Device Control Protocol (MDCP)* in both the IETF and the ITU. MDCP had more success in the ITU, where it was renamed *H.GCP*. Eventually, a version of MGCP influenced by MDCP emerged as *MegacoP*. Ultimately, MegacoP and H.GCP were merged into the IETF *MEGACO (MEdia GAteway COntrol)* protocol [55]. The ITU adopted the same protocol under the name *H.248*. (See Fig. 8.9.)

The emergence of MEGACO essentially brought development of MGCP to an end within standards bodies. With no official MGCP standard, there's no way to change or extend the protocol in any coherent way. Vendors who have MGCP products are on their own with regards to adding proprietary features. MGCP may easily fragment into different versions as it's modified to serve each organization's purposes. The developers of MEGACO hope their protocol will become a universal standard that replaces MGCP and prevents this sort of fragmentation.

MEGACO has adopted many of MGCP's features. Like MGCP, MEGACO is a text-based protocol. Like MGCP, MEGACO uses SDP to describe properties of media streams. Its processing of media-stream signals and events is similar to that of MGCP. It also uses digit maps, just as in MGCP, that tell MGs how to collect digits.

MEGACO enhances MGCP's simple command and response model by adding *transactions*. Transactions allow an MGC and MG to exchange

Figure 8.9
The evolution of MEGACO.

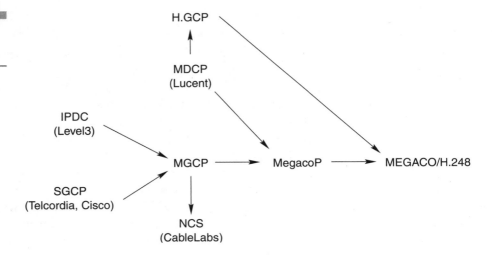

a series of messages, instead of just a single command and response, in order to carry out some activity. MEGACO messages may also contain multiple commands or responses. Since every command or response is associated with a unique transaction and every transaction has a unique ID, it's easy to sort these out and pass them on to the correct recipients.

Instead of MGCP's endpoints and connections, MEGACO models calls with terminations and contexts. A *termination* is the source or destination of a media stream—such as a DS0, an RTP stream, or an MP3 stream—passing through the MG. A termination has properties, such as media type, the signals that can be applied to it, and the events it can generate. Some terminations, such as a DS0, are permanent within the MG. These are usually represented by a port on the MG. Other terminations that represent flows on a packet network, such as an RTP stream, are *ephemeral*. Ephemeral terminations are created when needed, used, and then released.

A specific kind of termination and its properties can be defined in a MEGACO *package*. These are similar to MGCP packages except that adding a new package to MGCP requires a revision to the protocol. A new MEGACO package can be created merely by registering it with the *Internet Assigned Numbers Authority (IANA)*.

A *context* is a collection of all the terminations connected to each other in a call. A typical MEGACO two-party call is represented by a context with two terminations: a physical termination for a DS0 channel on a PSTN trunk, and an ephemeral termination for an RTP stream.

IP Telephony

Besides keeping track of the terminations in a call, contexts are also used for accounting and logging.

For connection management, MEGACO's termination/context model improves greatly on MGCP. This is particularly apparent with conference calls. In MEGACO, you need only have a context and add terminations to it to create a conference. If a party drops out of the conference, that termination is simply removed and nobody else is affected. In MGCP, you would have to allocate a conference bridge endpoint and then create separate connections to it from all the participating endpoints.

MEGACO's command set is virtually the same as MGCP's, though with different names, as shown in Fig. 8.10. The Move command, for

Figure 8.10
MEGACO commands and their MGCP equivalents.

MEGACO	MGCP
Add	CreateConnection
Modify	ModifyConnection
Subtract	DeleteConnection
Move	MoveConnection (proposed)
AuditCapabilities	No equivalent
AuditValue	AuditEndpoint, AuditConnection
ServiceChange	ReStartInProgress
Notify	Notify

which MGCP has no official equivalent, lets an MGC move a termination from one context to another. This simplifies the development of features such as Call Waiting, Call Transfer, and Call Hold. The AuditCapabilities command, which also has no MGCP equivalent, lets an MGC find out what packages an MG supports. Using this command, one can design MGs with varying capabilities and leave it up to the MGC to figure out which one to use in a given situation.

SIGTRAN AND SCTP In Fig. 8.5, the link from the SG to the MGC is labeled "SIGTRAN." This is a reference to the IETF `sigtran` working group, which is developing the *Stream Control Transport Protocol (SCTP)*[9] to handle transport of SS7 ISUP and TCAP messages on IP networks. Predecessors of SCTP included *Real-Time UDP (RUDP)* and the *Multiprotocol*

---

[9]Originally the *Simple Control Transport Protocol.*

Datagram Transport Protocol (MDTP). From these, `sigtran` selected MDTP and renamed it as SCTP.

SCTP runs directly on top of IP, sitting just below any application that uses its facilities. Such an application would consist of an *upper layer protocol (ULP)* and a *ULP adaptation layer (UAL)*. The UAL connects the native protocol of the application (for example, SS7 ISUP) to the services supplied by SCTP. UALs are not part of SCTP. There are now UAL definitions for SS7 ISUP, SCCP, and MTP, and for ISDN Q921.

SCTP's main purpose is to remedy limitations of TCP and UDP that make neither of them entirely suitable for carrying SS7 traffic (or other protocols that have the same requirements as SS7). Without going into all the gory details, suffice it to say that these relate to matters such as control over timers (available in UDP, but not TCP), reliable delivery (available in TCP, but not UDP), and fault management. These are all important matters. They're also complicated and shot through and through with subtle issues that would take many pages to discuss. See `http://www.ietf.org` if you're really interested.

8.5 Routing and Translation for IP Telephony

In a fully deployed IP telephony network, whether private or on the Internet, there could be many gateways. The end-to-end processing of a call may easily involve at least two gateways. The calling party end of the call will find the originating gateway on its own (for example, a call originating on the PSTN and going to an IP network via a trunking gateway). Finding the appropriate terminating gateway is another matter. How does one know which of many gateways operated by a variety of service providers is the right one to use for a call? In the PSTN, routing and translation use sophisticated algorithms that coordinate the efforts of many network elements. A comparable infrastructure is needed to support gateway location for IP telephony.

Routing protocols are a complex subject of their own, and it is not my intent to discuss them in depth here. However, it's worth knowing that many of the problems of gateway location are similar to the problems already faced by the Internet in routing data traffic from one *administrative domain* to another. An administrative domain—for example, a company or a university—takes care of routing within its own boundaries. However, to get inside those boundaries one first has to find an entry point. This is the function of the *Border Gateway Protocol (BGP)* [27].[10]

BGP distinguishes two types of traffic that may be carried within an administrative domain, or *Autonomous System (AS)*:

Local traffic stays entirely within the AS, originates within the AS and then leaves it, or arrives from another AS to terminate within the AS.

Transit traffic comes in from one AS and leaves via another.

BGP further distinguishes three types of AS, depending on the traffic they carry and their connections to other ASs:

Stub AS connects to only one other AS and carries only local traffic.

Multihomed AS connects to more than one other AS but does not carry any transit traffic.

Transit AS connects to more than one other AS and carries both local and transit traffic.

Each AS contains at least one host designated to be its *BGP speaker.* A BGP speaker provides information needed to reach networks contained within its AS. If it's speaking on behalf of a transit AS, it also provides information about other networks that can be reached through it. The BGP protocol provides a means for BGP speakers to exchange information about what networks can be reached from which AS.

The IP telephony equivalent of BGP is the *Telephony Routing Information Protocol (TRIP)*, now being developed in the IETF `iptel` working group. Besides figuring out whether or not a gateway can take a call to its destination, TRIP must consider other factors, such as current load and capacity on the gateways in question and the IP telephony protocols they support. TRIP is modeled after BGP, but includes features that allow it to find routes using additional factors.

Another routing issue that must be addressed within an IP telephony network is the translation of PSTN phone numbers to IP addresses from which one might find out the characteristics of the terminals and users to which those phone numbers correspond. Within the IETF, the `enum` working group is charged with defining techniques for doing this.

---

[10]BGP is the second interdomain routing protocol to be used in the Internet. The first, the *Exterior Gateway Protocol (EGP)*, severely limited the layout of the Internet in such a way that today's multibackbone architecture could not be supported. BGP is now in its fourth version, BGP-4. BGP is regarded as one of the more complex of the Internet protocols. That should give some sense of how complicated the gateway location problem is.

While their work is not yet finished, enough has been done to say something about the direction in which they are going.

The architecture proposed by `enum` is based on the Domain Name System (DNS) (see Section 6.2). DNS is how addresses such as `www.some-company.com` are translated into the numeric IP addresses like 192.168.0.55 that identify locations on the Internet. DNS defines both a database structure and a protocol to update and query this database. Since there are obviously far too many hosts on the Internet for any one database to keep track of them all (though this is how the Internet worked originally), queries to one DNS server can be redirected to another. This can be done as many times as necessary for a given query until, finally, a server is found with the needed information.

The `enum` working group proposes extending DNS to support entries that map E.164-style phone numbers to Internet URIs. This would be done by establishing a DNS domain—`e164.arpa`—for phone numbers and converting phone numbers to names in that domain by reversing their digits and putting a period between each digit. For example, 1-512-555-1234 would be `4.3.2.1.5.5.5.2.1.5.1.e164.arpa`. DNS analyzes names from right to left, so once it sees `e164.arpa`, it knows that phone number digits will come next and where it should go to look up entries in that domain. Each digit or group of contiguous digits can be assigned to a different naming authority, or *zone*, as it's known within DNS.

For example, the digit `1` could be assigned to a North American naming authority whose directory might hold entries for NPAs (for example, `2.1.5`) that point to DNS directories overseen by the organizations responsible for them. Those entries might point to directories for specific central offices (for example, `5.5.2`) which would hold records that map individual stations (for example, `4.3.2.1`) to URIs. Each URI would indicate the capabilities of the terminal it represents and could also be used to look up the IP address of that terminal. More details may be found in RFC 2916 [52].

This will work fine if the terminal is a PC or some other kind of equipment that has an IP address. But most phone numbers correspond to phones that are attached to the PSTN, and these most certainly do not have IP addresses. Even with convergence, it will be a long time, if ever, before phones can be expected to have IP addresses. In this situation, DNS lookup of a phone number should lead to a gateway that can carry the call from the Internet to the PSTN. That gets you right back to the problem that TRIP addresses: How do you get to the right gateway for a call?

Just as important as what `enum` is trying to do is what it is not trying to do:

- It is not the job of `enum` to design a means for locating telephony gateways. That's the job of TRIP.
- It is not the job of `enum` to develop protocols for mobile phone roaming or other routing services.
- A query to convert a phone number to a URI may yield more than one result. It is not the job of `enum` to specify means to resolve these. That's the job of the services making the queries.

8.6 For More Information

Other books that provide material on IP telephony from several different perspectives include Refs. [17], [14], and [5].

QoS is an entire discipline of its own. The reader who would like to learn just a bit more than was presented here should see Ref. [14]. A good survey of codec technology can be found in Ref. [11]. Those who would like to learn a lot more about Internet QoS protocols should go to the IETF Web site (http://www.ietf.org) to research RTP, RTCP, and RSVP. One should pay particular attention to the efforts of the `intserv`, `diffserv`, and `rsvp` working groups. ATM brings a lot to bear on QoS problems and has significantly influenced work in the IETF. See the ATM Forum's Web site (http://www.atmforum.org) for more.

The ultimate authority on H.323 is the ITU-T Recommendations ([60], [61], [62]).

All of the referenced Internet protocols (MGCP, MEGACO, BGP, TRIP, and so forth) are covered in their respective RFCs, which may be found at the IETF Web site.

CHAPTER 9

Session Initiation Protocol (SIP)

When to the sessions of sweet silent thought
  I summon up remembrance of things past,
  I sigh the lack of many a thing I sought,
  And with old woes new wail my dear times' waste.
      —*William Shakespeare*

9.1 Background

So far, I've focused on IP telephony infrastructure and haven't said much about IP telephony services. H.323 is an exception, though it blurs the lines between services and infrastructure. It's not a single protocol but rather an entire suite of protocols that covers everything from codecs to call and conference control, in one tightly coupled and vertically integrated package. The advantage is that, by controlling so many aspects of IP telephony, it's easier to ensure that H.323-based systems work together with each other.

On the other hand, the price you pay is that you tie yourself to a single family of technologies. For a mature technology, this may not be a problem, since the best solutions are likely to have been discovered and incorporated into standards. For a field as young as IP telephony though, nearly every problem and solution is the subject of ongoing debate. Old solutions are constantly revised or discarded in favor of new ones. In many situations, there is more than one proposed solution with no clear favorite having yet emerged. Even the criteria for judging one solution better than another are debated.

In this environment, many feel it's better to use a wide variety of protocols, each addressing a different aspect of the problem space. This is, by and large, the philosophy behind the IETF approach to IP telephony. The advantage is that one may choose from among many competing technologies and move to newer and, one would expect, better ones as they emerge. The disadvantage is that interoperability may suffer while matters are sorting themselves out. Those who advocate a "separation of powers" approach obviously feel that the advantages of flexibility outweigh those of interoperability at this stage.

SIP is an important piece of this modular approach to IP telephony protocols. A key part of any communication system is finding potential call participants and contacting them. The problem is made even more interesting if you assume participants may move from place to place, changing their locations and the addressable equipment they are using. Add to this the notion that calls need not be restricted to a single voice stream but may involve multiple streams of various media. Then consider that many—even thousands—of participants might be involved in a call, joining and leaving in a constantly changing topology. Put all this together and there's obviously a need for some sort of protocol to deal with generalized *sessions*. SIP fills this role.

SIP is an IETF standards track application layer protocol for establishing, modifying, and tearing down sessions whose participants are connected—directly or via a gateway—to a network (often, but not always, an IP network). SIP's main purpose is to help session originators deliver invitations to potential participants wherever they are. Having stated what SIP does, it's also important to be clear about what SIP does not do:

- SIP is not a session management protocol, though it provides a framework within which to implement session management.
- SIP is not a session description protocol, though some of its messages carry session description information, most often in the form specified by the separate *Session Description Protocol (SDP)*.
- SIP is not a conference control protocol. It does not handle things such as voting and polling, microphone passing, or chairperson selection.
- SIP is not a resource reservation protocol. Though the body of a SIP invitation includes information that can be used to reserve resources, SIP has nothing to do with this and doesn't even care whether resources are reserved or not.
- SIP has nothing to do with QoS. It has no facilities for giving feedback on or for adjusting the quality of a session in progress.

SIP was designed to solve only a few problems and to work with as broad a spectrum as possible of existing and yet to be developed IP telephony protocols. Even with its intentionally limited scope, SIP can be applied to many problems.[1] Some of this power results from the fact that, like HTTP, SIP messages can carry just about anything in their body, including messages of other protocols.

SIP places no restrictions on the kinds of sessions to which it can be applied. They can be pure IP phone calls, calls that use both the PSTN and an IP network, chat sessions, video feeds, collaborative applications, or any other kind of joint multimedia activity. SIP uses a separate standard, the *Session Description Protocol (SDP)* [38]—which isn't a protocol at all—to describe sessions.[2]

---

[1] There are some who think it's in danger of being applied to too many problems.

[2] SIP doesn't have to use SDP. In theory, it can use any session description technique that's understood by all the parties involved. In practice, SDP is almost certainly what you'll encounter.

SIP first gained attention in the mid-1990s with the advent of the *Mbone*, an experimental network that used the Internet to exchange multimedia content, such as broadcasts of space shuttle launches, academic meetings, and presentations. The Mbone used SIP to invite participants to listen in on or join a broadcast. As research on IP telephony increased, people saw that similar protocols would be needed for call setup and management. Because joining participants in a phone call is much like setting up a simple multimedia conference, SIP was seen to be an obvious candidate for this role. Formal work on SIP as an IETF standard began in the multiparty Multimedia Session Control (`mmusic`) working group and was later moved into its own `sip` working group. In March 1999, SIP became a Proposed Standard of the IETF [41].

SIP provides four basic functions:

User location Translating from a user's name to their current network address. Besides the obvious translations—for example of an e-mail address or E.164 phone number to an IP address—this function also includes user mobility features, i.e., keeping track of a user's address as the user moves to different locations in the network.

Feature negotiation Ensuring that all participants in a session agree on the features to be supported among them, since not all may have the same capabilities.

Call management Adding, dropping, transferring, or placing participants in a session on hold.

Feature modification Changing the features of a session while the session is in progress. For example, adding a video channel to a session that started with only a voice channel.

In providing these functions, SIP's designers worked from several fundamental assumptions:

- *SIP should be scalable.* It should allow many users on a given session. It should allow a user to participate in many sessions at once. SIP should work well on wide-area networks. One way SIP does this is by encouraging—though not mandating—fast, stateless servers in the core network and slower, stateful servers as you move closer to the edge.

- *SIP should reuse as many existing protocols and protocol design concepts as possible rather than inventing new ones.* Examples of such reuse in SIP would include modeling SIP after HTTP, using URLs for addressing, and using SDP to convey session information.

- *SIP should maximize interoperability.* It should be easy to tie SIP functions to existing protocols and applications, such as e-mail and Web browsers. SIP does this by adhering to the modular philosophy common to Internet protocols, addressing a specific set of functions and leaving it to other protocols to handle other issues. Its reuse of existing Internet technology, like URLs, further eases the task of integrating SIP with other applications. Compare this with the vertically integrated approach of H.323, which specifies a complete architecture that includes everything from codecs to gateways.

In the next sections, we'll see how these guidelines have been applied in the design of SIP's core capabilities. The first section presents the architecture of SIP: who talks to whom and for what purpose. The next sections describe the structure and purpose of the various SIP messages, including a brief overview of SDP, the primary means for describing SIP sessions. After that are some examples of how SIP is used in several services. The last two sections of this chapter discuss SIP's relation to other telephony protocols in both IP and traditional phone networks.

9.2 SIP Architecture

> While SIP typically is used over UDP or TCP, it could, without technical changes, be run over . . . carrier pigeons, frame relay, ATM AAL5, or X.25, in rough order of desirability.
> —Henning Schulzrinne, Columbia University

The software that runs at the endpoints of a SIP session is called a *user agent*. User agents have two parts: a client part—the *User Agent Client (UAC)*—that sends SIP request messages, and a server part—the *User Agent Server (UAS)*—that processes requests and sends back responses. (See Fig. 9.1.)

The UAC is sometimes compared to the calling party in a phone call and the UAS to the called party. This is a bit misleading. The UAC does initiate SIP sessions (calls) and the UAS does respond to those (answer, reject, and so forth). However, once a session is set up, the called party's UAC may send SIP request messages of its own to be handled by some UAS. In a multiparty session, a complex web of requests and responses may be woven among the participants. The key principle here is that all endpoints in a SIP session need to have a complete user agent with UAC and UAS aspects.

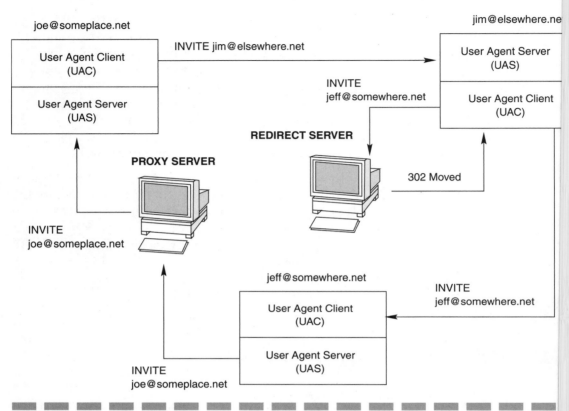

Figure 9.1
SIP architecture: User Agents and SIP servers.

If the session's originator knows where to find all the participa[nts to] be invited, the originator's user agent can contact the other user a[gents] directly. However, if the originator doesn't know where to fin[d] the invitees, a SIP server of some sort, either a redirect server or a p[roxy] server, will have to be used. The originator may know the inv[itees'] phone numbers or e-mail addresses, but not the addresses o[f the] machines at which they are receiving SIP invitations. SIP servers pr[ovide] name resolution and location services to user agents.

A session originator can get location information from a SIP *re[direct] server* before sending an invitation to its destination. The informati[on it] gets back may lead directly to the invitee, or it may lead to anothe[r SIP] server. Another possibility is for the originator to send an invitatio[n to a] *proxy server.* Proxy servers forward invitations towards invitees. They forward an invitation directly to the invitee, or they may forward it to another proxy server, which may forward it to another proxy server, and so on until it gets to the invitee. This is roughly comparable to switching in the PSTN, though the analogy probably shouldn't be pushed too far. When a request goes through a chain of proxy servers like this, the corresponding response(s) return through the same chain in reverse order.

It's important to note that the sender of a SIP request, whether it's the UAC that created the request or a proxy server that's forwarding it, doesn't need to know if the recipient is a UAS or a proxy server. The only difference between the two is that a SIP server can't accept or reject a request, while a UAS can. To a session originator, a proxy server appears and behaves just like an invitee. To an invitee, a proxy server appears and behaves just like an originator.

SIP has two kinds of proxy servers. *Stateless proxies* forward SIP request messages and promptly forget about them. Stateless proxies are fast, but they lack the smarts needed to build certain SIP services. They're usually placed in the core of a network that uses SIP. There they can move SIP messages along as quickly as possible.

Stateful proxies remember the state of a request after they forward it. This is useful when the proxy *forks* an invitation, i.e., forwards it to more than one location. When a user agent sends an invitation, it expects to get back only one response. When a stateful proxy forks an invitation, the proxy has to keep track of responses as they come back, associating them with the original request and returning only one response to the originator. Stateful proxies don't keep track of call state. That's the job of user agents. Once a proxy processes all the responses to a forked request and sends a single response back to the originator, it forgets about the transaction. Stateful proxies aren't as fast as stateless proxies because they have more work to do. Thus, stateful proxies usually go near the edge of a network that uses SIP, where they won't have as much traffic passing through.

SIP doesn't specify how location services work. A location service could be an LDAP directory. It could be a general purpose database. It could even be a simple text file. Regardless of how a location service works, users still must register themselves with SIP servers to appear in their location services. SIP does define the protocol by which users register themselves with SIP servers (and thus, with their location services).

Unlike conferencing systems that use a central registry to keep track of sessions and their participants, SIP uses a "light-weight session model." User agents at the endpoints of a session do most of the work needed to

keep track of and manage sessions. The only role SIP servers play is h
ing user agents get in touch with other user agents. Once that's b
done, SIP servers have little more to do other than relay subsequent r
sages for a session.³ This is a subject of some controversy in the netwo
ing world, namely, how much service intelligence should be at the e
(i.e., in endpoint equipment such as phones and PCs) and how m
should be in the core (switches in the PSTN or proxy/redirect ser
with SIP)? At one extreme, the traditional PSTN put little or no ser
intelligence at the edge and concentrated intelligence in the core. At
other extreme, some advocates of SIP feel strongly that service int
gence should go at the edge with the core network doing little n
than shuffling data around on behalf of edge devices. Bypassing
many fascinating discussions that proceed from these two viewpo
suffice it to say that it is this author's opinion that, as so often in
working controversies, a middle ground will prevail. What seems cer
is that there will in the future be more intelligence at the edge t
there used to be. What also seems certain is that core elements will
tinue to provide services that are difficult or impractical to offer on
edge (directory services being a particularly obvious example).

SIP is, by design, loosely coupled to the other protocols with whic
works. For example, SIP is independent of the underlying transport pr
col. UDP is the most common transport for SIP, but TCP can also be u
SIP can also work with other protocols, such as ATM or X.25. Bec
both SIP and TCP have "sessions," it's important to keep the two conc
separate. When SIP runs on top of TCP, all SIP sessions are comple
independent of the TCP sessions in which they reside. If the endpoin
a SIP session keep track of its SIP session ID, that session can survive
loss of its TCP session, even across reboots of participating machines.

The strict separation of responsibilities between SIP and other IP
phony protocols can lead to situations one wouldn't see in a tightly
pled system like the PSTN. For example, because SIP doesn't reserve
work resources, one might issue an invitation, have it accepted, and
discover there are no resources to support it. The result might be th
SIP-enabled phone would ring and be answered, but no voice w
flow. To get around this problem, you could add information to the
tation that tells the called party's CPE not to alert until the ne

---

³It could be argued that stateful proxy servers have a larger role to play in session ma
ment. If one regards user location as part of session management, that's true. Nevert
once a user has been located, even a stateful proxy server drops out of the session ma
ment picture.

resources have been obtained. Or you could let the CPE alert anyway and let the call go on with whatever quality it could get. Neither solution requires QoS and resource reservation to be handled in any particular way (say, RSVP).

9.3 SIP Messages

9.3.1 SIP Request Messages

Like HTTP, SIP is a text-based protocol with request and response messages. A SIP *request message* consists of a request line followed by header lines and an optional body. The format of the *request line* is:

SIP-Method Request-URI SIP-Version

where:

- *SIP-Method* is INVITE, ACK. OPTIONS, BYE, CANCEL, REGISTER, or a SIP extension method (see Sec. 10.2 for a description of extension methods).
- *Request-URI* is a SIP URL or some other URI that indicates the user or service to which the request is being sent. SIP URLs have the prefix `sip:` followed by a SIP address. SIP addresses look like e-mail addresses, for example, `sip:fred@someplace.com`. SIP can also use phone number URLs. For example, `tel:15125551234@voipservice.com` identifies a phone that can be reached via the gateway at `voipservice.com`. If it gets a non-SIP URL, the SIP software can query DNS for the corresponding SIP URL. Because they work just like any other URL, SIP URLS can be put in Web pages to provide a click-to-dial service, similar to the `mailto:` URL. Of course, the Web browser would have to recognize SIP URLs and know what to do with them for this to work.
- *SIP-Version* is the version of the SIP protocol being used by the message, say, SIP/2.0.

Following the request line, there may be one or more optional *header lines*. Just as in HTTP, each header line is of the form:

keyword: value

SIP reuses many HTTP keywords. The order of header lines in a request is not important except that header lines carried end-to-[end] must appear after those carried across only one hop.

The optional *body* comes after the header lines. Every SIP mess[age] except BYE can have a body. INVITE, ACK, and OPTIONS requests [use] the body for session descriptions. If present, the body is separated f[rom] the header lines by a blank line, and the header section must incl[ude] the following header lines:

- `Content-Length`: The length of the body in bytes.
- `Content-Type`: The MIME type of the body's content.[4] The most common value here is `application/sdp`, used when the body contains an SDP session description.

INVITE method   A SIP session originator sends an INVITE reque[st to] all potential participants. For reasons that should be obvious, th[is is] the most important of the SIP request methods. INVITE requ[ests] may be compared with SS7 ISUP's IAM message or the ISDN Q[.931] SETUP message. INVITE requests may be sent while a session i[s in] progress to modify its characteristics. The message body o[f an] INVITE request contains a description of the session (usually a[n] SDP message). Every SIP network element (i.e., proxy server, red[irect] server, UAC, and UAS) must support the INVITE method.

ACK method   A SIP session originator sends an ACK to let the re[cipi]ents of an INVITE know their final response to it has been rece[ived.] (Against every normal expectation, SIP classifies ACK as a request[, not] as a response.) If the ACK has a body, it contains a final descriptio[n of] the session as negotiated between the originator and the invitee. E[very] SIP network element must support the ACK method.

OPTIONS method   User agents send OPTIONS requests to SIP se[rvers] to find out their capabilities. Every SIP network element must [sup]port the OPTIONS method.

---

[4]*MIME (Multi-purpose Internet Mail Extensions)* is used to identify file types on the In[ternet.] Web servers and browsers use MIME types to figure out what to do with the file[s they] exchange. A *MIME type* has two parts, a type and a subtype, separated by a slash (/[). The] *type* names a broad file category such as `text`, `image`, `audio`, `video`, or `applica`[tion.] The subtype names a specific type of file within that general category, su[ch as] `text/plain`, `text/html`, `image/jpeg`, `audio/G.722.1`, `video/mpeg`, and `app`[lica]`tion/pdf`. A browser that gets a `text/html` file will probably display it as a Web pa[ge. An] `image/jpeg` file may be displayed as a JPEG image. A `video/mpeg` file may be dis[played] as a movie by an MPEG player. Official MIME types are registered with IANA. More [infor]mation on MIME and MIME types may be found in Refs. [33] and [34].

BYE method   A UAC sends a BYE request when it wants to leave a session. If a session has more than one participant, the session stays up until its last participant leaves. BYE requests may be compared with SS7 ISUP's REL and RLC messages and to ISDN Q931's DISC message. SIP proxy servers forward BYE requests the same way they do INVITEs. Proxy servers must support BYE. Redirect servers and UASs should support it, but it's not mandatory.

CANCEL method   A UAC sends a CANCEL request to terminate an earlier request that is still pending. The `Call-ID`, `To`, `From`, and `CSeq` header lines of the CANCEL request are the same as in the request being canceled. In theory, any pending request may be canceled. In practice, only INVITE is.

REGISTER method   User agents send REGISTER requests to tell a SIP server where they are or to update their service profile. If they're registering their current location, the `To` header contains the address being registered.

9.3.2 SIP Response Messages

A SIP *response message* consists of a response line followed by header lines and an optional body. The format of the *response line* is:

SIP-Version Status-Code Reason-Phrase

where:

- *SIP-Version* is the version of the protocol being used by the message, say, SIP/2.0.
- *Status-Code* is a 3 digit value.
- *Reason-Phrase* is a short text string that explains *Status-Code*.

SIP has six kinds of response status code, indicated by a code's first digit:

1xx   Information responses. A request has been received and is being processed, but it's not done yet. Further 1xx responses may follow as long as the request has not yet been completed. Information responses are comparable to *call progress tones* in the PSTN.

2xx   Success responses. The request has been successfully processed.

3xx   Redirection responses. Further action must be taken before the request can be completed.

4xx Client error responses. The request had an error or could no[t be] processed by the server that received it.

5xx Server error responses. The request was valid but the server fa[iled] while processing it.

6xx Global failure responses. The request could not be processed [by] any server.

Figure 9.2 shows examples of status codes from each of t[he] categories.

INFORMATION	REDIRECTION	SERVER ERROR
100 Trying	300 Multiple choices	500 Internal server error
180 Ringing	301 Moved permanently	501 Not implemented
181 Call being forwarded	302 Moved temporarily	502 Bad gateway
		503 Service unavailable
		504 Gateway timeout
		505 SIP version not suppor[ted]
SUCCESS	**CLIENT ERROR**	**GLOBAL FAILURE**
200 OK	400 Bad request	600 Busy everywhere
	401 Unauthorized	603 Decline
	402 Payment required	604 Does not exist anywhe[re]
	403 Forbidden	606 Not acceptable
	404 Not found	
	415 Unsupported media type	
	486 Busy here	

Figure 9.2
A partial catalog of SIP response-message status codes.

9.4 Session Description Protocol (SDP)

Before you can set up a session, the participants must agree on [the] media they will use to communicate with each other. To reach suc[h] agreement, they must have some way to describe to each other the n[um]ber and types of connections proposed. The most common vehicl[e for] carrying such information is the Session Description Protocol (SDP). SDP isn't a protocol at all, since there is no such thing as an "SDP sage." Instead, SDP is language for describing the connections in a session and their properties.

SDP defines a session as a set of *media streams*. A media stream travels on some medium and exists for some period of time. Different streams in the same session may exist at different times, so a session need not be continuously active. An SDP description contains the following information about a session:

- The name of the session and its purpose.
- The time(s) during which the session will be active.
- The media out of which the session is built and the information needed to tap into those media, including:
  - *—Media type.* Video, audio, and so forth.
  - *—Transport protocol.* RTP on UDP on IP, and so forth.
  - *—Media format.* G.711 voice, MPEG video, and so forth.
  - *—Multicast/unicast information.* For multicast sessions, the destination address and port of the multicast stream. For unicast sessions, the address and port to which data is sent and the address and port at which data is to be received.[5]

A session description may also include other information, such as the bandwidth needed, the person responsible for the session, and pointers (URIs) to locations that contain information an invitee can use in deciding whether to join or not.

An SDP description has two sections: a session-level part and a media stream part. The *session-level part* contains information about the entire session and all its media streams. The *media stream part* has zero or more media stream descriptions. A *media stream description* provides information about a single media stream. SDP descriptions are text based, consisting of a sequence of fields, each field a line of the form:

type= value

where *type* is a single character and *value* is a string the value of which depends on *type*. Furthermore, SDP specifies the order in which types must appear. This makes SDP easy to parse.

The following are the fields that go in the session-level part of an SDP description, in the order in which they must appear. An asterisk marks

---

[5]This is the default. Some media may use the address and port to specify a control channel.

optional fields. Note that the mandatory t= field appears in the mi[ddle] of a bunch of optional fields:

v=*protocol version*
o=*owner/creator and session identifier*
s=*session name*
i=*session information*
u=*URI of description*
e=*e-mail address*
p=*phone number*
c=*connection information*
b=*bandwidth information*
t=*time the session is active*
r=*repeat times*
z=*time zone adjustments*
k=*encryption key*
a=*session attributes*

Each media stream description in the media stream part uses the [fol]lowing fields, in the order shown (optional fields again marked b[y] asterisk):

m=*media and transport address*
i=*media title*
c=*connection information*
b=*bandwidth information*
k=*encryption key*
a=*session attributes*

If a message contains more than one SDP session description, the [start] of each one may be found by looking for the mandatory version field with which it starts.

Rather than work through SDP field by field (see Ref. [38] for t[hat], let's look at an example that uses some of the most commonly [used] fields. Here is an SDP description of a session with two media stream[s]:

```
v=0
o=jsmith 2873289751 2873289751 IN IP4 somemachine.nowhere.org
s=Multimedia Session
i=An example of an SDP session
```

```
u=http://www.nowhere.org/sessioninfo.html
e=jsmith@nowhere.org (John Smith)
c=IN IP4 192.168.0.1
t=2873397496 2873404696
a=recvonly
m=audio 49170 RTP/AVP 15
a=sendrecv
m=video 51372 RTP/AVP 32
```

The fields in this description may be interpreted as follows:

- `v=` The description begins with a mandatory version field. At this writing, its value must be zero (0).
- `o=` This field packs in lots of information. The first value (`jsmith`) is the login ID of the session originator on the machine from which the session will originate. If that machine doesn't support login IDs, this value is a minus sign (-). The second value (`2873289751`) is a numeric string selected in such a way that it can be used to produce a globally unique identifier for the session. While SDP does not specify how this value is to be selected, it suggests using the current time in *Network Time Protocol (NTP)* [25] format. The third value (`2873289751`) is a number that's used to keep track of different announcements for the same session (for example, because of changes in the requirements for a session). SDP mandates nothing about this value other than that it increase with each modified announcement of the same session. The SDP standard again suggests using an NTP time value. The fourth value (`IN`) is a string that indicates what kind of network will be used for the session, the Internet in this case. The fifth value (`IP4`) indicates what kind of address will follow. `IP4` means an IPv4 address. The last value is the globally unique address of the machine from which the session is being created. Here it's shown as `somemachine.nowhere.org`. It could also be a numeric IP address (say, 192.168.0.53).
- `s=` A mandatory field in which the originator assigns a name to the session.
- `i=` An optional field that provides information about the session. There aren't any rules for how to use this field. A common use is to indicate what the session will be about, much like the subject line of an e-mail.
- `u=` An optional field that gives the URI of a location to which invitees can go for more information about the session.
- `e=` An optional field that contains the e-mail address of the session originator. SDP also defines an optional `p=` field for the originator's phone number.

c= A field that provides information that invitees need to connect t
the session. This field is best described as semioptional. An SDP
description must provide the information somewhere. The c= field
may appear in the session-level part, in which case it applies to all
media streams. If there is no c= field in the session-level part, then
every media stream description must include its own c= field. If c
fields appear in both the session-level and media stream part, a
media stream's own c= field takes precedence over the one in the
session-level part. The first two values of this field—IN IP4 in the
example—have the same meaning they do in the s= field: network
type and address type, respectively. The rest of this field gives the
address (192.168.0.1) to which an invitee should connect in ord
to join the session.

t= The start and stop time for the session, given in NTP format. Th
are both usually zero (0) when SIP is used for IP telephony, since tl
is no preset start and stop time for a phone call.

a= Session attributes. The session-level part ends with zero or more
of these. These apply to all media streams in the session (unless the
are overridden by a= fields for the media streams themselves). The
example session has only one attribute, recvonly. This
means that, unless otherwise indicated, when people join this
session, they should be in receive-only mode. They can listen, but
talk.

m= A media stream description that specifies the media type, port,
transport protocol, and format(s) to be used for that stream. There
may be more than one media stream description, each with its ow
m= field and optional fields following it that apply only to that me
stream.

- *Media type.* SDP defines five media types: audio, video,
  application, data, and control. The difference between
  application and data may not be obvious. In SDP, an
  application stream generates data for display to participants
  during the course of a session. A data stream provides bulk
  transfer of data that is not usually displayed (for example, an
  executable program file). A control stream carries conference
  control messages for the session.
- *Port.* The port to which the media stream will be sent, 49170 f
  audio and 51372 for video in the example.

- *Transport protocol.* For a session where the `c=` field is `IP4`, the transport protocol may be either `RTP/AVP` for RTP using the Audio/Video profile on UDP, or `udp` for UDP by itself.[6]
- *Media formats.* The final values in the media stream field are media formats. For an `RTP/AVP` stream, these are numbers that represent different formats as defined in RFC 1890 [31]. The values shown in the example (`15` and `32`) represent G.728 audio and MPEG video.

`a=sendrecv` An attribute that applies only to the audio stream and overrides the session-level `a=recvonly` attribute. The audio stream in this session will send and receive data, but the video stream will receive only, since it has no attribute that states otherwise.

9.5 Call Signaling with SIP

9.5.1 Simple Two-Party Call

Every SIP session starts with an INVITE message. The service in Fig. 9.3 demonstrates INVITE in its simplest form: a two-party session in which both user agents talk directly to each other. Before turning to more complex SIP services, we'll look at the messages used by this simple service in some detail. Figures 9.4 and 9.5 show the state machines that the UAC and UAS use to handle this and other INVITE requests.

Mr. Smith, the session originator (calling party) begins by supplying the SIP URL of the called party, Ms. Jones, to his user agent. His user agent asks DNS for the corresponding IP address and then sends this INVITE request to that location:

```
INVITE sip:JillJones@BigCo.com SIP/2.0
Via: SIP/2.0/UDP host1.LittleCo.com
From: "Mr. Smith" <sip:BobSmith@LittleCo.com>
To: "Ms. Jones" <sip:JillJones@BigCo.com>
Call-ID: 111111@host1.LittleCo.com
CSeq: 1 INVITE
Contact: <sip:BobSmith@LittleCo.com>
Subject: Can we make this SIP thing work?
Content-Type: application/sdp
```

---

[6]As of this writing, these are the only two transport protocols defined for this field. Other protocols may be added by registering them with IANA.

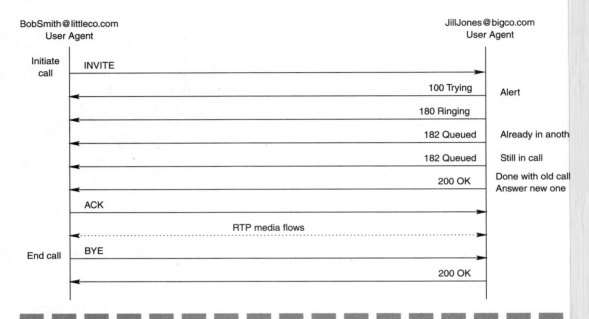

Figure 9.3
A two-party SIP call (user agent to user agent with no SIP servers involved).

```
Content-Length: ...

v=0
o=bobsmith 2890844526 2890844526 IN IP4 host1.LittleCo.com
s=Two-party call
c=IN IP4 100.101.102.103
t=0 0
m=audio 49170 RTP/AVP 4 9
```

The request line at the beginning of the message contains the method (`INVITE`), the SIP URL of the user to whom the request is be sent (`sip:JillJones@BigCo.com`) and the SIP version being used Mr. Smith's end (`SIP/2.0`). Because no port is specified in the SIP U (for example, `sip:JillJones@BigCo.com:12345`), the default po 5060—will be used to make the connection.

After the request line are several header lines with more informat about the proposed session. The `Via` header gives the path taken by request so far. As a message travels from one SIP server to another, e adds another `Via` header to it. `Via` ensures requests don't end up lo ing through a set of SIP servers forever. If a server sees its own URL a where in the `Via` lines of a request message, it knows the request is lo

Session Initiation Protocol (SIP)

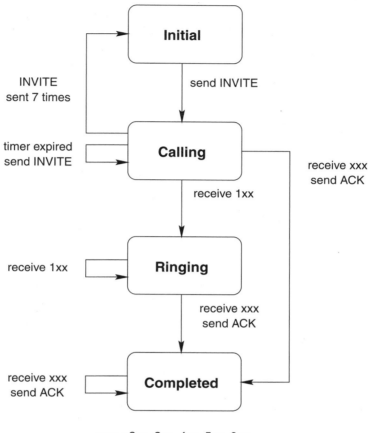

Figure 9.4
INVITE request state machine for User Agent Client (calling party).

ing. Similarly, it can tell it's about to loop a request if the host it's forwarding to already appears in the `Via` lines.

Besides `Via`, SIP has other mechanisms to short-circuit loops and catch other routing errors. Messages may have a `Max-Forwards` header line that limits the number of proxies through which a message will pass. The `Expires` header line limits the time a server can spend searching for a session invitee.

Besides detecting loops, `Via` ensures that responses return on the same route as the requests to which they correspond. Since the invitation in this example goes directly from inviter to invitee, it has only one `Via` header. The route taken by one SIP request has no effect on the route taken by subsequent requests, even requests for the same session.

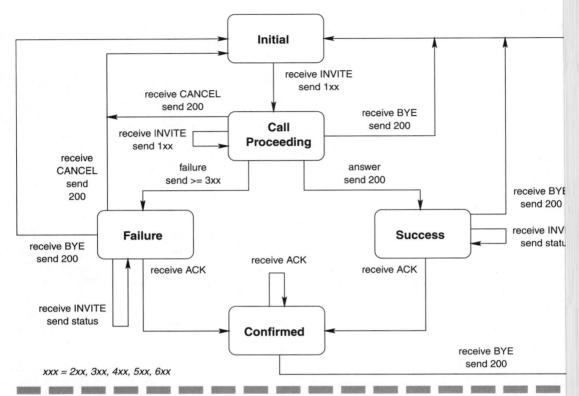

Figure 9.5
INVITE request state machine for User Agent Server (called party).

Responses return along the same path as requests, but each reque[st is] routed independently of any other.

The `From` header identifies the session originator. It contains [the] originator's SIP URL or other URI. It may also have the origina[tor's] name ("`Mr. Smith`"). `From` has much the same function as Caller I[D in] the PSTN (though it's perhaps more susceptible to mischief because [it's] provided by the originator himself). Likewise, the `To` header ident[ifies] the invitee, the *logical call destination* for the session. Like the `From` he[ader,] it contains a SIP URL or other URI and may also have a name ("[Mr.] `Jones`"). SIP requests and responses must have both `To` and `F[rom]` headers.

The `Call-ID` header uniquely identifies a specific invitation. It's [con]structed by combining the originating host name (`host1.Litt[le` `Co.com`) with a string of characters that's not been previously used any session originating at that host (111111). The @ sign that joins these two values has nothing to do with e-mail; it's just a syntactic convention SIP uses. Every session participant will use this call ID in SIP messages pertaining to that session.

The `CSeq` (command sequence) header contains a SIP method name and a number. Request messages echo the method they are requesting here. The number helps participants keep track of the order in which requests are sent so that responses can be matched to them. A session's first INVITE message can use any number (as long as it's less than 2, 147, 483, 648). Other request messages increase this number by one if their method, headers, or body differ from those in a previous request message for the same session. The only exceptions are ACK and CANCEL requests, which use the same sequence number as the INVITE request to which they refer. A retransmitted request uses the same `CSeq` values as the request it is repeating. Response messages echo the method name and sequence number of the request to which they are responding.

INVITE requests must have a `Contact` header containing the URI (normally a SIP URL) of the location from which the request was sent. The called party sends its request messages for the session originator to this address. In this way, the called party can send requests like BYE directly to the originator, bypassing any proxy servers in between. The `Contact` header is also used when a user agent registers itself with a SIP redirect server, as we'll see in Sec. 9.5.4.

The `Subject` header is optional. When provided, it's comparable to the same header in e-mail, a place to put a brief description of the session's purpose or topic. Software on Ms. Jones's end can display this to her so she can decide whether or not to accept the invitation.

We've seen `Content-Type` and `Content-Length` before. Here, the request has a body with an SDP description of the session to which Ms. Jones is being invited. In a real SIP message, `Content-Length` would be the number of bytes in that body. The reader may fill in the correct value as an exercise (albeit, a pointless one).

A blank line separates the headers from the body. The body is an SDP description of the proposed session. As always, it begins with the mandatory `v=0` field. This session description is similar to the one in Sec. 9.4. The originator's login ID is `bsmith` on `host1.LittleCo.com`, the machine from which the invitation is being sent. This session will use the Internet, and the originating machine uses an IP address. Time stamps are given that allow one to generate a unique session ID and keep track of the order in which messages for this session are sent. The session is given a name (`Two-party call`). Ms. Jones may join the session at IP address 100.101.102.103. The start and stop times for the sion are both zero (0), which is normal for a phone call that typi begins as soon as it is accepted and lasts until both parties have dis nected. The media field shows that Mr. Smith can receive an RTP a stream on port 49170 using either G.723 (4) or G.722 (9) encoding.

SIP messages may contain many kinds of address. To review, her the different addresses from the INVITE message example:

- The INVITE line contains the address of the location to which th request is sent.

- The `To` header contains the logical address of the party to whom the request is directed. This may not be the same as the address in the INVITE line. For example, when a proxy server forwards an invitation, the INVITE line contains the address to which it is forwarded, while the `To` header contains the address used by the originator (probably that of the first proxy to get the invitation).

- The `Via` headers hold the addresses of hosts through which the request has passed. Responses go back to the last host in this chair regardless of what's in the `From` header line.

- The `From` header line holds the logical source of the request. Sinc a request can be forwarded through many proxy servers before arriving at its destination, this identifies the party that initiated t request.

- The SDP description contains addresses that specify where the session originator wants media streams to be sent. As with `Via`, these may not be the same as the `From` address.

- The `Call-ID` looks like an address but it isn't one.

Once Ms. Jones's UAS in the domain `LittleCo.com` gets the in tion, it sends back a `100 Trying` response that lets Mr. Smith knov invitation has been received:

```
SIP/2.0 100 Trying
Via: SIP/2.0/UDP host1.LittleCo.com
From: "Mr. Smith" <sip:BobSmith@LittleCo.com>
To: "Ms. Jones" <sip:JillJones@BigCo.com>
Call-ID: 111111@host1.LittleCo.com
CSeq: 1 INVITE
Content-Length: 0
```

The UAS sends `100 Trying` immediately rather than wait to res] until it's contacted Ms. Jones, because that could take some tin might have to locate her first. Or the call might be queued somev

In any case, the calling party gets back this message to let it know things are proceeding normally. If it didn't get a response back, the originator's UAC would send the invitation every 0.5 seconds until it did get a response. This is how SIP can be reliable, even though it usually runs on top of the unreliable UDP. The advantage is that it's generally easier to adjust transmission timers in SIP than in a reliable transport protocol, such as TCP.

The first line of the response contains the SIP version used by the called party's user agent software, the response status code (`100`), and a short text explanation of that code (`Trying`). The rest of the message mostly echoes information from the original INVITE request. The `From` header still refers to Mr. Smith and the `To` header still refers to Ms. Jones, even though this messages goes from Ms. Jones to Mr. Smith. The `CSeq` header is the same as in the invitation to which this message is responding.

The most common response at this stage of a call is `100 Trying`, but there's nothing to keep the called party's UAS from sending as many `100` responses as it wants to tell the calling party what's going on. Since the UAC doesn't care what's in the text portion of the first line, the UAS can put other messages there besides `Trying`, for example, `100 Looking up number`...`100 Looking for carrier`...`100 Got carrier SBC`...`100 Dialing`...and so forth.

After sending the `100 Trying` response, Ms. Jones's UAS needs to find her and tell her she has a session invitation. SIP doesn't say how the UAS is supposed to do this. Once it's been done, the UAS sends back a `180 Ringing` response:

```
SIP/2.0 180 Ringing
Via: SIP/2.0/UDP host1.LittleCo.com
From: "Mr. Smith" <sip:BobSmith@LittleCo.com>
To: "Ms. Jones" <sip:JillJones@BigCo.com>
Call-ID: 111111@host1.LittleCo.com
CSeq: 1 INVITE
Content-Length: 0
```

There's not much to be said about this message, other than that it also echoes the original INVITE message's `CSeq` values. That's because `100 Trying` and `180 Ringing` are both provisional responses that leave the INVITE request still pending from the perspective of Mr Smith's UAC.

To make things more interesting, let's assume Ms. Jones is in the middle of another call and decides to queue Mr. Smith's call rather than reject it. In so doing, she or her software can supply information to Mr. Smith about what's going on, information that's returned in a `182 Queued` response message:

```
SIP/2.0 182 Queued, On another call, be with you shortly
Via: SIP/2.0/UDP host1.LittleCo.com
From: "Mr. Smith" <sip:BobSmith@LittleCo.com>
To: "Ms. Jones" <sip:JillJones@BigCo.com>
Call-ID: 111111@host1.LittleCo.com
CSeq: 1 INVITE
Content-Length: 0
```

The `CSeq` remains unchanged because this is still a provisi[onal] response to the original invitation. Ms. Jones can send more than [one] `182 Queued` response:

```
SIP/2.0 182 Queued, Almost finished
Via: SIP/2.0/UDP host1.LittleCo.com
From: "Mr. Smith" <sip:BobSmith@LittleCo.com>
To: "Ms. Jones" <sip:JillJones@BigCo.com>
Call-ID: 111111@host1.LittleCo.com
CSeq: 1 INVITE
Content-Length: 0
```

Finally, Ms. Jones responds to the invitation and her UAS sends a [200] OK response to Mr. Smith:

```
SIP/2.0 200 OK
Via: SIP/2.0/UDP host1.LittleCo.com
From: "Mr. Smith" <sip:BobSmith@LittleCo.com>
To: "Ms. Jones" <sip:JillJones@BigCo.com>
Call-ID: 111111@host1.LittleCo.com
CSeq: 1 INVITE
Content-Type: application/sdp
Content-Length: ...

v=0
o=jilljones 2891802492 2891802492 IN IP4 somehost.BigCo.com
s=Let's talk
c=IN IP4 200.201.202.203
m=audio 5004 RTP/AVP 9
```

This body of this response contains an SDP description of the se[ssion] as accepted by Ms. Jones. The most important thing here is the m[edia] field in which Ms. Jones indicates she can receive a G.722 encoded audio stream on port 5004 of `somehost.BigCo.com`.

Mr. Smith's UAC acknowledges Ms. Jones's acceptance of the ca[ll by] sending an ACK request:[7]

```
ACK sip:JillJones@BigCo.comSIP/2.0
Via: SIP/2.0/UDP host1.LittleCo.com
From: "Mr. Smith" <sip:BobSmith@LittleCo.com>
To: "Ms. Jones" <sip:JillJones@BigCo.com>
```

---

[7] That's not a typo. In SIP, ACK is a request, not a response.

```
Call-ID: 111111@host1.LittleCo.com
CSeq: 1 ACK
Content-Length: 0
```

The `CSeq` number remains 1 since this message still pertains to the INVITE request with that sequence number, but the `CSeq` method name is now ACK. There is no response to an ACK request. If there were, the response would presumably have to be acknowledged, and that acknowledgment responded to, and that response acknowledged, and ... you get the idea. ACKs are sent only in response to a response to an INVITE request (yes, like most things about SIP ACKs, that sounds a bit odd). If an INVITE response is not ACKed, it's retransmitted until its ACK arrives.

Mr. Smith and Ms. Jones are now connected to each other in a single RTP session, with Mr. Smith receiving on port 49170 of `host1.LittleCo.com` and Ms. Jones receiving on port 5004 of `somehost.BigCo.com`. Eventually one or the other of them hangs up and their UAS sends a BYE request. Here it's Mr. Smith who hangs up:

```
BYE sip:JillJones@BigCo.com SIP/2.0
Via: SIP/2.0/UDP host1.LittleCo.com
From: "Mr. Smith" <sip:BobSmith@LittleCo.com>
To: "Ms. Jones" <sip:JillJones@BigCo.com>
Call-ID: 111111@host1.LittleCo.com
CSeq: 2 BYE
Content-Length: 0
```

The main thing worth noting is that the `CSeq` number goes up by one, since this is the second independent request to have been generated within the session. The party receiving the BYE request acknowledges it with a `200 OK` response:

```
SIP/2.0 200 OK
Via: SIP/2.0/UDP host1.LittleCo.com
From: "Mr. Smith" <sip:BobSmith@LittleCo.com>
To: "Ms. Jones" <sip:JillJones@BigCo.com>
Call-ID: 111111@host1.LittleCo.com
CSeq: 2 BYE
Content-Length: 0
```

9.5.2 Two-Party Call with Redirect Server

The scenario shown in Fig. 9.6 is a bit more complicated. The first thing it adds is a SIP redirect server. Just as in the previous example, this call starts off with an INVITE request to Ms. Jones from Mr. Smith. However, this time Ms. Jones's SIP URL leads to a SIP redirect server operated

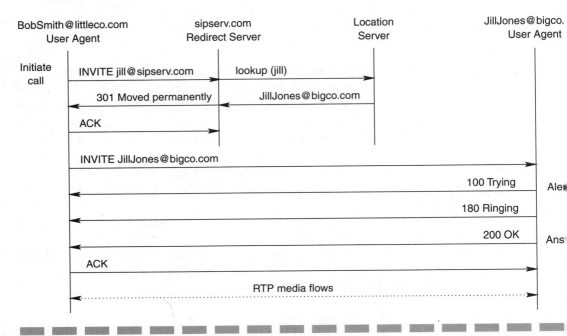

Figure 9.6
A two-party SIP call with redirect server.

by her SIP service provider. From now on, I'll show the complete te a message only if it introduces a significant new feature. The mi bits of this message would be largely the same as in the earlier exa Here's the message:

```
INVITE sip:jill@sipserv.com SIP/2.0
Via: SIP/2.0/UDP host1.LittleCo.com
From: "Mr. Smith" <sip:BobSmith@LittleCo.com>
To: "Ms. Jones" <sip:jill@sipserv.com>
Call-ID: 222222@host1.LittleCo.com
CSeq: 1 INVITE
. . .
```

The redirect server at `sipserv.com` is supposed to know wher Jones's user agent really is at any given time. SIP doesn't specify ho redirect server does that. It just assumes there's a location service th redirect server can query for Ms. Jones's current location. The loc service could be an LDAP directory. It could be a mobile network tion register. However it's done, the redirect server obtains the SIP at which Ms. Jones can now be contacted and returns it to Mr. S user agent in the `Contact` header of a `301 Moved permane` response message. Mr. Smith's user agent can use that information to send the invitation to the correct location:

```
SIP/2.0 301 Moved permanently
Contact: sip:JillJones@BigCo.com
Via: SIP/2.0/UDP host1.LittleCo.com
From: "Mr. Smith" <sip:BobSmith@LittleCo.com>
To: "Ms. Jones" <sip:jill@sipserv.com>
Call-ID: 222222@host1.LittleCo.com
CSeq: 1 INVITE
Content-Length: 0
```

After Mr. Smith's user agent gets the redirect server's response, it returns an ACK. Once that's taken care of, his user agent sends the invitation again, but this time to the Contact address, sip:JillJones@BigCo.com. The call proceeds from here just as it did in the previous example. Note that the To and Call-ID headers of the new invitation keep the values they had in the original invitation while the CSeq value goes up by one because this is a new invitation, not a continuation of the previous one:

```
INVITE sip:JillJones@BigCo.com SIP/2.0
Via: SIP/2.0/UDP host1.LittleCo.com
From: "Mr. Smith" <sip:BobSmith@LittleCo.com>
To: "Ms. Jones" <sip:jill@sipserv.com>
Call-ID: 222222@host1.LittleCo.com
CSeq: 2 INVITE
. . .
```

The 301 Moved permanently response sent by the redirect server means that Ms. Jones will be staying at her current address for a while. If Mr. Smith wants to bypass the redirect server from now on and go directly to Ms. Jones's current address, he should update any address books used by his SIP software with that address.

The redirect server might also have sent back a 302 Moved temporarily response:

```
SIP/2.0 302 Moved temporarily
Contact: sip:JillJones@BigCo.com
. . .
```

This tells Mr. Smith's user agent to send the invitation to the address in the Contact header, just as it did with the 301 response. However, this address will be valid only for this call and shouldn't be saved for future calls. That's not to say Ms. Jones might not stay there for some time. However, Mr. Smith's user agent will have to keep checking with the redirect server each time he calls her to find out where she is.

9.5.3 Two-Party Call with Proxy Server

The next example is another basic call, this time using a SIP p
server, as shown in Fig. 9.7. The initial invitation from Mr. Smith l
the same as in the other examples:

```
INVITE sip:jill@sipserv.com SIP/2.0
Via: SIP/2.0/UDP host1.LittleCo.com
From: "Mr. Smith" <sip:BobSmith@LittleCo.com>
To: "Ms. Jones" <sip:jill@sipserv.com>
Call-ID: 333333@host1.LittleCo.com
CSeq: 1 INVITE
  . . .
```

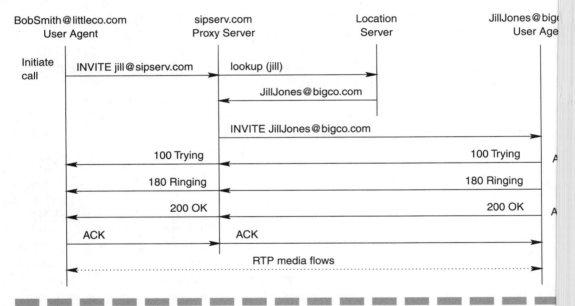

Figure 9.7
A two-party SIP call with proxy server.

This time Ms. Jones's SIP URL leads to a SIP proxy server operat
her SIP service provider. The proxy server looks up Ms. Jones's
address using a location service, just as the redirect server di
instead of sending Ms. Jones's address back to Mr. Smith in a `301`
response, the proxy server forwards his invitation directly to Ms.
The proxy server changes the SIP URL in the INVITE line. It also
`Via` line with its own address and a branch parameter, a unique

Session Initiation Protocol (SIP)

that correlates requests with their responses when the proxy forwards a request to more than one location. The branch parameter is superfluous here (but required nonetheless):

```
INVITE sip:JillJones@BigCo.com SIP/2.0
Via: SIP/2.0/UDP sip-proxy.sipserv.com;branch=a7yu45z87dfj
Via: SIP/2.0/UDP host1.LittleCo.com
From: "Mr. Smith" <sip:BobSmith@LittleCo.com>
To: "Ms. Jones" <sip:jill@sipserv.com>
Call-ID: 333333@host1.LittleCo.com
CSeq: 1 INVITE
. . .
```

The rest of this call proceeds largely as it did in the other examples, except that now requests and responses pass through the proxy server. For example, when Ms. Jones's user agent gets the invitation, it returns `100 Trying` to the proxy server:

```
SIP/2.0 100 Trying
Via: SIP/2.0/UDP sip-proxy.sipserv.com;branch=a7yu45z87dfj
Via: SIP/2.0/UDP host1.LittleCo.com
From: "Mr. Smith" <sip:BobSmith@LittleCo.com>
To: "Ms. Jones" <sip:jill@sipserv.com>
Call-ID: 333333@host1.LittleCo.com
CSeq: 1 INVITE
Content-Length: 0
```

The extra `Via` line for the proxy server is the only thing that would set this response apart from the corresponding response in the redirect server scenario. The proxy server checks this line for its own address. If it's there, the proxy server strips that line off and forwards its response to the address it finds in the next `Via` line of the response it received:

```
SIP/2.0 100 Trying
Via: SIP/2.0/UDP host1.LittleCo.com
From: "Mr. Smith" <sip:BobSmith@LittleCo.com>
To: "Ms. Jones" <sip:jill@sipserv.com>
Call-ID: 333333@host1.LittleCo.com
CSeq: 1 INVITE
Content-Length: 0
```

This is the same message Mr. Smith's user agent would have received if it had been talking directly to Ms. Jones's end. To the user agents on the ends of a session, a proxy server looks just like a user agent.[8]

More than one proxy may sit between the participants in a SIP call. These proxies are essentially invisible to their user agents, which behave

---

[8]Well, on the called party end, there are those extra `Via` lines that proxies add to the header of request messages. But since user agents don't use `Via` information, it's a moot point.

just as though they were talking directly to each other. For example
proxy to which Mr. Smith sends his invitation might look up Ms. J
in its location service and get the address of another proxy. It do
know that's a proxy and not Ms. Jones's user agent, but it doesn't m
It forwards the invitation just as it would if it were sending it to
Jones's user agent. This second proxy in the chain does the same th
as the first proxy: It looks up Ms. Jones in its location service and
wards the request to the address it gets back (with a new `Via` line a
for itself). This is repeated as often as necessary to get the request to
Jones's user agent. Any responses flow back through the proxy cha
reverse order, with each proxy stripping off its own `Via` line and
warding the message to the address in the `Via` line just below its
Once a session has been established through, proxy servers are n
the path along which media flows pass from one endpoint of a se
to another.

9.5.4 Registration

I haven't said much about the location services that SIP redirec
proxy servers use to find SIP users. SIP doesn't say how they're
implemented. But the information has to get in there somehow. If
SIP server has its own implementation of a location service, how d
users get a record of themselves in there so the server can find t
Every location service may have its own unique interface. It's not re
able to expect user agents to know how to interact with every po
location service.

SIP's solution to this problem is the REGISTER request. The follo
message registers Ms. Jones at the `sipserv.com` server (which cou
either a redirect or a proxy server). After the server processes this re
any invitations it receives for `sip:jill@sipserv.com`, which
value in the `From` and `To` headers of the REGISTER request,[9] w
redirected or proxied (depending on the type of server) to the add
the `Contact` header:

   sip:JillJones@sip1.BigCo.com.

---

[9]When the party registering is the same as the party being registered, the `From`
headers are the same. If Party A registers on behalf of Party B, Party A's address
the `From` header and Party B's address goes in the `To` header.

There's no expiration given, so it will last until another registration overrides it:

```
REGISTER sip:sipserv.com SIP/2.0
Via: SIP/2.0/UDP somehost.BigCo.com
From: sip:jill@sipserv.com
To: sip:jill@sipserv.com
Call-ID: 444444@somehost.BigCo.com
CSeq: 1 REGISTER
Contact: <sip:JillJones@sip1.BigCo.com;transport=udp>
Content-Length: 0
```

The server returns a `200 OK` response that lists the current registrations it has for Ms. Jones. Here it returns only the registration she just made:

```
SIP/2.0 200 OK
Via: SIP/2.0/UDP somehost.BigCo.com
From: sip:jill@sipserv.com
To: sip:jill@sipserv.com
Call-ID: 444444@somehost.BigCo.com
CSeq: 1 REGISTER
Contact: <sip:JillJones@sip1.BigCo.com;transport=udp>
Content-Length: 0
```

A registration's life can be limited by adding an `Expires` header. Here's a registration that lasts for one hour:

```
REGISTER sip:sipserv.com SIP/2.0
Via: SIP/2.0/UDP somehost.BigCo.com
From: sip:jill@sipserv.com
To: sip:jill@sipserv.com
Call-ID: 555555@someotherhost.BigCo.com
CSeq: 1 REGISTER
Contact: <sip:JillJones@sip2.BigCo.com;transport=udp>
Expires: 3600
Content-Length: 0
```

9.5.5 Forked Invitations

The last services in this chapter use *forked invitations*. Forked invitations allow a proxy server to contact an invitee who might be at any one of several locations, even though the session originator gave only one address in the original invitation. Service logic and registration data on the proxy send the invitation to multiple locations, evaluate the results, and return a single response to the originator. To the originator, forking is transparent (other than, perhaps, taking longer to set up the session).

We'll first look at a sequential search that forwards invitations to location after another. Then we'll look at a parallel search, in which tations are sent to all locations at once.

SEQUENTIAL SEARCH Figure 9.8 shows that an invitation for into a sequential search starts out like any other proxied SIP invit

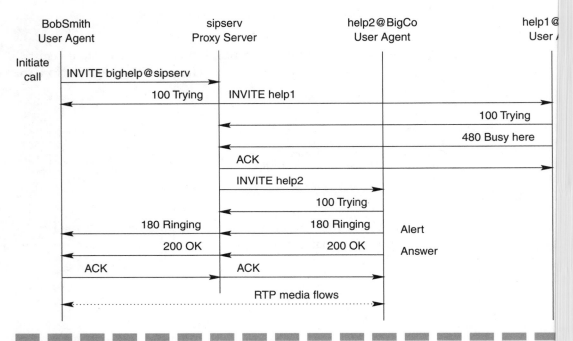

Figure 9.8
Forked INVITE with sequential search.

with the originator sending an invitation via a proxy server. The tion uses a SIP URL known to the proxy server:

```
INVITE sip:bighelp@sipserv.com SIP/2.0
Via: SIP/2.0/UDP host1.LittleCo.com
From: "Mr. Smith" <sip:BobSmith@LittleCo.com>
To: "Help Desk" <sip:bighelp@sipserv.com>
CSeq: 1 INVITE
 . . .
```

The proxy server looks up `bighelp` in its location service (not in the diagram) and finds it should try several locations in seq

The proxy forwards the invitation to the first of these, `help1@BigCo`, using the `help1` address in the INVITE line and the original `bighelp@sipserv` address in the `To` header. The proxy server adds itself to the `Via` headers, together with a `branch` parameter, a unique value that's used to correlate forked requests and their responses:

```
INVITE sip:help1@BigCo SIP/2.0
Via: SIP/2.0/UDP server1.sipserver.com; branch=3d8a8745ab4.1
Via: SIP/2.0/UDP host1.LittleCo.com
From: "Mr. Smith" <sip:BobSmith@LittleCo>
To: "Help Desk" <sip:bighelp@sipserv>
CSeq: 1 INVITE
. . .
```

The proxy server sends a `100 Trying` response to let the originator know call setup is proceeding. The UAS at `help1@BigCo` responds to its invitation with `100 Trying`, but the proxy doesn't forward that because it's already sent one of those to the originator.

Once the UAS for `help1` has a chance to process the invitation, it finds the party is busy at its location and sends back a `480 Busy here` response. This tells the proxy to try the next address for `help@BigCo`. It ACKs `help1`'s response and sends the invitation to `help2@BigCo`. Again the INVITE line uses the true address of the recipient and the `To` header uses the address from the original invitation. The `branch` parameter is different because this is a different invitation:

```
INVITE sip:help2@BigCo SIP/2.0
Via: SIP/2.0/UDP server1.sipserver.com; branch=3d8a8745ab4.2
Via: SIP/2.0/UDP host1.LittleCo.com
From: "Mr. Smith" <sip:BobSmith@LittleCo>
To: "Help Desk" <sip:bighelp@sipserv>
CSeq: 1 INVITE
. . .
```

The UAS for `help2` sends back `100 Trying`, which the proxy server again keeps to itself. This time, the UAS finds its party unoccupied and sends back a `180 Ringing` response. The proxy server sends this back to the originator. The person at location `help2` answers the call and the UAS there sends a `200 OK` response. From here on, the call behaves just like a normal proxied call: The proxy forwards the response, the originator ACKs the response, and the two parties start talking.

PARALLEL SEARCH Figure 9.9 shows an invitation forking into a parallel search. Again, it starts out like any other proxied SIP invitation with the originator sending an invitation via a proxy server:

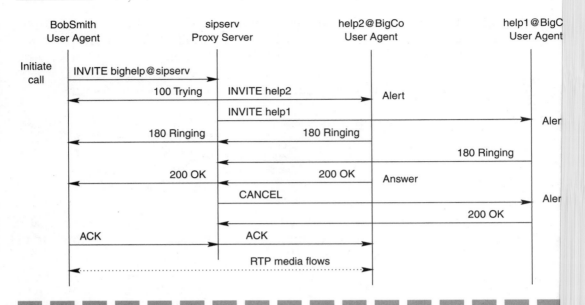

Figure 9.9
Forked INVITE with parallel search.

```
INVITE sip:bighelp@sipserv SIP/2.0
Via: SIP/2.0/UDP host1.LittleCo.com
From: "Mr. Smith" <sip:BobSmith@LittleCo>
To: "Help Desk" <sip:bighelp@sipserv>
CSeq: 1 INVITE
. . .
```

The proxy server looks up `bighelp` in its location service (not sh and finds it may be at any of several locations that should be tr parallel. The proxy server returns a `100 Trying` response to l originator know call setup is proceeding, and then forwards the tion to all indicated destinations. The `branch` parameter in th header is similar to that in the sequential fork. As in that sce the true address goes in the INVITE line and the ori `bighelp@sipserv` address goes in the `To` header, producing:

```
INVITE sip:help1@BigCo SIP/2.0
Via: SIP/2.0/UDP host1.LittleCo.com
Via: SIP/2.0/UDP server1.sipserver.com; branch=3d8a8745ab4.1
From: "Mr. Smith" <sip:BobSmith@LittleCo>
To: "Help Desk" <sip:bighelp@sipserv>
CSeq: 1 INVITE
. . .
```

and:

```
INVITE sip:help2@BigCo SIP/2.0
Via: SIP/2.0/UDP host1.LittleCo.com
Via: SIP/2.0/UDP server1.sipserver.com; branch=3d8a8745ab4.2
From: "Mr. Smith" <sip:BobSmith@LittleCo>
To: "Help Desk" <sip:bighelp@sipserv>
CSeq: 1 INVITE
. . .
```

Both UASs alert their parties and send back `180 Ringing` responses. The proxy server forwards one of these back to the originator.[10] The party at `help2` answers the call, so its UAS sends a `200 OK` response. The proxy forwards this to the originator. The proxy sends a CANCEL request to `help1` to stop the call attempt there. Then `help1` responds with `200 OK` (not ACKed because only the response to INVITE is ACKed). Finally, the originator ACKs the answer from `help2` and everything proceeds like a normal two-party call. If the originator were to hang up before anyone answered, with forked invitations still pending, his CANCEL would be sent to every UAS that hadn't yet responded.

9.6 SIP and H.323

Many people think of SIP as the IETF's answer to H.323. While there is overlap between the two, SIP was not developed as an alternative to H.323. But because SIP came out of the Internet culture and H.323 has its roots in the ITU, and because SIP and H.323 tackle similar problems, it's not surprising that SIP and H.323 have come to be seen as rivals.

When IP telephony was first being developed, H.323 quickly leaped to the forefront. Even though it had been designed for LANs, it was quickly adopted for IP telephony simply because it was more mature than the alternatives. Nevertheless, experience has shown that H.323 is far from an ideal solution. While H.323 is not likely to go away, it appears SIP will have a much more important role to play as a vehicle for building convergent network services on the Internet.

SIP is a simpler protocol than H.323.[11] The documents describing the base H.323 protocols take up more than 700 pages—and that does not

---

[10] According to RFC 2543, the proxy could forward all the `180 Ringing` responses if it so desired.

[11] Though it's arguable whether it could anymore be called a "simple" protocol, as new features continue to be added.

include ASN.1 and PER, both essential elements of H.323. The cor[e]
protocol document, by contrast, is about 150 pages. In part, th[is is]
because SIP tries to do much less than H.323. As stated previously, [H.323]
is a vertically integrated suite of protocols that addresses a broad r[ange]
of IP telephony issues, including such things as codecs, terminal reg[istra]tion, call control, address translation, admission control, and call a[utho]rization. In many cases, there is no clear separation of responsib[ility]
between these H.323 protocol elements. It's not uncommon for a se[ssion]
to require interactions among a number of them. SIP, on the other [hand,]
was designed to do nothing more than support session setup (a si[zable]
topic in itself) and relies on other, unspecified, protocols and ap[plica]tions to take care of everything else. SIP's modularity lets it work [well]
with H.323: A session originator can send a normal SIP invitatio[n,]
use SIP's user location features to find the called party and fin[d]
their capabilities, including H.323, and then use SIP's redirect facilit[y to]
take the call to a URL that refers to an H.323 endpoint and let H.3[23 set]
up the call.

SIP reuses existing Internet technology, for example, URLs, M[IME,]
and DNS. This makes SIP smaller. At the same time, SIP can
easily be integrated with existing Internet applications. Because [SIP's]
syntax is closely modeled on that of HTTP, knowledge of that [com]mon protocol provides a good foundation for learning SIP. W[hile]
H.323's designers intended to reuse existing protocols like Q9[31, in]
practice the version of Q931 in H.323 isn't nearly as much like the [orig]inal as planned. Thus, the hoped-for integration of H.323 and [ISDN]
never occurred.

Even if H.323 covered less territory, it would still be more compl[ex]
than SIP, because it's a binary protocol based on ASN.1 while SI[P is a]
text-based protocol. A more complete discussion of the differ[ences]
between these two styles of protocol design may be found in Se[ction]
2.4.2, but the highlights can be summarized as follows:

- Software that works with text-based protocols is generally less expensive to develop and easier to debug. Software that works wi[th] binary protocols can be considerably more expensive to develop (H.323 uses ASN.1 PER encoding, development tools for which ru[n] on the order of tens of thousands of dollars) and harder to debu[g] (the binary messages can't be easily read without being passed through debugging tools).

- While it's been claimed that binary protocols take up fewer byte[s] than text protocols, in practice this is often not so. For example,

H.323 messages using ASN.1 PER encoding can be longer than their SIP equivalents. Furthermore, space efficiency may not be an important criterion for protocols that exchange only a few intermittent messages, a good description of many call signaling protocols.

H.323's complexity can also be a problem when you consider devices with limited processing power and memory, and stringent power requirements, i.e., wireless phones, PDAs, and the like. In this sort of equipment, H.323's sheer computational needs make it a less than optimal solution.

Both SIP and H.323 can be extended to support applications not directly addressed in their original specifications. H.323 uses something it calls *nonStandardParams*, protocol elements that are spread throughout in specific locations. SIP's extension facility is more general and has fewer restrictions. (I cover SIP extensions at more length in the next chapter.)

Scalability wasn't an initial concern of H.323's designers because it was originally designed to be used on LANS, which are inherently of limited scale. SIP was designed for the Internet, so it had to keep scalability in mind from the beginning. This contrast is particularly noticeable with regard to servers. SIP proxy servers can be either stateless or stateful, with stateful servers at the edge, where traffic is lower, and stateless servers in the core, where traffic is high. H.323 gatekeepers, which carry out many of the same functions as SIP proxy servers, must be stateful, retaining call state for the entire duration of a call.

H.323 conference handling features are another potential bottleneck. In H.323, a central network element—the MCU—handles all aspects of even the smallest conferences. If the user providing the MCU function leaves a call, the entire conference ceases to exist. SIP distributes conference management among all participants, reducing the potential for bottlenecks and ensuring all users have the basic facilities needed to support a conference while it is in progress.

To summarize, H.323 is a first-generation IP telephony technology. It has served and will continue to serve as an important part of the IP telephony landscape. Nevertheless, its origins in the LAN environment and its initial focus on video conferencing, plus its complex design, led designers to shift their attention to more flexible and easier to implement protocols like SIP. While H.323 is adequate for phone-to-phone and PC-to-phone services, its future role is likely to be as a basic access technology for IP telephony.

9.7 Interworking SIP

For now, one should expect that H.323, MGCP, MEGACO, and SIP
have to coexist. Because SIP was designed to be a modular compone
a larger IP telephony solution, it is well suited to work with these
other protocols. Though there is some overlap of functionality bet
SIP and these protocols, getting SIP to play nicely with them is n
overwhelming challenge.

I touched briefly on the coexistence of SIP and H.323 in the pre
section. Turning to what I call, for lack of any other established
gateway layer protocols, MGCP and MEGACO by themselves aren't en
to build a complete IP telephony system. Both were designed to
the needs of a master (MGC) controlling devices attached to a slave
They still need something to handle MGC-to-MGC communicatio
is well suited to take on that role because, to SIP, an MGC looks jus
any other SIP user agent. It just happens to be a user agent with
connections.

An MG under the control of an MGC doesn't have to know t
call in which it's involved was established through a SIP convers
between one MGC and another MGC. It's hard to see how an MG
know SIP was involved, since it only speaks MGCP or MEGACO t
MGC. That leaves the MGC as the only network element that nee
speak both SIP and MGCP/MEGACO.

It's true that MGCP and MEGACO can be used to control
phone, and this puts them in competition with SIP to some d
However, MGCP and MEGACO assume the devices under their co
are equivalent to a dumb POTS phone with only a switch-hool
twelve buttons. MGCP and MEGACO don't let you do anyth
phone like that can't understand. That opens the door for SIP to
the picture, if your goal is to surpass the capabilities of POTS.[12]

Figure 9.10 shows SIP interworking with the PSTN through a ga
in support of a call from a SIP originator to a called party in the
Calls from the PSTN to a SIP-called party would be similar t
example. The SIP originator's user agent contacts the called
through a proxy that interacts with the gateway. The gateway co
SIP and SS7 messages. As usual, the call begins with the SIP origi
user agent sending an INVITE message and getting a 100 T
response back. The proxy uses a location service (not shown) to fi

---

[12]If it isn't, then why are you reading this?

Session Initiation Protocol (SIP)

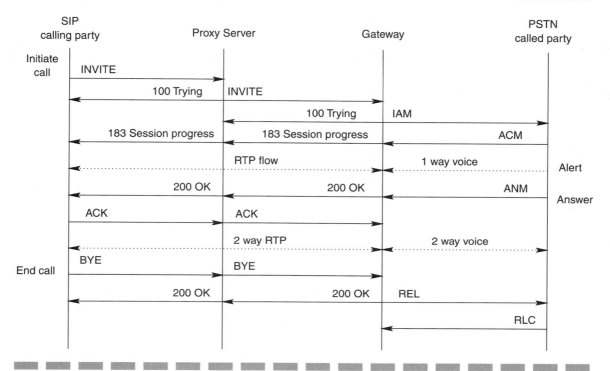

Figure 9.10
Call from SIP originator to PSTN.

appropriate gateway host and forwards the invitation to it. The gateway initiates SS7 call origination signaling in the PSTN with an IAM message.

Once the called party has been notified, that terminal rings and an ACM message flows back to the gateway. Rather than returning `180 Ringing` to the SIP originator, it sends `183 Session progress`, because the ringback comes directly from the PSTN, not from a called-party user agent. The PSTN has a path on which it plays ringback back through the gateway. The originator hears this on an RTP media between it and the gateway.

When the called party answers, the PSTN sends an ANM message to the gateway. The gateway turns this into a `200 OK` response, which goes to the proxy and from there to the originator. There is now a two-way voice path between the called party and the gateway, and a two-way RTP flow between the originator and the gateway. The gateway converts the PSTN-encoded voice into whatever format the originator needs.

Tearing down the call proceeds as one might expect. Assume the inator hangs up first, causing that user agent to send a SIP BYE me to the proxy. The proxy forwards this to the gateway, which resp with 200 OK, which the proxy sends back to the originator. Meanv the gateway sends an SS7 REL message toward the called party and dles the RLC message that's returned.

Even more ambitious uses of SIP to interwork with the PSTN been and are being standardized by the `pint` and `spirits` wo groups of the IETF. Those efforts, plus other applications of SIP, ar subject of the next chapter.

9.8 For More Information

At this writing, the best sources of information about SIP are o Henning Schulzrinne has an excellent Web site (`http://` `cs.columbia.edu/sip`) that collects in one place the best Web ences on SIP. The final authorities are the IETF RFCs for SIP [41 SDP [38]. Other valuable information may also be found at th working group's Web site at the IETF (`http://www.ietf.org`), i ing examples of SIP call flows for basic PSTN-like services and PST interworking [46], and Centrex-like [47] services.

CHAPTER 10

Going Further with SIP

Plant a kernel of wheat and you reap a pint; plant a pint and you reap a bushel.

—*Anthony Norvell*

10.1 Introduction

In the previous chapter, we looked at some basic SIP capabilities an[d ser]vices. In this chapter we'll look at some things you can do with SI[P that] go beyond the basics. SIP was designed as an extensible prot[ocol.] Many—but not all—of the services in this chapter use extensions t[o SIP.] Some of these have been standardized (for example, those introduce[d by] the PINT protocol). Others are, at this writing, being discussed wit[hin the] IETF (for example, the REFER method and the SPIRITS protocol[) and] have yet to be standardized. However, SIP has facilities that allo[w the] endpoints of a session to discover and negotiate features among t[hem]selves. Once features are negotiated, the endpoints can use any they [have] in common, whether standardized or not.

Nevertheless, even though the SIP protocol is standardized, and [even] though there are standardized means for extending it for findin[g and] negotiating those extensions among session participants, there's no [guar]antee those participants will interoperate successfully in provid[ing a] given service. This is because, while SIP provides a framework w[ithin] which to build services, the services remain largely unstandar[dized.] Because there is usually more than one way to implement a give[n ser]vice, there's lots of room for misunderstanding among participants. [This] goes also for many services covered in this chapter. In most cases, [what's] shown is "a" way to implement a service, not "the" way to do it.

This is not SIP's fault. It's occurred within traditional phone net[works] as well, as evidenced by the longstanding feature interaction pr[oblem] that causes one service to interfere with the operation of anoth[er. It's] important to keep in mind that SIP is not a panacea for the per[ceived] ills of the PSTN. While SIP makes it easier to experiment wit[h and] develop new services, something that is hard to do in the PSTN, d[esign]ing SIP services so they'll work across platforms, service provider[s, and] national borders remains a challenging problem.

10.2 Extending SIP

SIP can be extended by adding new request methods, new header[s, new] body types, or new parameters in existing headers. SIP does not re[quire] that every implementation support all or even any extensions. Ho[wever,] every implementation should support the ability of session partic[ipants] to discover and negotiate features among themselves.

10.2.1 Adding Request Methods

New methods can be added at any time without changing the core SIP protocol. If a user agent gets a request message with a method name it doesn't recognize, it just sends back a `405 Method not allowed` response with an `Allow` header line naming the methods it does recognize:

```
METHOD1 sip:jill@sipserv.com SIP/2.0
CSeq: 1 METHOD1
. . .
```

and:

```
SIP/2.0 405 Method not allowed
Allow: INVITE,ACK,OPTIONS,BYE,CANCEL,REGISTER,METHOD2
CSeq: 1 METHOD1
. . .
```

10.2.2 Adding Headers and Parameters

New headers and header parameters may be added to SIP at any time. Recipients ignore those they don't understand. Session participants don't negotiate headers by themselves. Instead they negotiate *features*, named services and service frameworks that include, in their definition, any new SIP headers and header parameters. Features are identified by their unique case-insensitive *option names*.

Standardized features may register their option names with IANA. Registered option names use the prefix `org.iana.sip`. The feature named `foo` would have the full option name `org.iana.sip.foo`, if registered with IANA. Unregistered option names use the reverse domain name of the organization responsible for that feature. If `foo` were developed at BarCorp and remained unregistered, its full option name would be `com.bar.foo`. BarCorp—and any other SIP feature developer—must ensure that all option names within its domain are unique. Features developed within the IETF and documented in an RFC use the option name prefix `org.ietf.rfc.N`, where N is the number of the RFC.[1]

---

[1] Nothing guarantees that the endpoints of a session will have the same understanding of what a feature provides and how it is to be used. The option naming mechanism merely provides a means by which endpoints can come to some sort of agreement. If they've all done their homework and have a common view of what's in a given feature, they should behave sensibly. Making sure they do that is the job of the designers and developers of SIP-enabled software.

Below is an invitation that uses feature negotiation. The Re[quire] header names the features the invitee must support for the sess[ion to] proceed. Here the session originator tells the invitee that it needs th[e fea]ture defined by BarCorp with the option name billing (or Bil[ling] or BiLlInG, or any other such permutation, since option names [don't] depend on case). The billing feature defines SIP extensions the [origi]nator uses, presumably the nonstandard Payment header and its a[ssoci]ated values, us_dollar and euro, that appear next in the invitati[on. If] the invitee understands the billing feature, it should be able to [inter]pret everything in this invitation:

```
INVITE sip:jill@sipserv.com SIP/2.0
Require: com.bar.billing
Payment: us_dollar, euro
CSeq: 1 INVITE
. . .
```

If the invitee understands the billing feature, it handles the i[nvita]tion as usual, perhaps sending back a response defined b[y the] com.bar.billing feature. On the other hand, if the invitee d[oesn't] support the com.bar.billing feature, it sends back a 420 [Bad] extension response with an Unsupported header that names th[e fea]ture(s) it doesn't recognize:

```
SIP/2.0 420 Bad extension
Unsupported: com.sipserv.billing
```

10.2.3 OPTIONS Requests

An OPTIONS request, which is one of SIP's set of six core method[s, can] be sent before setting up a session to find out what capabilities a[nother] SIP user supports. The Accept header in an OPTIONS message tel[ls the] recipient what MIME type it may use in the response body. If th[ere is] no Accept header, application/sdp is assumed:

```
OPTIONS sip:wizard@oz.com SIP/2.0
Via: ...
From: "Dorothy Gale" <sip:dorothy@kansas.com>
To: Humbug <sip:wizard@oz.com>
Call-ID: 123@kansas.com
Accept: application/sdp
CSeq: 1 OPTIONS
```

The response to an OPTIONS request includes methods recog[nized] by the recipient (Allow header), compression methods it ca[n

Going Further with SIP

(`Accept-Encoding`), ISO 639 language codes for human languages understood on its end (`Accept-Language`), features it knows about (`Supported`), and any or all media types it supports (SDP description). The q value in the `Accept-Language` header indicates relative preferences when more than one language is understood. Its default value is 1. Here, the user known as "Wizard" prefers English but will accept German and Danish as well (in that order). The SDP body shows that the Wizard can use G.721 and G.722 audio and H.261 and H.263 video:

```
SIP/2.0 200 OK
Via: ...
From: "Dorothy Gale" <sip:dorothy@kansas.com>
To: Humbug <sip:wizard@oz.com>
Call-ID: 123@kansas.com
Allow: INVITE,ACK,OPTIONS,BYE,CANCEL,REGISTER,METHODX
Accept-Encoding: compress, gzip
Accept-Language: en, de;q=0.8, da;q=07
Supported: com.sipserv.billing
CSeq: 1 OPTIONS
Content-Length: ...
Content-Type: application/sdp

v=0
m=audio 0 RTP/AVP 2 9
m=video RTP/AVP 31 34
```

10.3 PINT

10.3.1 History

PSTN/Internet Internetworking (PINT), now standardized in RFC 2848 [51], is a defined set of SIP extensions with which applications can send requests from the Internet to the PSTN, asking the PSTN to execute one of several services on their behalf. Work on PINT began towards the end of 1996, as several prototypes demonstrating hybrid PSTN/Internet services emerged from R&D organizations. Many of these used IN as a building block. Realizing that these prototypes formed a practical basis for building a standard that supported PSTN/Internet convergence, the `pint` working group was chartered by the IETF in mid-1997 to develop just such a standard.

The first task the group faced was to decide on the scope of their work. They soon decided to focus on means by which Internet applications might request and enrich services in telephone networks. While

they recognized that requests going the other way—from phon[e]
works to the Internet—were also important, they felt that tryi[ng to]
cover both would be overly ambitious and increase the risk of a[ccom]
plishing nothing by attempting too much at the outset. They fu[rther]
constrained their problem space by assuming that for the services [they]
considered, voice would be carried entirely within a phone net[work.]
Thus PINT was not to be a VoIP protocol. PINT would use the Int[ernet]
only for nonvoice interactions.

PINT became an IETF standards track protocol in June 2000, [with]
the publication of RFC 2848 [51]. With their work nearly finished, [they]
agreed the `pint` working group would dissolve once some final m[atters]
were taken care of, leaving the definition of other PINT-like service [and]
protocols to future designers.

10.3.2 The PINT Milestone Services

To further increase the focus of their work, the `pint` working g[roup]
decided to concentrate on a core set of services. As with their decisi[on to]
focus only on services in which the Internet sends requests to the [PSTN,]
they felt that if they tried to design a protocol that supported every [con]
ceivable such service, they might end up with nothing at all. By re[strict]
ing themselves to a small set of services, they hoped to set a sta[rting]
point on which to base future extensions.

The four so-called "PINT milestone services" they chose are as foll[ows:]

Click-to-dial-back A user makes a request through an IP h[ost to]
place a phone call in the PSTN. For this to work, the user must [have]
both voice (PSTN via phone) and data (IP network via PC or [other]
device) access. Oft-cited examples of this service include shoppin[g and]
customer service on the Web. Users could go to Web sites to make [pur]
chases or get support and click on a button to set up a p[hone call]
between themselves and customer representatives. The PSTN m[ight]
vary its treatment of such a call by, for example, selecting a call c[enter]
or call center agent to handle the call, depending on the time o[f day]
or day of week or other factors.

Click-to-fax A user requests that a fax be sent to a particular nu[mber.]
This might be useful when somebody has a fax machine bu[t no]
Internet access. The example often given for this service is a We[b site]
through which one can make reservations at hotels around the w[orld.]
Even today, it's more likely a hotel will have a fax machine tha[n]

Internet connection. The user could go to the reservation page, fill out a form, and click to have an image of the form faxed to the hotel.

Click-to-fax-back A user requests that a fax be returned from some location. This would be useful for getting a response back from the Internet-less hotel in the previous example.

Voice access to Web content This service makes designated information on the Web available as audible content that can be delivered to a phone rather than a data terminal, for example, giving Web access to blind persons.

10.3.3 The PINT Architecture and Protocol

The next task of the PINT working group was to establish a framework for the protocol they were to develop. The prototyping work that preceded PINT showed that HTTP-like protocols were a good technology for PINT services. Among candidates suggested for PINT, SIP stood out because it was modeled after HTTP, it occupied a similar problem space, and it could be extended in the ways PINT needed. Since it had a reasonable track record of its own already, SIP was chosen as the basis for PINT.

Figure 10.1 shows the PINT architecture. A *PINT client* (which also happens to be a SIP client) connects to a network of *PINT servers* (which also happen to be SIP servers) in the "PINT cloud." PINT servers relay PINT requests to *PINT gateways* (which act like a SIP UAS). PINT gateways sit at the edge of the PINT cloud and interact with *executive systems* at the edge of a phone network. Executive systems interact with other elements of the phone network to carry out PINT requests.

PINT doesn't care how a gateway interacts with an executive system or how an executive system interacts with a phone network. PINT's only concerns are how PINT clients and servers interact and how requests are routed to a PINT gateway. PINT doesn't say much about what happens on the PSTN side. In practice, the executive system would likely reside on an IN service node or SCP in the PSTN, or on a PBX or telephony server in a private phone network.

PINT client software may run on a user's PC or on a Web server with which that user is interacting. A PINT services starts when a PINT client sends a PINT invitation to a PINT server. The SDP describes the media session that is to result. In a normal SIP session, at least some of the session's media traffic would flow over the Internet or some other IP network. With PINT, media traffic—voice or fax—flows exclusively on a

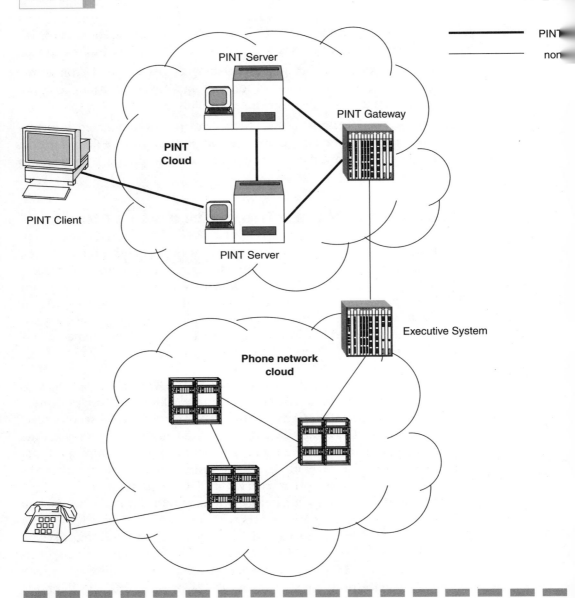

Figure 10.1
The PINT architecture.

Going Further with SIP

phone network. Some of PINT's extensions to SIP and SDP handle the ways it uses phone networks:

- PINT adds a new SDP network type, TN, for phone networks, and a new address type, RFC2543, for phone addresses. RFC2543 addresses are described in RFC 2806 [50].[2]
- PINT adds three new SDP protocol transport keywords—voice, fax, and pager—and their format types and attributes.
- PINT adds new SDP attributes for passing information to phone networks.
- PINT adds new SDP attributes that let it carry content as part of a multipart MIME payload.
- PINT messages can have multipart MIME payloads in their body. This lets them carry both SDP descriptions and content such as voice, paging, or fax data.

PINT designates a subset of the SIP URLs as *PINT URLs*. PINT doesn't mandate their use, but PINT implementations are encouraged to use them. The user part of a PINT URL identifies the requested PINT service:

- R2C for Request-to-call
- R2F for Request-to-fax
- R2FB for Request-to-fax-back
- R2HC for Request-to-hear-content

Anybody who uses the PINT extensions to support other PINT-like services must use names other than these in the user part of any PINT/SIP URLs they define. The host part of a PINT URL gives the domain name of the PINT service provider. A PINT URL may also use the optional tsp parameter to suggest a phone network for the service (this may be useful if a PINT gateway connects to more than one phone network). Here are some examples of PINT URLs:

```
sip:R2C@pintservices.com
sip:R2F@pintservices.com;tsp=fredstelco.com
sip:R2HC@pint.bigco.com;tsp=mainpbx.bigco.com
```

---

[2]They used to be described in RFC 2543, but then they were moved to RFC 2806, hence the seeming discrepancy in the name of this address type.

Just because a PINT server accepts a PINT invitation, there's no [guar]antee the request will succeed. Once it gets to a PINT gateway and [from] there to a phone network, there's nothing more a PINT server ca[n do.] There has to be some way for a PINT server to tell clients what's [hap]pened to requests they've made. Otherwise, the client software wo[n't be] able to give its user feedback about what might happen next. PINT [adds] three new SIP methods that clients and servers use to track pe[nding] requests:

SUBSCRIBE is sent when a PINT client wants to start monitoring t[he] status of a request.

UNSUBSCRIBE is sent when a PINT client no longer wants to moni[tor] the status of a request.

NOTIFY is sent to tell a subscribed PINT client about the status of [a] request.

Finally, PINT makes mandatory several elements for discovering [and] negotiating features that SIP leaves optional:

- PINT servers must be able to include `Warning` headers in respons[es] to tell clients about features that are not supported, such as text to speech or fax. PINT clients must recognize this header.

- PINT clients must use the `Require` header to tell PINT servers what PINT extensions to SIP they support. They should use the option name:

  `org.ietf.sip.subscribe`

  if they support SUBSCRIBE, UNSUBSCRIBE, and NOTIFY monit[or]ing and:

  `org.ietf.sdp.require`

  if they support the SDP `require` field.

10.3.4 PINT Applications

Just looking at the PINT extensions to SIP doesn't reveal much ab[out] how it might be used in practice. Some examples of PINT servi[ces,] based on those in RFC 2848 [51], should make it more clear how PI[NT] can be applied.

Going Further with SIP

CLICK-TO-CALL The first service is a click-to-call that requests a specific agent in a call center:

```
INVITE sip:R2C@pint.oz.com SIP/2.0
Via: ...
From: "Dorothy Gale" <sip:dorothy@kansas.com>
To: sip:tin.woodman@oz.com
Call-ID: 123@kansas.com
CSeq: 1 INVITE
Subject: Sale on emeralds
Content-type: application/sdp
Content-Length: ...

v=0
o=dorothy 2353687637 2353687637 IN IP4 tornado.kansas.com
s=R2C
i=Emerald sale
e=dorothy@kansas.com
c=TN RFC2543 +1-316-555-1234
t=current-time 0
m=audio 1 voice -
```

The PINT URL in the request line says this is a request for the click-to-call service (`R2C`) directed to the PINT server at `pint.oz.com`. The network type on the `c=` line is `TN` (rather than `IN`, as in the SIP examples) and the address type is `RFC2543` (rather than `IP4`). The `c=` line ends with phone number at which the requester can be reached during this session.

The media type `audio` on the `m=` line should be familiar, but the transport protocol `voice` is new to PINT. The other media types used in PINT are `text`, `image`, and `application`. The port number on the `m=` line is 1. Phones don't have ports, so you can put anything here. The usual value is 1. The format field is a dash (-). This means the requester will let the executive system decide how to encode the media stream. This is normal in PINT requests. The IP network is asking a phone network to do something on its behalf and has little or nothing to say about how the phone network should handle calls. So it just lets the phone network makes these sorts of decisions.

CLICK-TO-FAX-BACK The next example is a request to get a fax:

```
INVITE sip:R2FB@pint.oz.com SIP/2.0
Via: ...
From: "Dorothy Gale" <sip:dorothy@kansas.com>
To: sip:+9999-800-555-4321@oz.com
Call-ID: 456@kansas.com
CSeq: 2 INVITE
```

```
Content-type: application/sdp
Content-Length: ...

v=0
o=dorothy 2353687660 2353687660 IN IP4 tornado.kansas.com
s=R2FB
e=dorothy@kansas.com
t=current-time 0
m=application 1 fax URI
c=TN RFC2543 +1-316-555-1234
a=fmtp:URI uri:http://localstore/Products/Emeralds/9876.html
```

The PINT URL in the request line shows this is a request fo click-to-fax-back (`R2FB`) service on the same PINT server, `pint.oz` used in the click-to-call example. The network type is once agai rather than `IP`, and the address type is `RFC2543`. The request line with the phone number of the machine to which the fax sh be sent.

The `m=` line shows an `application` media type sent using the transport protocol. Data with the `application` media type is displ during the course of a session rather than being stored somewhere. makes sense, because this data will go directly to a fax machine. port number is `1`, because, like phones, fax machines don't have p The format field is `URI` this time, not a a dash as in the click-to example. When the format field is something other than a dash, the mat should be a valid MIME subtype of the media type.[3] With `application` media type, `URI` means that before it's faxed, the c should be processed by an application that knows how to turn a l into an image.

When the format field is something other than a dash, PINT : needs an `a=fmtp:` line to tell it where to find the data. The `a=fmt` line in this example shows the fax is stored on `localstore`, a locat within the PINT server's IP network (but not one that's necessar known outside that network).

SENDING CONTENT TO A PAGER The last example is a PII request to send a text message to a pager:

```
INVITE sip:R2F@pint.oz.com SIP/2.0
Via: ...
From: "Dorothy Gale" <sip:dorothy@kansas.com>
To: sip:R2F@pint.oz.com
Call-ID: 789@kansas.com
```

---

[3]See [34] for more on the various MIME subtypes and their uses.

```
CSeq: 3 INVITE
Content-Type: multi-part/related; boundary=--next
----next
Content-Type: application/sdp
Content-Length: ...

v=0
o=dorothy 2353687680 2353687680 IN IP4 tornado.kansas.com
s=R2F
e=dorothy@kansas.com
t=current-time 0
m=text 1 pager plain
c= TN RFC2543 +9999-800-555-9876
a=fmtp:plain spr:2@53655768
----next
Content-Type: text/plain
Content-ID: 2@53655768
Content-Length: ...
You big humbug! I said Kansas, not Arkansas!
----next--
```

The PINT URL in the request line shows a request for the click-to-fax (R2F) service on `pint.oz.com`.[4] The body is multipart MIME, as indicated by the header `Content-Type: multi-part/related`. Each part of the body that follows then has its own `Content-Type`.

The first part has `Content-Type: application/sdp`. The network type on the `c=` line of this part is `TN`, and the address type is `RFC2543`. That line ends with the phone number of the pager to which the message is to be sent. The `m=` line shows a `text` media type that will be sent using the `pager` protocol. A `text` media type with a `plain` format field means the data is just a stream of characters that can be sent "as is" without other formatting. Pagers don't have ports, so the port number is `1`. The format field isn't a dash, so there has to be an `a=fmtp:` line that says where to find the data. Here, `spr:` (stands for "sub-part reference") means the data is in another section of the multipart MIME body. After `spr:` is the content ID—`2@53655768`—of the part of the body with the data. The next MIME part has that content ID, so this is where you would find the message to be sent with type `text/plain`.

---

[4]Even though we've requested click-to-fax, we don't have to send data to a fax. R2F is just a convenient abbreviation for a PINT service that sends data to a display device. The means of transport is given in the SDP body. It could be `voice`, `fax`, or `pager` (though it's not clear what `voice` would mean for click-to-fax).

SUBSCRIBE, UNSUBSCRIBE, AND NOTIFY[5] If you pres[s a] click-to-call button on a Web page, you'd probably like to know i[f the] phone network can't place your call rather than just sit by the p[hone] waiting for something to happen. Many PINT services will have t[o give] users feedback like this about their requests. To handle that, PINT i[ntro]duces three new SIP methods: SUBSCRIBE, NOTIFY, and UN[SUB]SCRIBE.

PINT clients send SUBSCRIBE requests to a PINT server when [they] want to know the outcome of a PINT service invitation. They [send] UNSUBSCRIBE when they are no longer interested in hearing any [more.] As PINT servers get information about subscribed invitations, they [send] NOTIFY messages to their clients.

A PINT client will send a SUBSCRIBE message to the server that [orig]inally handled the invitation being subscribed or to a server, the a[ddress] of which appeared in the `Contact` header of a subsequent respon[se to] that invitation. SUBSCRIBE messages include the SDP of the PINT [invi]tation being subscribed.[6] If the subscriber is the same as the send[er of] the original invitation, the SUBSCRIBE message's call ID may b[e the] same as the call ID in that invitation. If anyone else sends a SUBSC[RIBE] for that invitation, they need to use a new call ID. Either way, the `o=` line in the SUBSCRIBE message is the same as in the original in[vita]tion. SUBSCRIBE messages may use a `Contact` header to tell the s[erver] where to send NOTIFY messages. SUBSCRIBE messages should ha[ve an] `Expires` header so the server will know when to stop monitorin[g if it] never gets an UNSUBSCRIBE.

PINT servers send a NOTIFY message whenever there's a chan[ge in] the state of a service that's being monitored. NOTIFY messages co[ntain] a (possibly) modified SDP description of the session in which the se[rvice] is running. This SDP may include more accurate information abou[t the] service, for example when execution began and ended. The `i=` field [may] describe the status of the service in the phone network (for exam[ple "3] of 5 pages sent" or "Ringing"). NOTIFY messages may also have a `W`[arn]`ing` header that describes any problems the PINT cloud had whe[n try]ing to invoke the service (for example, "No answer after 10 rings").

---

[5]This section describes SUBSCRIBE, UNSUBSCRIBE, and NOTIFY as they were d[efined] for PINT. Since then, proposals have been made to further extend SUBSCRIBE and [NOTI]FY as described in [56].

[6]If the original invitation had a multipart body, only the SDP part is sent in th[e SUB]SCRIBE request.

10.4 Music on Hold

Reference [57] shows how Centrex and PBX features can be built on top of SIP. This and the next example are based on scenarios in that document. Many of the services in Ref. [57] use the SIP extension method REFER. REFER tells its recipient to contact a third party using information provided by the method. REFER uses a new SIP header, Refer-To, the value of which is the URL of the third party to be contacted. Refer-To must appear in every REFER request. REFER requests must also use the new Referred-By header, which contains the URL of the request originator and other information (for example, a digital signature) about them. If the referral succeeds, the recipient sends back a 200 OK response. Music on Hold is an interesting application of REFER because it shows how to put a user on hold, how to refer them to another media source, and how to reestablish the held call. (See Fig. 10.2.)

The session starts as a normal SIP call, with User A calling User B and the two establishing an RTP voice path between themselves:

```
INVITE sip:UserB@bplace.com SIP/2.0
Via: ...
From: UserA <sip:UserA@aplace.com>
To: UserB <sip:UserB@bplace.com>
Contact: <sip:UserA@aplace.com>
Call-ID: 123@aplace.com
CSeq: 1 INVITE
Content-Type: application/sdp
Content-Length: ...

v=0
o=UserA 2890844526 2890844526 IN IP4 client.aplace.com
s=Session SDP
c=IN IP4 100.101.102.103
t=current-time 0
m=audio 49170 RTP/AVP 0
```

At some point, one of the parties—User B here—puts the other on hold with music. B sends a REFER request that tells the music server to contact User A at the URL in the Refer-To header lines. This request has a call ID that's different from the one for the original A to B call:

```
REFER sip:UserA@aplace.com SIP/2.0
Via: ...
From: UserB <sip:UserB@bplace.com>
To: "Music Server" <sip:music@server.com>
Call-ID: 456@bplace.com
CSeq: 1 REFER
Refer-To: <sip:UserA@here.com>
```

Figure 10.2
Message flows for Music on Hold feature.

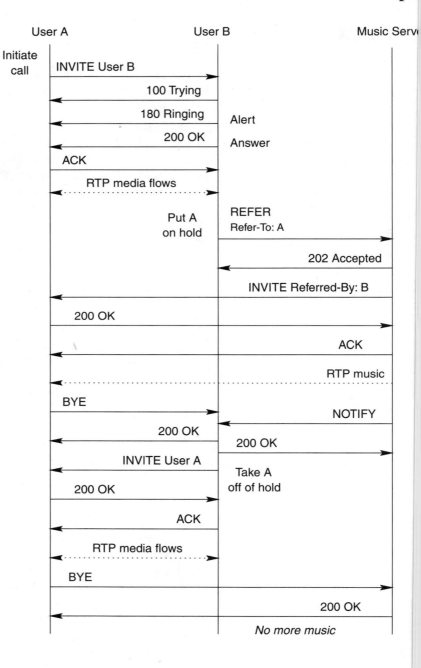

Going Further with SIP

```
Referred-By: <sip:UserB@there.com>
Content-Length: 0
```

The music server accepts the referral:

```
SIP/2.0 202 Accepted
Via: ...
From: UserB <sip:UserB@bplace.com>
To: "Music Server" <sip:music@server.com>
Call-ID: 456@bplace.com
```

The music server sends an invitation to User A to join a session at which music is being played. The music session replaces the one between A and B, so it gets a new call ID. The `Referred-By` header tells User A that User B initiated this invitation. The SDP tells User A how to connect to the music:

```
INVITE sip:UserA@aplace.com SIP/2.0
Via: ...
From: <sip:music@server.com>
To: User A <sip:UserA@aplace.com>
Call-ID: 789@server.com
CSeq: 1 INVITE
Referred-By: <sip:UserB@bplace.com>
Contact: <sip:music@server.com>
Content-Type: application/sdp
Content-Length: ...

v=0
o=MusicServer 2890844526 2890844526 IN IP4 music.server.com
s=Music Server SDP
c=IN IP4 50.60.70.80
t=current-time 0
m=audio 49170 RTP/AVP 0
```

User A responds with a `200 OK` message. The body of this message says that A has accepted the session as described by the music server and will take the music as a receive-only transmission:

```
SIP/2.0 200 OK
...
Call-ID: 789@server.com
CSeq: 1 INVITE
Content-Type: application/sdp
Content-Length: ...

v=0
o=UserA 2890844526 2890844526 IN IP4 client.aplace.com
s=Music Server SDP
c=IN IP4 100.101.102.103
t=current-time 0
m=audio 49170 RTP/AVP 0
a=recvonly
```

The music server ACKs this response and starts a one-way fl[ow of] music to A.

Next, User A sends a BYE request to User B. The `Replaces` h[eader] tells B that the music session is replacing the call formerly in pr[ogress] between A and B:

```
BYE sip:UserB@bplace.com SIP/2.0
Via: ...
...
Call-ID: 123@aplace.com
CSeq: 1 BYE
Content-Length: 0
```

User B also gets a NOTIFY message from the music server. This [mes]sage is not part of the core SIP protocol, but was introduced by [57] as an extension to SIP (see Sec. 10.3.5). As used here, it takes advanta[ge of] a further extension to NOTIFY, an `Event` header (described in [57]) from which B learns that A successfully transferred to the m[usic] on-hold server.[7] The message body of MIME type `application/sip` contains the `200 OK` response (without the SDP body) that A sent t[o the] music server's invitation. From this, B learns the call ID of the m[usic] on-hold session:

```
NOTIFY sip:UserB@bplace.com SIP/2.0
Via: ...
...
Call-ID: 456@bplace.com
CSeq: 1 NOTIFY
Event: refer
Content-Type: application/sip
Content-Length: ...
SIP/2.0 200 OK
Via: ...
From: <sip:music@server.com>
To: UserA <sip:UserA@aplace.com>
Call-ID: 789@server.com
CSeq: 1 INVITE
Contact: <sip:UserA@here.com>
Content-Type: application/sdp
Content-Length: ...
```

User B sends `200 OK` responses to the BYE and NOTIFY mes[sages,] and A remains on hold listening to cheesy music. If A is lucky, B s[oon]

---

[7]You may have noticed that B never sent a SUBSCRIBE request before getting this [NOTI]FY. The best I can tell, this is an error in [57]. But I could be wrong. I thought of a[dding] my own NOTIFY to clean things up, but decided to leave things as they are bec[ause it] highlights the need to standardize SIP service flows to ensure interoperability.

Going Further with SIP

or later sends a new INVITE request that takes A off hold. This invitation replaces A's session with the music server:

```
INVITE sip:UserA@there.com SIP/2.0
Via: ...
From: UserB <sip:UserB@bplace.com>
To: UserA <sip:UserA@aplace.com>
Call-ID: 321@bplace.com
CSeq: 1 INVITE
Contact: User B <sip:UserB@bplace.com>
Content-Type: application/sdp
Content-Length: ...

v=0
o=UserB 2890844527 2890844527 IN IP4 client.bplace.com s=Session SDP
c=IN IP4 110.111.112.113
t=current-time 0
m=audio 3456 RTP/AVP 0
```

A accepts the invitation, sends a BYE to the music server, and Users A and B are once again talking to each other:

```
BYE sip:music@server.com SIP/2.0
Via: ...
From: UserA <sip:UserA@aplace.com>
To: <sip:music@server.com>
Call-ID: 789@server.com
CSeq: 1 BYE
Content-Length: 0
```

10.5 Call Forward Busy Line (CFBL)

The *Call Forward Busy Line (CFBL)* service forwards calls to a designated number whenever the called party is busy. It's an essential building block of PSTN services, such as voicemail and various follow-me/find-me offerings. A SIP version of CFBL can be built with a stateful proxy server, as shown in Fig. 10.3.

As always, the call starts with an INVITE request:

```
INVITE sip:UserB@bplace.com SIP/2.0
Via: ...
From: UserA <sip:UserA@aplace.com>
To: UserB <sip:UserB@bplace
Contact: <sip:UserA@aplace.com>
Call-ID: 123@aplace.com
CSeq: 1 INVITE
Content-Type: application/sdp
Content-Length: ...

...
```

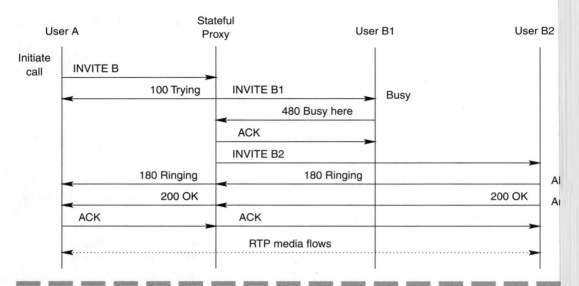

Figure 10.3
Call Forward Busy Line with SIP stateful proxy.

The proxy server looks up the SIP URL of the called party and it should try one location first and, if that's busy, try another. The proxy server forwards the invitation to the first location, inserting its address in the request line, and sends a `100 Trying` response to the calling party:

```
INVITE sip:UserB1@bplace.com SIP/2.0
Via: ...
From: UserA <sip:UserA@aplace.com>
To: UserB <sip:UserB@bplace
Contact: <sip:UserA@aplace.com>
Call-ID: 123@aplace.com
CSeq: 1 INVITE
Content-Type: application/sdp
Content-Length: ...

...
```

If the first location is busy, it returns a `480 Busy here`. The proxy server forwards the invitation to the second location, this time putting the second location's address in the request line:

```
INVITE sip:UserB2@bplace.com SIP/2.0
Via: ...
From: UserA <sip:UserA@aplace.com>
```

Going Further with SIP

```
To: UserB <sip:UserB@bplace
Contact: <sip:UserA@aplace.com>
Call-ID: 123@aplace.com
CSeq: 1 INVITE
Content-Type: application/sdp
Content-Length: ...
```

...

The rest of Fig. 10.3 shows the second location ringing and answering as usual. There's not much sophisticated SIP-craft here. What makes it work is service logic in the proxy server.

10.6 SPIRITS

With the PINT effort well underway, the *Services in the PSTN/IN Requesting Internet Service (SPIRITS)* effort was kicked off in late 1999 to develop protocols for services with messages that flow opposite to PINT's, i.e., those in which a phone network sends a request to an IP network or tells an IP network when an event has occurred. Among the circumstances that might cause the PSTN to ask the Internet to do something on its behalf are the following:

- A voicemail message has been left.
- An incoming call has arrived.
- A subscriber has subscribed or unsubscribed to a PSTN service.

As with PINT, the initial work in SPIRITS is being driven by emerging services. After reviewing several of these, the IETF `spirits` workgroup settled on the *Internet Call Waiting (ICW)* service as their starting point. Many variations on ICW are possible, but the core service works like this:

1. A user makes a dial-up connection to the Internet, thus tying up the regular phone.
2. Somebody calls that person. If the user didn't have ICW, the caller would get a busy signal or go to wherever the Call Forward Busy Line feature sent them (probably voicemail). With ICW, the user sees a pop-up window on the PC that says a new call has arrived. If calling party name and number are available, those may be displayed also.
3. From the PC, the user chooses how to handle the call. Some of the options include:

- End the Internet session and take the call on the phone line th
  had been in use.
- Take the call over the Internet using Voice over IP (VoIP).
- Reject the call, perhaps playing a message to the calling party.
- Play a prerecorded message to the calling party and disconnect
  the call.
- Forward the call to voicemail.
- Forward the call to another number.

More elaborate implementations of ICW handle incoming calls
matically, based on such things as calling number, time of day, an
of week.

After analyzing various ICW implementations, the `spirits` wo
group developed a SPIRITS architecture proposal [49] (Fig. 10.4)

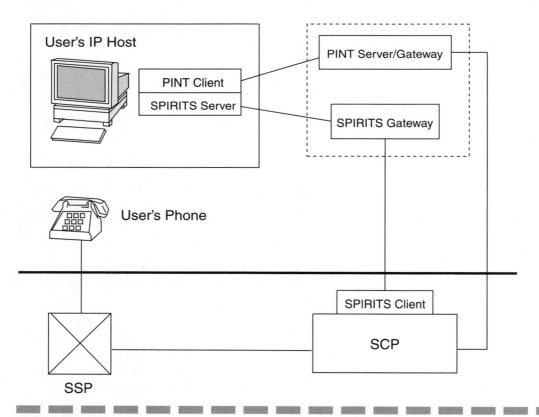

Figure 10.4
The proposed SPIRITS architecture.

coexisting SPIRITS and PINT elements. In this architecture, SPIRITS users have both phones and IP hosts (usually PCs). These IP hosts have both PINT client and SPIRITS server software. PINT clients initiate PINT requests, which go to PINT gateways (possibly) via PINT servers. From there, these requests go to a phone network to be processed as described in Section 10.3. SPIRITS servers receive and process PSTN-generated SPIRITS requests, which are generated by SPIRITS clients at the frontier of a phone network. A SPIRITS gateway sits between a SPIRITS client and a SPIRITS server, shuttling their messages back and forth between each other.

The ICW service has PINT and SPIRITS aspects. PINT clients initiate ICW sessions, presumably when a user first dials up to connect with the Internet. This makes sense (even though it's not one of the standardized PINT services), because requests to activate ICW flow from an IP network to a phone network. Information about calls to an ICW user appear as SPIRITS messages requesting instructions on what to do with them. These requests flow from an SSP to an SCP acting as a SPIRITS client, and from there to a SPIRITS server via a SPIRITS gateway. The SPIRITS client interprets responses from a SPIRITS server and supplies appropriate instructions to the SCP with which it is associated. That SCP directs the SSP to dispose of the call accordingly.

Not surprisingly, given the relationship between PINT and SPIRITS, SIP was selected as the basis for the SPIRITS protocol. Though that protocol has yet to be defined, it's likely to have the following characteristics:

- A typical message from a SPIRITS client will be of the form:

    event-notification parameter-list$_{DP}$

    where an *event-notification* corresponds to a DP in the IN BCSM (see Section 7.7.2) and the list of parameters varies with each DP. It's been proposed that these DPs include those in both wireline and wireless IN.

- There are two ways a phone network could be told it is to generate a SPIRITS message upon reaching a DP:
    1. The phone network's service management systems could set the behavior statically.
    2. The SIP SUBSCRIBE/NOTIFY mechanism could (with extensions yet to be defined) set the behavior dynamically. Just because an event hasn't been subscribed to doesn't mean

notifications for it won't be received (notification may have be
set statically within the phone network).[8]

- IN messages use ASN.1 binary encoding, while SIP is a text-based
  protocol. Thus, conversions of DP information to SPIRITS messag
  parameters will have to be defined carefully. The SPIRITS client
  will make the conversion between the two formats.

- Before a SPIRITS service can be invoked, the IP host acting as its
  SPIRITS server will have to register itself, and that information w
  have to be passed to the relevant SPIRITS client.

- Much of the work in defining the SPIRITS protocol will be
  figuring out how to represent the call instructions returned to a
  phone network from a SPIRITS server. For example, with ICW, th
  possible call dispositions that must be represented include:
    - —Accept the call.
    - —Reject the call for some supplied reason.
    - —Send the call to another location.

SPIRITS is in an early stage of its design. For that reason, any
seriously interested in this protocol should monitor the `spirits`
group's activities by visiting their IETF Web site and joining the rel
e-mail discussion groups.

10.7 For More Information

RFC 2543 [41] describes the SIP extension and feature negotiation m
nisms. Reference [45] provides guidelines for deciding whether an e
sion to SIP is the correct solution to a given problem. The PINT e
sions to SIP are described in RFC 2848 [51] along with many
service examples. PINT is not a complex protocol, but there are
subtleties I merely alluded to or ignored completely.

The `sip` working group's Web site at the IETF (`http://v
ietf.org`) has several documents describing proposed extensions t
and ways SIP could be used to provide services. References [46] and
document possible SIP call flows for some of these. These docun
are not definitive in their current form. They show how services
be implemented, but there's nothing (yet) to keep them from being

---

[8]Which could be a solution to the unsubscribed NOTIFY problem raised in the
on-hold example.

in other ways. Reference [57] describes the REFER method, Ref. [53] describes the INFO method, and Ref. [56] proposes further extensions to SUBSCRIBE and NOTIFY beyond what PINT supports.

The ongoing work of the `spirits` working group may be monitored at their IETF Web site. Documents there describe pre-SPIRITS implementations of the ICW service [54], the proposed SPIRITS architecture [49], and requirements for the SPIRITS protocol [48].

CHAPTER 11

Web-oriented Service Technology

I at least have so much to do in unravelling certain human lots, and seeing how they were woven and interwoven, that all the light I can command must be concentrated on this particular web, and not dispersed over that tempting range of relevancies called the universe.

—*George Eliot*

11.1 Services and Service Logic

So far I've concentrated on protocols that support convergent ne[twork] services. I haven't said much (if anything) about how messages of [these] protocols come to be created and exchanged in the first place. [That's] what service logic does.

Figure 11.1 shows a simple model of service logic and the prot[ocol] beneath it. Messages flow in and out of an engine that handles pr[otocol]

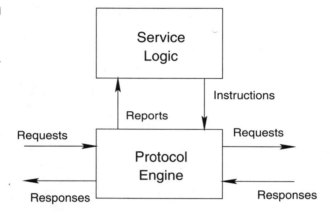

Figure 11.1
A model for service logic.

details. Some of these messages cause the protocol engine to [report] events to service logic. Service logic processes these events and [sends] instructions back, telling the protocol engine what messages it [should] send in response.

Before implementing this model, several questions must be ans[wered.]

- *Should service logic run on servers or on adjuncts to those servers?*
  Putting logic on an adjunct may improve the server's reliability, since a service crash would affect only the adjunct and not the [server.] entire server. Adjuncts can also improve performance by offloa[ding] the server. Using more than one adjunct may help even more. O[n] the other hand, it may be easier to build a system with logic tha[t] runs on the server. With logic running on adjuncts, there has to [be] some way for servers and adjuncts to converse. This could be a n[ew] protocol. Or it could be RPCs and distributed object technologi[es] such as CORBA and DCOM (see Chap. 5). Either way, it can be a [lot] of work to build.

- *How does the protocol engine know when to send a report to service logic and what to put in that report? And how much should one let service logic influence the behavior of the protocol engine?* It's probably not a good idea to have the protocol engine know too much about specific services, but it has to know something if it's going to interact with service logic. At the very least, it has to know when to send event reports to service logic.
- *How much access should service logic have to network resources?* The more things a service can touch, the more can go wrong. But the fewer things a service can touch, the fewer things it can do.

Different networks and services will answer these questions in different ways. The result is many different technologies for building service logic. Some of these give the service programmer much greater access to server and network resources than others.

The Internet has already confronted many of these issues. Ten years ago, developing an Internet service took skills possessed by very few. The World Wide Web made it easier to put things on the Internet, but you still had to learn HTML (at the very least). While HTML is a simple language, it was still a barrier for many people. But it was simple enough that graphically oriented editors were soon developed that allowed even novices to build sophisticated Web pages.

The voice service world has yet to catch up, but it's increasing the pace. This chapter covers several voice service development tools—CPL, SIP CGI, SIP servlets, and VoiceXML—that were strongly influenced by Web development. All bear strong similarities to software already used on the Web. Because they resemble software with which many programmers are already familiar, these tools could greatly expand the number of potential voice service developers.

11.2 Call Processing Language (CPL)

What It Is and What It's For

The *Call Processing Language (CPL)* [44] is an XML-based scripting language for building convergent network phone services. CPL programmers can write scripts on their own from scratch, or they can work with service creation tools that collect information and generate the required CPL. Once written, CPL scripts are loaded onto CPL serv[ers to] be executed.

Like AIN, CPL emphasizes ease of development and network re[liabil]ity at the expense of flexibility by making available only a limited [set of] functions. Because of these restrictions, a CPL server can make [some] checks on scripts submitted to it. That means it can accept scripts [from] users who might be otherwise untrusted, thus making it possib[le for] more people to create services than otherwise:

- A CPL server can verify that it is willing and able to perform all [of] the functions in a script before it executes that script. This is [especially] important because a failed script might cause a user no longer t[o] receive calls.

- A CPL server can verify that a script will finish its work in some finite time. CPL has two restrictions that support this:

  1. CPL scripts can use external resources for only a fixed time. T[his] keeps scripts from asking external resources over which they have no control to do something on their behalf and then waiting forever for that to happen.
  2. There's no way to write a program loop in CPL. That means there's no way to write a script that runs forever.[1] Scripts that never finish could monopolize a server while cutting off use[rs] from other activities.

- CPL's limited set of features prevent scripts from wandering at [will] through other data or resources on the server and doing damag[e to] the server or other users.

Besides making CPL safe to run on servers, CPL's designers war[ted to] make it as portable and easy to use as possible:

- Because CPL is based on XML, scripts can be moved without change from one machine to another and from one service creation tool to another. Service writers can focus on scripts, not the environment in which they run. And because XML is a text-based language, CPL scripts can be read without much trou[ble] by people and machines.

---

[1] The best example of an endless loop appears in a classic joke (at least it's a classi[c among] programmers): Why did the programmer get stuck in the shower? He followed t[he direc]tions on the shampoo: "Lather, rinse, repeat."

Web-oriented Service Technology

- CPL assumes its runtime environment has only a basic set of capabilities. That means a CPL script can be executed any place with those capabilities and doesn't depend on specific protocols and technologies. For example, a CPL server could be implemented to work with SIP, H.323, or Q931.
- Because XML is designed for transport across the Internet, moving CPL scripts onto servers is a simple matter. Scripts can even be embedded into Web pages.
- CPL is not a full programming language. That makes it easier for novice service authors to learn.

CPL models a service as a decision graph with *decision nodes* and *action nodes*, as shown in Fig. 11.2. One of those nodes is the *start node*. This is

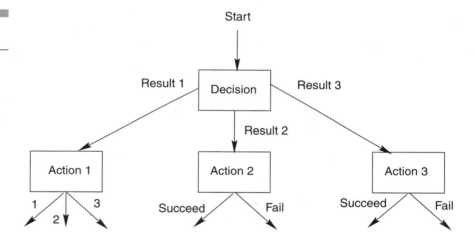

Figure 11.2
Decision graph.

the same model AIN uses. Executing a node produces results. These results determine which of the arrows leading from a node is taken to the next node in the graph. At the bottom of the graph are nodes with no result arrows. A script ends once one of these is reached.

While it's conceivable the arrows in such a decision graph could go both up and down, CPL allows only down arrows. Once a script passes through a decision or action node, it can't return. Thus, you're guaranteed to reach the end of the graph sooner or later. That's why CPL scripts can't run forever. And because the longest path through the graph can't be infinitely long (unless you write an infinitely long CPL script), the server can calculate how long a service might be expected to

run in the worst case. Servers can reject scripts they think might
too long.

11.2.2 A Simple Script: Call Redirection

The CPL script in Fig. 11.3 sends all incoming calls to the phone nu
it serves to another number. The first two lines are the XML pre

Figure 11.3
Unconditional call redirection script.

```
<?xml version="1.0" ?>
<!DOCTYPE cpl PUBLIC"-//IETF//DTD RFCxxxx CPL 1.0//EN" "cpl.dtd
<cpl>
   <incoming>
      <location url="sip:smu@sip-phone.cplservice.org">
         <redirect />
      </location>
   </incoming>
</cpl>
```

that says this is an XML document and its DTD may be found
`DOCTYPE` line. A `<cpl>` tag surrounds the CPL script.

The first tag in the script, an `<incoming>` node, is a *top-level*
There are two kinds of top-level action in CPE—incoming call an
going call:

- Incoming calls are handled by actions inside a script's
  `<incoming>` node.

- Outgoing calls are handled by actions inside a script's
  `<outgoing>` node.

A script may handle incoming, outgoing, or both incoming an
going calls.

Every script has a list of locations (usually URLs) associated
The actions of the script are directed at those locations. Outgoing
start with a list of locations to which they will send calls. Inc
nodes have an empty location list. *Location nodes*, like `<location`
and remove locations from this list. The `<location>` node in F
adds `sip:smu@sip-phone.cplservice.org` to the script's lo
list. Location nodes don't have results. CPL assumes they always s
Inside a location node are other nodes, the actions of which are a
to the current location list. In Fig. 11.3, there's only one node to
the location node applies: a redirection (`<redirect>`) node. The r

tion node tells the CPL server to send an incoming call to the current list of locations. This was just set to `sip:smu@sip-phone.cplservice.org`, so all incoming calls intercepted by this script will go to that location.

11.2.3 A More Complicated Script: Call Screening

The script in Figure 11.4 shows a call screening service that checks calling party identity and time. The `<subaction>` node at the beginning contains a bit of CPL code that's used throughout the script. Rather than repeat that code, it goes here with the name `voicemail`. There'll be more to say about it later.

As in the previous example, this script handles only incoming calls. The first node in the top-level `<incoming>` action, `<address-switch>`, is a CPL *switch node*. CPL uses switch nodes to make decisions. A switch node contains a list of conditions to be checked and actions to take if one of them matches. The server checks these conditions in the order in which they appear in the switch node. The first one matched has its action taken. Switch nodes may have an optional `<otherwise>` condition, the action of which is taken if no other condition matches. If no condition matches and there is no `<otherwise>` tag, the CPL server takes some default action (see Ref. [44] for the default actions to be taken in various situations).

In Figure 11.4, the `<address-switch>` looks at the user part (`subfield="user"`) of the call's originating address (`field="origin"`). If it's an exact match with `anonymous`, the script rejects the call by executing a `<reject>` node. The `status` attribute on this node gives the rejection type: `busy`, `notfound`, `reject`, or `error`. Protocol-specific status values may also be used for this attribute. The `<reject>` node terminates a CPL script, so it has no results. Its `reason` attribute gives an explanation for the rejection.

If the calling party isn't `anonymous`, the script goes into the `<otherwise>` branch, where it makes another check on the user part of the call's originating address. If it's an exact match with `mom`, the script proceeds into a `<time-switch>`. If the current time is between 11 a.m. and 1 p.m. on any day from Monday through Friday, the script sends Mom to the called party's voicemail (perhaps he's at lunch?). Otherwise, the script sends Mom directly to the called party's SIP phone by adding the URL for that phone—`sip:smu@sip-phone.cplservice.org`—to

Figure 11.4
Call screening script.

```xml
<?xml version="1.0" ?>
<!DOCTYPE cpl PUBLIC"-//IETF//DTD RFCxxxx CPL 1.0//EN" "cpl.dtd"
<cpl>
  <subaction id="voicemail">
    <location url="sip:smu@voicemail.cplservice.org">
      <proxy />
    </location>
  </subaction>
  <incoming>
    <address-switch field="origin" subfield="user">
      <address is="anonymous">
        <reject status="reject"
          reason="I don't accept anonymous calls"/>
      </address>
      <otherwise>
        <address-switch field="origin" subfield="user">
          <address is="mom">
            <time-switch>
              <time timeofday="1130-1300" day="1-5">
                <sub ref="voicemail"/>
              </time>
              <otherwise>
                  <location url="sip:smu@sip-phone.cplservice.or
                    <proxy timeout="30">
                      <busy>
                        <sub ref="voicemail"/>
                      </busy>
                      <noanswer>
                        <sub ref="voicemail"/>
                      </noanswer>
                    </proxy>
              </otherwise>
            </time-switch>
          </address>
          <otherwise>
            <sub ref="voicemail"/>
          </otherwise>
        </address-switch>
      </otherwise>
    </address-switch>
  </incoming>
</cpl>
```

the script's (until now) empty location list. The `<proxy>` node im
ately after the `<location>` node sends the call that triggered this
to the address(es) in the location list. The `timeout` attribute say
many seconds the script should wait after placing the call before
ing it won't be answered. If omitted, the script uses a default time
20 seconds. Once the `<proxy>` node has been executed, the scrip
for its result (or timeout). If the called party is busy, the script e
the nodes inside the `<busy>` result tag. If there's no answer bef

timeout, the script executes the nodes inside the <noanswer> tag.[2] If the call goes through, neither result is triggered and the script uses a default behavior, which in this case is to terminate.

Returning to the <proxy> results, both <busy> and <noanswer> send the call to voicemail by invoking the voicemail subaction defined at the beginning of the script. This subaction sets a location for voicemail—sip:smu@voicemail.cplservice.org—and forwards the call to it with a <proxy> node. The scope of a <location> node extends only as far as the nodes it surrounds, so the location node the script executed before has no effect on the <proxy> for voicemail. The voicemail <proxy> has no result tags, so the script assumes a call to voicemail will succeed.

That takes care of anonymous calls and calls from Mom. The last <otherwise> node in Fig. 11.4 sends all other callers to voicemail, again using the voicemail subaction.

11.3 SIP CGI and SIP Servlets

SIP CGI and SIP servlets are another way to create convergent network services using Web tools. Like CPL scripts, they run on a server. Unlike CPL, which can be used with any signaling protocol, SIP CGI and SIP servlets were both designed to be used with SIP.

SIP CGI is modeled on regular HTTP CGI. SIP CGI programs can be written in any language, just like regular CGI programs. SIP servlets are written in Java, just like other servlets. The only difference between SIP servlets and regular servlets is that SIP servlets invoke telephony operations with the SIP servlet API.

Because SIP CGI programs are so flexible, they can't be checked as CPL scripts can. SIP servlets are somewhere between SIP CGI and CPL in safety. They have the flexibility of a general-purpose programming language, introducing the possibility of problems, such as infinite loops. On the other hand, they run in a Java Virtual Machine and so gain all the protections that provides.

SIP CGI works just like regular CGI in that every request to execute a SIP CGI program spawns a new process. Like regular servlets, SIP servlets don't have this performance liability because they all run in a single

---

[2]Other <proxy> results are possible, but these are the most common.

process. CPL is also more efficient than SIP CGI from this pr[ocess] spawning perspective, since it usually runs in the same process [as the] server.

SIP CGI and SIP servlets are meant for different purposes than [CPL.] CPL has limited capabilities and was designed for end users who w[ant to] create their own (simple) services. SIP CGI and SIP servlets supp[ort all] the features of a general-purpose programming language and ca[n pro]vide access to many communications capabilities. They can do [more] than CPL, but they can also go wrong in more ways than CPL.

11.3.1 SIP CGI

SIP CGI is about 90% the same as regular HTTP CGI. Thus, many [of the] skills, tools, and components for regular CGI development can be [used] with SIP CGI also. There are, however, some important differ[ences] between regular CGI and SIP CGI:

- HTTP CGI programs generate only responses—typically Web pages—to requests. Because SIP CGI programs execute on SIP servers and some services may send further SIP requests from th[e] server, SIP CGI programs generate both requests and responses.

- An HTTP CGI program is usually named by the URL in the request. SIP CGI may use other elements of a request to identify the program to be invoked, for example, the To: and From: field[s] or the method name in the SIP URL.

- Some SIP services use timers (for example, "call forward no answ[er").] This is something SIP CGI provides that's not part of HTTP CG[I.]

- Unlike Web servers, SIP servers can fork requests to multiple recipients.

- A regular CGI program gets a request, responds to it, and goes a[way.] A SIP server can generate many related requests, each with its o[wn] response. Thus, a SIP CGI program may need to persist over a se[ries] of messages to handle responses to requests it's generated. A response may generate other requests with their own responses, and so on. To stay as consistent with regular CGI as possible, SI[P] CGI programs get a request, do their work (generate requests and/or responses), and terminate, just like a normal CGI progra[m.] However, SIP CGI programs can generate a unique token to kee[p] track of their state. When a later message arrives for that service[, the] same program is invoked. But this time the token generated ear[lier] is passed to it. The program uses that token to figure out where it should resume executing. When the program reaches a point where it knows there will be no more messages, it terminates without generating a token.

When a SIP message arrives at a SIP CGI server, the server decides if it should be handled by a SIP CGI program. If not, the message is handled like any other SIP message. Otherwise, the message goes to the appropriate SIP CGI program. That program executes and returns one or more messages that tell the server how to proceed.

METAVARIABLES The server passes information to the program in *metavariables*. On many systems, these are implemented as environment variables, but they don't have to be. SIP CGI specifies about 20 mandatory metavariables that every implementation must support. Among these are the following:

CONTENT_LENGTH The number of bytes in the body of the SIP message that triggered execution of a SIP CGI program.

REMOTE_ADDR The IP address of the client that sent the message. This may not be the same as the originating user agent client or user agent server.

REQUEST_METHOD If the message that triggers a SIP CGI program is a SIP request, this holds the name of the request method.

REQUEST_URI If the message that triggers a SIP CGI program is a SIP request, this contains the SIP URL in the request line.

REQUEST_TOKEN When a SIP CGI program sends a proxied request, it can provide a unique token that identifies responses to that request in later invocations of the program. When those responses arrive, the server sets this metavariable to equal the token of the request with which the responses are associated.

RESPONSE_STATUS If the message that triggers a SIP CGI program is a SIP response, this contains the numeric response code.

RESPONSE_REASON If the message that triggers a SIP CGI program is a SIP response, this contains the reason phrase that follows the numeric response code.

RESPONSE_TOKEN If the message that triggers a SIP CGI program is a SIP response, the server supplies a unique token that associates that response with later messages pertaining to it. This token is passed to the SIP CGI program when it is called again to handle those messages.

MESSAGES SIP CGI programs read the body of the reque[st]
response message that triggered them on the standard input strea[m of]
the system on which they run (for example, `stdin` for UNIX sys[tems].)
They write their results on the standard output (`stdout` for UNIX[...) The]
output of a SIP CGI program is one or more *messages*. A messag[e has]
exactly one *action line* that tells the server what to do, plus zero or [more]
CGI headers, zero or more SIP headers, and an optional SIP [mes-]
sage body.

Action lines SIP CGI defines five action lines:

Status Tells the server to send a SIP response message toward t[he]
user agent client. The format of this action line is the same as [that]
of the first line of the response to be sent:

```
SIP/2.0³ status-code reason-phrase
```

`CGI-PROXY-REQUEST` Tells the server to forward a request to [the]
SIP-URL. If the SIP CGI program was triggered by a request [mes-]
sage, that request is forwarded. If the program was triggered b[y a]
response to an earlier request, that request is forwarded. The f[ormat]
of this action line is as follows:

```
CGI-PROXY-REQUEST SIP-URL SIP/2.0
```

`CGI-FORWARD-RESPONSE` Tells the server to forward a respons[e to]
its final destination. The *response-name* is either a response [name]
token provided by the server in the `RESPONSE_TOKEN` metava[riable]
or the string "`this`." The latter may be used only if the progr[am]
was triggered by a response message. If that's so, the triggerin[g]
response message is forwarded. The format of this action line [is as]
follows:

```
CGI-FORWARD-RESPONSE response-name SIP/2.0
```

`CGI-SET-COOKIE` If a SIP CGI program runs again to handle [mes-]
sages related to earlier messages it processed or generated, it m[ay]
need to know something about what was going on during its [previ-]
ous execution. SIP CGI programs can generate a string token, [called]
a *cookie*, anytime they handle a new request or generate a pro[xied]
request. The `CGI-SET-COOKIE` action passes the cookie's valu[e]

---

[3]If another version of SIP is being used, this would change. The same comment a[pplies to]
other instances of this value in action lines.

the server. The server saves this value without examining or otherwise processing it. When the SIP CGI program runs again, the server makes the cookie available to it. The format of this action line is as follows:

```
CGI-SET-COOKIE cookie-token SIP/2.0
```

CGI-AGAIN If a program should be executed again to process requests and responses that are part of a single transaction, this action line specifies "yes." If there will be no further need to execute the program for later requests and responses, it specifies "no." The format of this action line is as follows:

```
CGI-AGAIN yes-or-no SIP/2.0
```

CGI headers CGI headers all start with the string "CGI-." They provide information a SIP CGI program wants to export to the server. SIP CGI defines two CGI headers:

CGI-Request-Token As described for the REQUEST_TOKEN metavariable, a SIP CGI program may specify a unique token that identifies later responses to a proxied request. The CGI-Request-Token header gives the value of this token:

```
CGI-Request-Token: token
```

CGI-Remove A program can remove headers from a SIP message by listing them here. The *SIP-header-list* is a comma-separated list of SIP header names the server should remove before sending the message:

```
CGI-Remove: SIP-header-list
```

PROCESSING SIP CGI PROGRAM OUTPUT If the output of a SIP CGI program does not include action lines for status, CGI-PROXY-REQUEST, or CGI-FORWARD-RESPONSE, the server takes its default action. If the program was triggered by a SIP request, the default action is for the server to see if its own domain is the same as the URI in the request line. If it isn't, the server proxies the request to the requested URI. Otherwise, it checks its registration database to see how requests from the originating user agent should be handled (proxy or redirect). If the SIP CGI program was triggered by a SIP response, the server handles it like any other SIP response.

If the output of a SIP CGI program includes a status, CGI-PR[OXY-]
REQUEST, or CGI-FORWARD-RESPONSE action line, the server pr[oceeds]
as described earlier for those action lines. In so doing, it mak[es the]
changes required by SIP in any messages it sends out (for exa[mple,]
adding a Via: header) and merges any SIP headers in the SI[P CGI]
program output into those messages. For example, the followi[ng SIP]
CGI program output causes a request to be proxied to dick[ens@]
author.org with a contact of editor@publisher.com. It [sets a]
request token and tells the server to run the program again to [handle]
responses:

```
CGI-PROXY-REQUEST sip:dickens@author.org SIP/2.0
CGI-Request-Token: ahdydsfymndfjkh87hqhhas
CGI-AGAIN: yes
Contact: sip:editor@publisher.com
```

11.3.2 SIP Servlets

SIP CGI programs can generate all the parts of a response message[, head-]
ers, status codes, and reason phrases. Servlets usually generate on[ly the]
body of an HTTP message. The SIP servlet API lets servlets wor[k with]
all parts of a SIP message. Like SIP CGI programs, a single SIP serv[let can]
handle a series of messages that together form a transaction. In[forma-]
tion is passed from one execution of a servlet to another by stori[ng it in]
a database or by storing it as Java static data in the servlet.

The SIP servlet API has about fifteen classes and interfaces. Th[e next]
few examples show how a few of these are used. The SIP servlet p[rogram]
in Fig. 11.5 rejects all calls (INVITE requests). Its first line import[s decla-]
rations of classes and interfaces defined by the SIP servlet API:

```
import org.ietf.sip.*;
```

The next line declares a new class, RejectServlet, derived fr[om the]
SipServletAdapter class in the SIP servlet API. The server cre[ates an]
instance of this class to run the servlet. SipServletAdapter p[rovides]
default implementations of all the methods a SIP servlet mus[t have.]
Most of these default implementations return values that tell the [server]
to do what it normally would do if a SIP servlet hadn't gotten in[volved.]
Most of the work of writing a SIP servlet is deciding which o[f the]
default implementations to redefine and then writing code for th[em.]

```
public class RejectServlet extends SipServletAdapter
```

Web-oriented Service Technology

Figure 11.5
A SIP servlet that rejects all calls.

```
import org.ietf.sip.*;

public class RejectServlet extends SipServletAdapter {

  protected int statusCode;
  protected String reasonPhrase;

  public void init(ServletConfig config) {
    super.init(config);
    try {
      statusCode = Integer.parseInt
                   (config.getInitParameter("status-code"));
      reasonPhrase = config.getInitParameter("reason-phrase");
    }
    catch (Exception e) {
      statusCode = SC_INTERNAL_SERVER_ERROR;
    }
  } // END init

  public boolean doInvite(SipRequest req) {
    SipResponse res = req.createResponse();
    res.setStatus(statusCode, reasonPhrase);
    res.send();
    return true;
  } // END doInvite
} // END class RejectServlet
```

The next two lines define some variables for SIP response-message status codes and reason phrases:

```
protected int statusCode;
protected String reasonPhrase;
```

The rest of the `RejectServlet` class redefines the `SipServletAdapter` methods `init` and `doInvite`. The first of these, `init`, performs any setup work when the servlet starts running. It's called automatically every time the server creates an instance of the servlet. The server passes `init` an instance of `ServletConfig` in the `config` argument. `ServletConfig` is part of the regular Java servlet API. It provides servlets with information about the environment within which they are running:

```
public void init(ServletConfig config)
```

The first thing this method does is invoke the `init` method of the class from which it is derived (`SipServletAdapter`) to take care of any high-level initializations:

```
super.init(config);
```

init then sets `statusCode` and `reasonPhrase` by calling getI
Parameter to get the values of servlet initialization parameters
the `config` object. Initialization parameters may be set in several
On many servers it's done by editing a configuration file. Plausible
tusCode and reasonPhrase values for this service would be 480
porarily Unavailable or 486 Busy Here. The try and catch
handles runtime errors that might occur while setting statusCod
reasonPhrase:

```
try {
  statusCode = Integer.parseInt
                  (config.getInitParameter("status-code"));
  reasonPhrase = config.getInitParameter("reason-phrase");
}
catch (Exception e) {
  statusCode = SC_INTERNAL_SERVER_ERROR;
}
```

The `doInvite` method handles SIP INVITE requests. It's pass
instance of `SipRequest` that represents the invitation:

```
public boolean doInvite(SipRequest req)
```

The servlet creates a response by invoking the request object's
ateResponse method. This method set header values in the
response object by copying their corresponding values from the r
object:

```
SipResponse res = req.createResponse();
```

The servlet sets the response object's status code and reason
and sends it on:

```
res.setStatus(statusCode, reasonPhrase);
res.send();
```

The final line of `doInvite` tells the server that the servl
changed the normal handling of an INVITE request. Thus, the
should not do what it normally would with the invitation (pr
redirect it):

```
return true;
```

Figure 11.6 shows a more complex service that logs information
a call's duration by processing an invitation and subsequent messa
the call it initiated. Once again, the class that implements this
LogCallDuration, extends the `SipServletAdapter` class:

```
public class LogCallDuration extends SipServletAdapter
```

Figure 11.6
A SIP servlet that logs call durations.

```java
import java.util.Hashtable;
import org.ietf.sip.*;

public class LogCallDuration extends SipServletAdapter {

  static Hashtable startTimes = new Hashtable();

  public void init(ServletConfig config) {
    super.init(config);
    // open database connection...
  } // END init

  public boolean doInvite(SipRequest req) {
    req.recordRoute(true);
    return false;
  } // END doInvite

  public boolean gotResponse(SipResponse res) {
    String callID = res.getCallID();
    String method = res.getMethod();
    if (res.getStatus()/100 == 2) {
      if (method.equals(INVITE)) {
        Long start = new Long(System.currentTimeMillis());
        startTimes.put(callID, start);
      }
      else if (method.equals(BYE)) {
        Long start = (Long) startTimes.remove(callID);
        if (start != null) {
          logCall(callID, res.getFrom(), res.getTo(),
                  start.longValue(), System.currentTimeMillis());
        }
      }
    }
    return false;
  } // END gotResponse

  void logCall(String callID, SipAddress from,
               SipAddress to, long start, long end) {
    // save call details to database...
  } // END logCall

  public void destroy() {
    // close database connection...
  } // END destroy

} // END class LogCallDuration
```

To keep track of call durations, the servlet needs to know when a call started. It stores those in `startTimes`, a table that matches durations with call IDs. This variable is static because instances of `LogCallDuration` may come and go, but the data must persist:

```
static Hashtable startTimes = new Hashtable();
```

The `init` method calls the parent class's `init` method, as us[ual,] then opens connections for writing log records to a database.[4] [Once] that's done, the servlet is ready to handle messages. The servlet do[es that] in its `doInvite` method:

```
public boolean doInvite(SipRequest req)
```

The server running this servlet needs to be on the path of any [subse-]quent messages relating to this invitation. A `Record-Route:` h[eader] added to the request ensures that:

```
req.recordRoute(true);
```

Finally, `doInvite` returns `false` to let the server know it s[hould] continue processing the message—now with an added `Record-R[oute]` header—as it normally would:

```
return false;
```

The `gotResponse` method is new. The server calls this method [when] it gets a response to an earlier request it sent. The first thin[g the] method does is extract the call ID and method name from the res[ponse:]

```
String callID = res.getCallID();
String method = res.getMethod();
```

It then gets the status code and checks if it's a `200 OK` response:

```
if (res.getStatus()/100 == 2)
```

If it is, the method checks if it's a response to an invitation an[d, if so,] records the current time as the start time for that call:

```
if (method.equals(INVITE)) {
  Long start = new Long(System.currentTimeMillis());
  startTimes.put(callID, start);
}
```

If the response is not a response to an invitation, it migh[t be a] response to a BYE request. If it is, a call has ended. The servlet ca[lculates] the duration and logs that in the database:

---

[4]We skip the details of how that's done. If you're interested, see any reference on [Java] Database Connectivity (JDBC) API. It is worth pointing out that, if this servl[et isn't] created anew for each message, the database connections remain open as long as it [is in exis-]tence. That may considerably reduce overhead, since opening database connection[s is] lots of work.

```
    else if (method.equals(BYE)) {
      Long start = (Long) startTimes.remove(callID);
      if (start != null) {
        logCall(callID, res.getFrom(), res.getTo(),
                start.longValue(), System.currentTimeMillis());
      }
    }
}
```

The servlet ignores any other responses. It finishes by telling the server to process the response as it normally would:

```
return false;
```

Everything else in this servlet is code (not shown) to write logs to the database and close database connections when the servlet is destroyed.

11.4 VoiceXML

VoiceXML is an XML-based scripting language for building *Automated Speech Recognition (ASR)* and *Interactive Voice Response (IVR)* applications. VoiceXML is being defined by the VoiceXML Forum (`http://www.voicexml.org`), an industry consortium of over 300 companies. VoiceXML grew out of several earlier efforts that were developing ways to provide voice access to the Internet. Among these were the *Phone Markup Language (PML)* developed by both AT&T and Lucent (though in different dialects), Motorola's *VoxML,* and IBM's *SpeechML.* The VoiceXML Forum was established in 1999 to blend these into a single standard. The result of that work was the VoiceXML 1.0 standard, released in March 2000. The W3C accepted VoiceXML 1.0 in May 2000 and has continued work on VoiceXML 2.0 in its own Voice Browser working group.

A VoiceXML script takes audio input (voice or phone keypad tones) and generates audio output (recorded or generated on the fly using *Text To Speech,* or *TTS,* technology). While VoiceXML was developed with regular phones in mind, it can also be used with VoIP. One of the main applications of VoiceXML is making Web content available through a voice interface, as shown in Fig. 11.7. Connecting the two is a natural development for both. It makes the phone more useful and the Internet more accessible. And, believe it or not, phones have some advantages over PCs:

- Phones are cheap, much cheaper than even the least expensive PC, especially when you compare mobile phones with mobile PCs.
- Phones are much smaller and lighter than even a laptop PC.

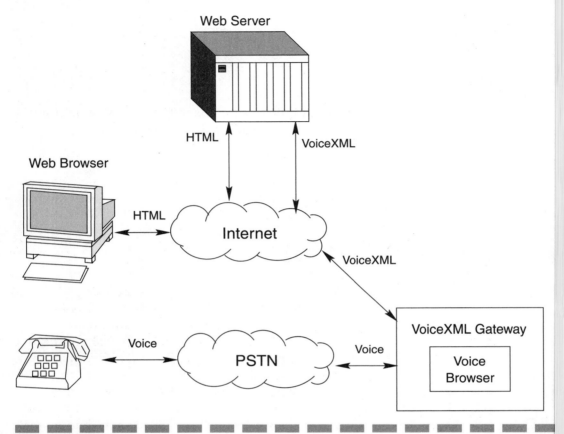

Figure 11.7
A VoiceXML gateway bridges the Internet and the PSTN for voice access to Web content.

- Batteries last much longer in a phone than in a laptop PC.
- Phones don't have to be booted up and they don't crash (so far).

WAP, the *Wireless Application Protocol,* is another way people have to blend the Internet with the phone. It is, however, a wireless sol as its name suggests, and does nothing for wireline phone users. It only with special WAP phones. These show only show a few cha at a time on a small screen. Entering data is no easy task with just or so buttons available on a phone keypad. The *i-mode* system, p in Japan, has been more successful, but has many of the sam itations as WAP. VoiceXML gives any phone Internet access in that takes advantages of the phone's strengths and acknowledges its weaknesses.

VoiceXML can be used in many ways, not just for Web access. It can be used for a variety of information retrieval applications, of which Web access is only one sort. Other ways in which VoiceXML could be used (with or without the Web) are:

- Getting directions for driving
- Getting weather reports
- Tracking packages
- Financial applications: banking, stock quotes and trading, account status
- User interfaces for enhanced phone services: voice dialing, reviewing and updating service data, setting up and managing teleconferences, unified messaging, getting e-mail from a phone, recording messages
- Corporate intranet applications, such as querying an inventory database

A VoiceXML application is made of one or more *documents*. By convention, VoiceXML document names end with *.vxml*. The format of a VoiceXML document is as follows:

```
<?xml version="1.0"?>
<vxml application="vxml-filename" version = "1.0">
  vxml instructions
</vxml>
```

The instructions inside the `<vxml>` tag are divided into forms marked by the `<form>` tag. Each form handles part of a VoiceXML dialog. Every form has a unique name given by its `id` attribute:

```
<form id="form-name">
  form contents
</form>
```

VoiceXML forms are similar to HTML forms, which are also used to interact with a user, albeit through a graphical rather than an audio user interface.

Forms normally have *fields*, which are bits of information to be obtained from a user. The form below prompts a user for a phone number:

```
<form id="getPhoneNumber1">
  <field name="phoneNumber">
```

```
    <prompt>Please tell us your phone number.</prompt>
    <grammar src="../grammars/phone.gram" type="application/x-jsgf"/>
   <help>Please say your ten digit phone number.</help>
  </field>
</form>
```

The `<field>` in this form names a *field item variable*, called ph Number; its value will be set when the form executes. Insid `<field>` tag are tags to be executed when setting phoneNumber first of these, `<prompt>`, provides a phrase to be spoken by a engine on the VoiceXML platform. The `<grammar>` tag that follow the script how to recognize a user's input. The `src` attribute sa grammar is in a file at the relative URL ../grammars/phone. The `type` attribute gives the MIME type of the grammar; here M type says the grammar is in *Java Speech API Grammar Format* Everything the user says is compared with this grammar. If ther match, i.e., the user speaks a 10-digit phone number, the `<field>` is left with phoneNumber set to the phone number just spoken. user says "help" instead of a phone number, the script speaks the p in the `<help>` tag and waits for the user to say something again can compare with the grammar.

VoiceXML has several predefined grammars for things like numbers. Using the built-in `phone` grammar, the script looks like t

```
     <form id="getPhoneNumber2">
       <field name="phoneNumber" type="phone">
         <prompt>Please tell us your phone number.</prompt>
         <help>Please say your ten digit phone number.</help>
       </field>
     </form>
```

Simple grammars may appear directly within the `<gramma` The VoiceXML below has a small JSGF grammar inside it. If th says "hamburger" or "burger," the `sandwich` field item variable is hamburger. If the user says "chicken" or "chicken sandwich," san is set to `chicken`:

```
<form id="getOrder">
  <field name="sandwich">
    <prompt>What kind of sandwich would you like?</prompt>
    <grammar type="application/x-jsgf">
      hamburger | burger {hamburger} | (chicken [sandwich]) {chicken}
    </grammar>
    <help>Do you want a hamburger or a chicken sandwich?</help>
  </field>
</form>
```

Figure 11.8 shows a complete VoiceXML script. The first few lines are standard boilerplate that says this is an XML document and, more specifically, a VoiceXML document:

```
<?xml version="1.0" ?>
<vxml version="1.0">
```

The next three tags—`<help>`, `<noinput>`, and `<nomatch>`—define default *event handlers* for the script. The `<help>` event handler is used when a user says "help." It speaks a message to the user:

```
<help>Sorry, there's no help here.</help>
```

The `<noinput>` event handler tells the script what to do if a user doesn't provide input, spoken or keypad. Here, it speaks a message and repeats the prompt that got no input (`<reprompt/>`):

```
<noinput>
  I couldn't hear what you said.
  <reprompt/>
</noinput>
```

The `<nomatch>` event handler tells the script what to do if the user says or enters something the script didn't expect:

```
<nomatch>
  I couldn't understand that.
  <reprompt/>
</nomatch>
```

The form that follows makes up the bulk of the script. The form tag in this script doesn't have an `id` attribute. That's not a problem. You need it only in scripts that refer to a form from elsewhere in the script. We could have left `id` off all the other examples as well.

The `<field>` tag that begins the form gets a sandwich order from the user. Most of the code inside that tag should already be familiar. It has its own `<help>` event handler. This event handler is used in place of the default `<help>` event handler at the beginning of the script. The `<filled>` tag is new. It tells the script what to do once it gets acceptable input from the user. Here it echoes back what the user ordered before proceeding to the next part of the form:

```
<field name="sandwich">
  <prompt>What kind of sandwich would you like?</prompt>
  <grammar type="application/x-jsgf">
    hamburger | burger {hamburger} | (chicken [sandwich]) {chicken}
```

```
<?xml version="1.0" ?>
<vxml version="1.0">

   <help>Sorry, there's no help here.</help>

   <noinput>
     I'm sorry. I didn't hear anything.
     <reprompt/>
   </noinput>

   <nomatch>
     I didn't get that.
     <reprompt/>
   </nomatch>

   <form>
     <field name="sandwich">
        <prompt>What kind of sandwich would you like?</prompt>
        <grammar type="application/x-jsgf">
           hamburger | burger {hamburger} | (chicken [sandwich]) {chicken}
        </grammar>
        <help>Do you want a hamburger or a chicken sandwich?</help>
        <filled>
           <prompt>
              You ordered <value expr="sandwich"/>.
           </prompt>
        </filled>
     </field>
     <var name="total"/>
     <field name="quantity" type="number">
        <prompt>How many do you want?</prompt>
        <filled>
           <if cond="sandwich=='hamburger'">
              <assign name="result" expr="Number(quantity) * 1.29"/>
              <prompt>
                 Your total is <value expr="total"/>
              </prompt>
           <else>
              <assign name="result" expr="Number(quantity) * 1.39"/>
              <prompt>
                 Your total is <value expr="total"/>
                 <audio src="http://www.voxwich.com/theme.wav"/>
              </prompt>
           </if>
        </filled>
     </field>
   </form>
</vxml>
```

Figure 11.8
A complete VoiceXML script.

Web-oriented Service Technology

```
        </grammar>
        <help>Do you want a hamburger or a chicken sandwich?</help>
        <filled>
          <prompt>
            You ordered <value expr="sandwich"/>.
          </prompt>
        </filled>
      </field>
```

The `<var>` tag between the two `<field>` tags declares a variable named `total` that's used in the next part of the script. The variable `total` goes in its own declaration because it's internal to the script, unlike a field item variable that's obtained from a user external to the script:

```
<var name="total"/>
```

The `<field>` that follows asks the user how many sandwiches he wants. Because this field has no event handlers of its own, it uses all the default event handlers. Like the phone number example, it uses one of VoiceXML's built-in grammars, the one for numbers. The `<filled>` section calculates a total amount for the order and speaks that back to the user. The `<audio>` tag inside the prompt plays a sound file after the total has been spoken. The `<if>` and `<else>` tags let the script calculate different amounts, depending on what kind of sandwich was ordered:

```
<field name="quantity" type="number">
  <prompt>How many do you want?</prompt>
  <filled>
   <if cond="sandwich=='hamburger'">
     <assign name="result" expr="Number(quantity) * 1.29"/>
     <prompt>
       Your total is <value expr="total"/>
     </prompt>
   <else>
     <assign name="result" expr="Number(quantity) * 1.39"/>
     <prompt>
       Your total is <value expr="total"/>
       <audio src="http://www.voxwich.com/theme.wav"/>
     </prompt>
   </if>
  </filled>
</field>
```

11.5 For More Information

Reference [84] describes the model of service logic used in this chapter. The same document also has high-level descriptions of CPL and SIP CGI.

CPL is by no means a finished piece of work. New constructs ar[e]
stantly being discussed and added, of which I covered only a fe[w]
the most recent information, see the `iptel` working group's page [o]
IETF Web site. The best current reference is the IETF draft that des[
CPL [44].

There's not much information on either SIP CGI or SIP se[
beyond the IETF drafts in which they are described ([43] and [39]
documentation for both is sketchy. One might wonder to what [
they are being implemented and used. Nevertheless, it's worth kn[
something of them because they show some ways convergent se[
could be implemented and deployed on the Web.

The VoiceXML 1.0 specification [91] can be obtained fro[m]
VoiceXML Forum's Web site (`http://www.voicexml.org`).

CHAPTER 12

JTAPI

Well, if I called the wrong number, why did you answer the phone?

—*James Thurber*

12.1 JTAPI and Computer Telephony Integration (CTI)

12.1.1 What Is CTI?

The *Java Telephony API (JTAPI)* is a Java API for developing *Comput[er Tele]phony Integration (CTI)* applications. CTI technology blends voice an[d data] processing for single-user and enterprise environments. CTI system[s] range from simple caller ID and address book applications on P[Cs all] the way to call centers that integrate many agents, computers, an[d tele]phony devices. Examples of CTI applications include:

- Call logging and tracking
- Auto-dialing
- Telephony with a graphical user interface on a PC
- Screen-pop software
- Call routing applications
- IVR systems
- Call centers
- Fax send and receive
- Voicemail

Although JTAPI was not designed for public networks, JTAPI a[pplica]tions may interact with public networks. Furthermore, as one of t[he ear]liest Java-based telephony APIs, JTAPI has had a strong influe[nce on] public network APIs such as Parlay and JAIN, which I cover in Ch[apters] and 14.

Figure 12.1 shows the elements of a simple CTI system. A work[station] and a phone are both connected to the PSTN through a CTI serve[r. The] CTI server and its software determine how the user will interac[t with] the PSTN. The figure shows an incoming call arriving at the CTI [server.] The server takes caller ID information from the call's signaling m[essage] and sends that to the workstation. Ringing signals go to the ph[one. A] CTI application in the workstation uses the caller ID to look up [infor]mation about the caller in a database. Once retrieved, that informa[tion]

---

[1] If this were a PC application, the CTI server would probably be a modem or o[ther tele]phony card in the PC itself.

JTAPI

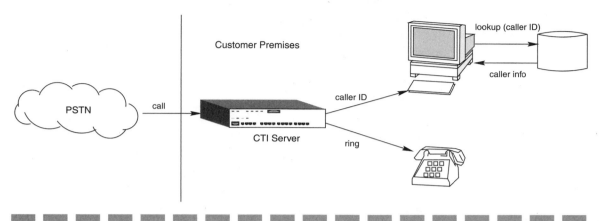

Figure 12.1
A simple CTI system.

displayed on the workstation. This, in a nutshell, is a *screen pop*, one of the most common CTI applications.

Call centers are the quintessential CTI application. Just about anything that can be done with CTI is done by call centers in one way or another. A call center is a system with which a group of agents makes calls to and takes calls from customers or clients of the organization operating the call center. Typical uses are inbound and outbound telemarketing and help desks. Among the functions a call center might support are:

- *Incoming call queuing.* A call center can intercept incoming calls and hand them on to agents according to an organization's policies. A common technique is to handle calls in *first come-first served* order. An organization that wanted to offer different levels of service to different customers might use *priority queuing* instead, giving certain (say, high-dollar, high-volume) customers faster access than they might get in a first come-first served system.

- *Call distribution.* A call center can send incoming calls to the agents best able to handle them. Different call centers may define "best able to handle" in different ways. A call center with an *Automatic Call Distributor (ACD)* routes calls to the agent that has gone without a call for the longest time. Another call center might route calls based on caller ID or other information provided by the caller. Many call centers use an IVR dialogue with the caller to get this information.

- *Agent monitoring.* A call center can keep track of the time spent responding to calls, the duration of each call, or other load and efficiency statistics to be used in managing it.

CTI (and in fact, all network service) APIs must provide two sets of capabilities:

- *Call processing* manipulates calls and connections within calls. Typical call processing activities include call setup and teardown, call transfer, adding and removing parties to a call, and forwarding calls. Call processing is also referred to as *call control* or *connection control*.
- *Media processing* covers the transport and processing of the information content in a call. This includes such things as playing sound files, voice transmission, speech recognition, and speech synthesis. *Voice processing* is a subset of media processing that applies specifically to handling human speech.

There are two ways in which a CTI application can be built. A *party call control* application works with only one telephony device, example, a PC with an attached phone. It's the basis of many PC CTI applications for voicemail, call logging, and feature management. *Third-party call control* gives control to a system other than the making and receiving a call. Third-party call control systems typically oversee many telephony devices. It forms the basis for most call applications.

12.1.2 APIs for CTI

Until recently, most CTI APIs worked only with one vendor's ment. Once a customer developed CTI applications for such a platform, equipment providers hoped they would think twice before moving another platform and abandoning all that carefully crafted software. What actually happened was that many customers opted not to any CTI software at all.

In the late 1980s and early 1990s, several organizations started to nonproprietary APIs for CTI. Among these were:

Computer-Supported Telecommunications Applications (
CSTA defines an API for PBXs. It was developed by several companies and PBX manufacturers and network providers at the invitation

JTAPI

British Telecom (BT) and Digital (DEC), and released as an *European Computer Manufacturer's Association (ECMA)* standard in 1991.

Telephony Services Applications Programming Interface (TSAPI) TSAPI is an implementation of CSTA by Novell and AT&T. It supports both first-party call control and third-party call control. TSAPI makes a PBX into a Novell Netware server and gives it third-party call control capabilities. Desktop phones are attached directly to the PBX, but desktop PCs can control phones by interacting with a TSAPI server attached to the PBX (Fig. 12.2).

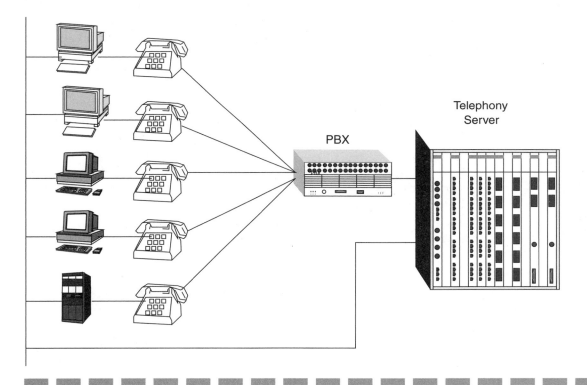

Figure 12.2
TSAPI architecture.

Telephony Applications Programming Interface (TAPI) TAPI is a Microsoft API for writing CTI applications that run on the Windows operating system. TAPI supports both first-party and third-party call control.

S.100 S.100 is a standard developed by the *Enterprise Computer Tel[ephony] Forum (ECTF)*. S.100 defines an operating-system-independent A[PI that] abstracts various types of call processing hardware. Application[s] use S.100 services to allocate, configure, and use this hardware.

JTAPI builds on this legacy and brings Java technology to th[e CTI] world. Work on JTAPI began in August 1996 with representatives [from] Intel, Lucent, Nortel, Novell, and Sun. Their goal was to de[sign a] portable, object-oriented CTI API that would be simple, scalable, e[xtensi]ble, easy to implement on top of existing CTI platforms, and capa[ble of] both first-party and third-party call control. They released version [1.1 of] JTAPI in October 1997. They were subsequently joined by com[panies] such as IBM, Dialogic, and Siemens. The enlarged team develope[d the] JTAPI 1.2 specification, which they released in February 1998. B[ecause] JTAPI was in many ways similar to the work being done by the [JCP,] that organization took responsibility for JTAPI, issuing JTAPI 1[.3, the] most recent version in July 1999.

JTAPI can be used to build the full range of CTI applic[ations.] Because it's based on Java, a JTAPI application is as portable as any [other] Java application and can be executed in any environment that pr[ovides] the telephony features it needs. Furthermore, JTAPI applications c[an use] everything else provided by Java: security; a large, ready-to-us[e class] library; and access to a wide range of computing techno[logy] that includes such things as multimedia, databases, directories, a[nd the] Internet.

12.1.3 The Service Provider Interface (SPI)

An essential piece of any CTI system, including one based on JT[API, is] the software that connects application software and the software [of the] API, with the hardware and software that make the API work. Thi[s is the] job of a *Service Provider Interface (SPI)*. To use a particular vendor'[s hard]ware or software with a CTI API, there must be an SPI that brid[ges the] two. For example, a telephony card for a PC may be able to make [outgo]ing calls. The manufacturer of that card will provide a set of lo[w-level] drivers—usually software that you install on your PC. If you have [a CTI] API with the method `makeOutgoingCall` and you want to wri[te pro]grams that make outgoing calls from your PC, you'll need the ap[propri]ate SPI software to get from `makeOutgoingCall` to the drive[rs that] actually tell your telephony card to make a call.

JTAPI

While the need for an SPI might be obvious once you give the matter a bit of thought, it's caused much confusion among developers new to JTAPI. With many—if not most—Java APIs, when you download code from Sun you get working implementations of the classes in that API. However, if you go to Sun's JTAPI Web site (`http://java.sun.`

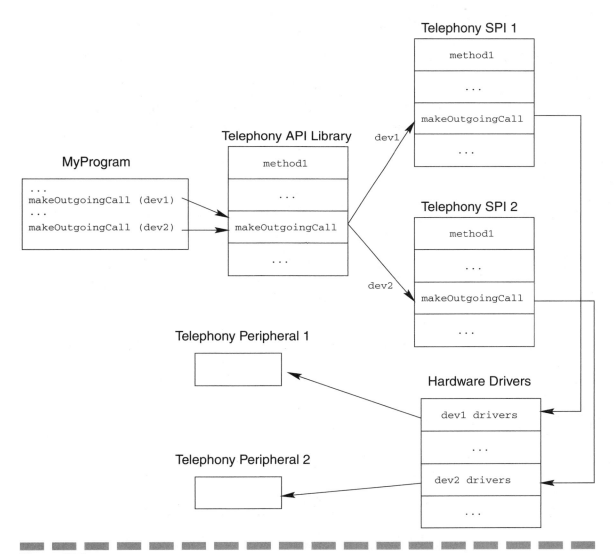

Figure 12.3
An SPI is the link between an API and hardware.

com/products/jtapi) and download the JTAPI code, all you'll g
definitions of JTAPI class interfaces, not implementations of those c
People who assume JTAPI is just like other Java APIs soon find
programs don't do much of anything. That's because they still ne
SPI that implements JTAPI and the hardware with which it int
(Fig. 12.3). It's important to understand this because it applies also
network service APIs discussed in Chaps. 13 and 14. As far as JT
concerned, the SPI for a JTAPI implementation could be a CTI API
as TSAPI or TAPI, or it could be something built entirely from scra

Figure 12.4 shows how this works. A JTAPI application is a Jav
gram that runs inside a Java Virtual Machine (JVM). It makes c
methods in the Java APIs, including calls to classes that impleme
JTAPI interfaces. The JTAPI implementation, through its SPI, make
that manipulate the telephony features of the underlying CTI platf

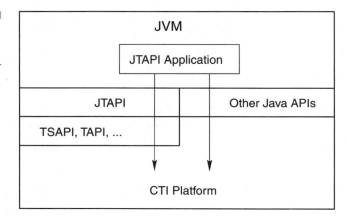

Figure 12.4
A JTAPI application stack.

12.2 JTAPI Core

The foundation of JTAPI is a set of core classes that must be pres
every implementation. To this core it adds optional extension class
provide specialized functions such as voice recording and pla
sophisticated call center capabilities, and wireless network su
JTAPI does not specify everything needed to build a phone s
For example, signaling protocols such as SS7 and ISDN, while p
the environment within which JTAPI applications run, are not
to them.

JTAPI's classes fall into four groups:

Call Control Call Control classes monitor and control call processing, both first party and third party.

Physical Device Control Physical Device Control classes monitor and control the user interface elements of a phone, such as keypad buttons, displays, lamps, hook-switch, ringer, speaker, and microphone.

Media Services Media Services classes manipulate and process a call's media streams, with such tasks as generating and detecting tones, processing faxes, text-to-speech translation, and speech recognition.

Administrative Services Administrative Services classes provide everything needed to start up, shut down, and manage a JTAPI-based system.

12.2.1 The JTAPI Call Model

Six of the core JTAPI classes make up the JTAPI call model:

`Provider` An instance of `Provider` represents the telephony system—for example, a PBX, a phone/fax card in a PC, or a VoIP platform—used by a JTAPI application.

`Terminal` Instances of `Terminal` represent the physical endpoints of a call (phones).

`Address` Instances of `Address` represent the logical endpoints of a call (phone numbers).

`Call` An instance of `Call` represents all the logical and physical entities that together represent the joining together of two or more endpoints (call parties).

`Connection` An instance of `Connection` represents the dynamic relation between an instance of `Call` and an instance of `Address` for the duration of that call. A two-party call has one `Call` object and two `Connection` objects. A multiparty call has one `Call` object and three or more `Connection` objects.

`TerminalConnection` An instance of `TerminalConnection` represents the dynamic relation between an instance of `Connection` and an instance of `Terminal` for the duration of a call.

When a JTAPI-based system starts up, it has no calls in pr(
JTAPI's *structural call model*, shown in Fig. 12.5, represents its state
time. The structural call model represents object relationship
change infrequently or not at all. It has no instances of Call, Co

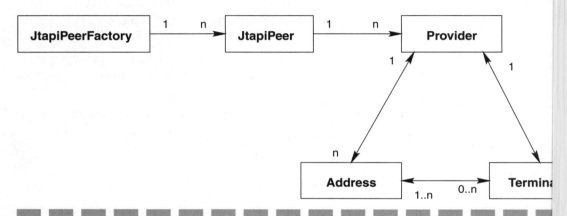

Figure 12.5
JTAPI's structural call model.

tion, or TerminalConnection. There's a JtapiPeerFactory
creates and tracks instances of JtapiPeer which create and
instances of Provider on behalf of JTAPI applications. A *peer* is
form-specific implementation of a Java interface or API (for ex
JTAPI), a *factory* is a singleton class (a class that has only a single in
that creates instances of peers, and a *provider* is an object that sup
specific service using the platform-specific capabilities of the pe
which it is associated. In JTAPI, providers represent a telephony e
such as a PBX or a telephony card in a PC. The factory-peer-p
pattern is common among Java APIs for network software, and yo
it again when we look at JAIN.

Associated with each instance of Provider are any num
Addresses and Terminals for which it is responsible. Each Ad
may refer to zero or more Terminals, and each Terminal has
one and perhaps more Addresses.

Once a JTAPI application starts processing calls, the JTAPI *d
call model* comes into play, as shown in Fig. 12.6. The dynamic call
is built on top of the relationships of the structural call model,
by dotted lines. The objects and relationships of the dynamic call
are as follows:

JTAPI

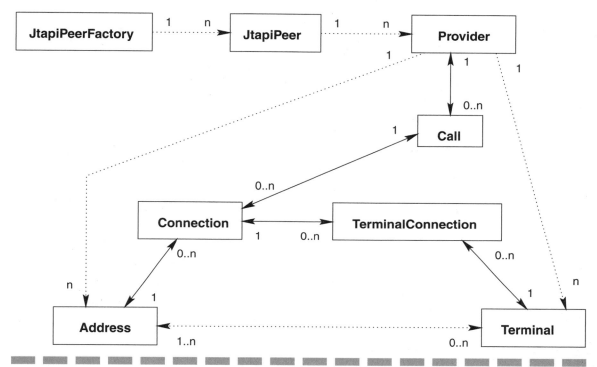

Figure 12.6
JTAPI's dynamic call model.

- A `Provider` creates `Call` objects as needed by a JTAPI application. Each `Call` is associated with just a single `Provider`.
- A `Call` is a collection of zero or more `Connections`.
- A `Connection` binds a single `Address` (a logical endpoint) to a single `Call`.
- A `TerminalConnection` binds a single `Connection` to a single `Terminal` (a physical endpoint). Note that a given `Connection` or `Terminal` may participate in more than one `TerminalConnection`.[2]

With `Connections` and `TerminalConnections`, JTAPI applications can see both the logical and the physical endpoints of a call. This is useful

---

[2] All parties outside the domain of an application's `Provider` have only a `Connection` and `Address` in this model. This makes sense because we can make calls to and receive calls from phone numbers not under a `Provider`'s control. In that case, we know something about the `Address`, but nothing about the `Terminal`.

in applications where dialing a single number causes more than
phone to ring, either all at once or one after another. It's also usef
dealing with phones that support more than one phone number.

To process a call, a JTAPI application gets an instance of an idle
from its `Provider` by invoking `Provider.createCall`. The J
core classes provide methods for the basic activities of call proce
making a call, answering a call, and ending a call. One might expe
these methods would appear in `Call`. However, JTAPI divides the
according to whether they deal with an entire call, one logical end
of the call, or a single physical endpoint of the call. That's why an
for example, is a method of `TerminalConnection`:

`Call.connect (Terminal origTerm, Address origA`
`String tn)` This method makes an outgoing call from an ori
ing terminal and address to a specified phone number.

`TerminalConnection.answer()` Applications that expect to r
phone calls monitor `Terminal` objects with event listeners (se
12.2.4). When a call arrives at a `Terminal`, it generates an eve
answer that call, an application invokes this method on the Te
nalConnection associated with the `Terminal` that generate
event.

`Connection.disconnect()` This method removes an `Address`
a call. All `Terminals` associated with that `Address` are also rer
from the call.

12.2.2 Core Object States

`Call`, `Connection`, and `TerminalConnection` objects change s
they do things or as things are done to them. These states indicate
happening in a call and what methods the objects associated wit
call can execute. Besides the states described in the following s
there's one more state both `Connection` and `TerminalConne`
have that's not shown in their state diagrams:

UNKNOWN A `Connection` or `TerminalConnection` may
the UNKNOWN state from any nonfinal state and move fror
any other state. When an object is in the UNKNOWN sta
`Provider` in whose domain it resides has not been able to fi
its state. Methods invoked on objects in this state will have

dictable results. An object might enter this state if the telephony equipment on which it depends has a problem that leaves the system unsure of the status of calls in progress.

CALL STATES `Call` objects have the state machine shown in Fig. 12.7.

Figure 12.7
Call states.

- **IDLE** When a `Call` object is created, it enters the IDLE state. It has no `Connections` associated with it. Applications may invoke its `connect` method to make an outgoing call.
- **ACTIVE** Once a `Call` has one or more `Connections`, it enters the ACTIVE state and becomes the subject of ongoing activities by its application.
- **INVALID** This is the final state of a `Call`. It has no `Connections` because they've all moved to their DISCONNECTED state. The name is perhaps a bit misleading, because this is a perfectly legal state in which to find a call. However, it may no longer be the subject of any more actions by its application.

CONNECTION STATES Figure 12.8 shows the state machine for `Connection` objects.

- **IDLE** Newly created `Connections` start in the IDLE state. You can't do much with an IDLE `NameConnection`. It usually represents a party that has just joined a call and quickly moves to one of the other `Connection` states.
- **INPROGRESS** This state applies only to `Connections` on the called end of a call. It indicates a call is being placed to the endpoint represent by the `Connection`.
- **ALERTING** This state applies only to `Connections` on the called end of a call. It indicates the `Address` at the called end is being alerted.
- **CONNECTED** A `Connection` moves into the CONNECTED state when it and its associated `Address` become an active part of a call. For the calling end, this happens when it initiates the call. For the called end, this happens when it answers.

Figure 12.8
Connection states.

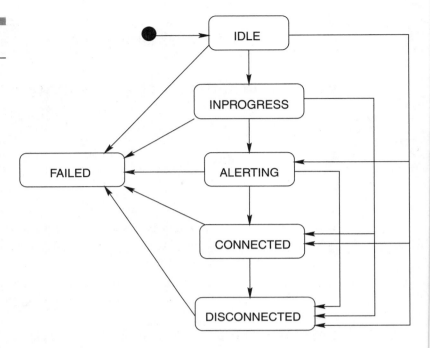

DISCONNECTED A Connection moves into the DISCONNE[CTED] state when it leaves a call. No methods may be invoked on a DIS[CON]NECTED Connection.

FAILED A Connection moves into the FAILED state if a call to it fails for some reason, for example, because the party it repr[esents] is busy.

TERMINALCONNECTION STATES Figure 12.9 shows th[e state] machine for TerminalConnection objects:

IDLE Just like Connection objects, newly created TerminalCo[nnec]tions start in the IDLE state.

RINGING A TerminalConnection moves to the RINGIN[G state] when its associated Terminal signals an incoming call.

ACTIVE A TerminalConnection moves to the RINGING stat[e when] its associated Terminal becomes an active part of a call, s[uch as] when a phone goes off hook. On the calling end, a TerminalCo[nnec]tion enters this state as soon as the call is made. On the called [end it] enters this state when the call is answered.

JTAPI

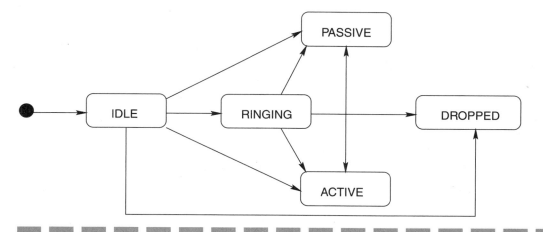

Figure 12.9
TerminalConnection states.

PASSIVE This state means the associated Terminal is part of a call but not now an active participant.

DROPPED This state means the associated Terminal was once part of a call but has left it permanently.

12.2.3 A Simple JTAPI Application

Figure 12.10 is a simple JTAPI application that makes an outgoing phone call. The interactions among its classes and objects are shown in Fig. 12.11. This application demonstrates the basic activities most JTAPI programs carry out. The first line of the program imports definitions of the JTAPI core interfaces. These are all part of the javax.telephony package. The second line imports the Foo company's implementation of the JTAPI interfaces:

```
import javax.telephony.*;
import com.foo.jtapi.*;
```

The class that implements this application is named OutCall. It has only one method: main. The first thing the program does is obtain an instance of FooJtapiPeer from JtapiPeerFactory by calling its

```
import javax.telephony.*;
import foo.com.jtapi.*;

public class OutCall {
  public static void main (String [] args) {
    FooJtapiPeer peer =
        (FooJtapiPeer)JtapiPeerFactory.getJtapiPeer
                         ("com.foo.jtapi.FooJtapiPeer");
    Provider provider = peer.getProvider ("OutcallProvider");
    Terminal origTerminal = provider.getTerminal ("PhoneA");
    Address origAddr = provider.getAddress ("5551111");
    Call call = provider.createCall();
    Connection[] conns = call.connect (origTerminal, origAddr, "5552222");
  }
}
```

Figure 12.10
A simple JTAPI outgoing call application.

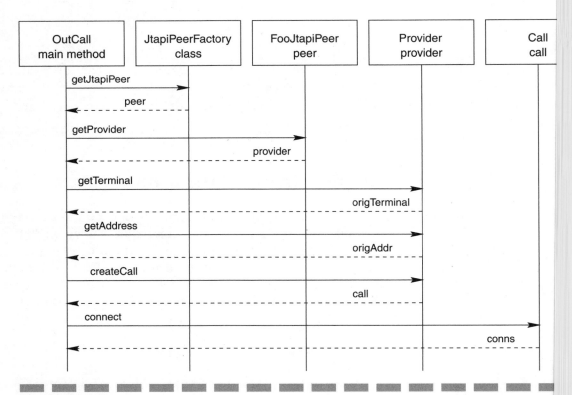

Figure 12.11
Object interactions for the `OutCall` program.

getJtapiPeer method, passing an argument that says it wants the Foo company's implementation of JtapiPeer:

```
FooJtapiPeer peer =
   (MyJtapiPeer)JtapiPeerFactory.getJtapiPeer("com.foo.jtapi.FooJtapiPeer");
```

Once the program has a JTAPI peer, it gets a Provider from the peer by supplying the name of the service it wants—the ability to make outgoing calls in this example—to the getProvider method:

```
Provider provider = peer.getProvider("OutcallProvider");
```

The program can perform call operations on any Address and Terminal known to provider. Of course, that begs the question of how provider learned about those in the first place. Typically, those relationships and their data are set up during system initialization.

To make an outgoing call, the program gets a Terminal object that represents the device from which the call is made:

```
Terminal origTerminal = provider.getTerminal ("PhoneA");
```

Next the program gets one of the Addresses associated with that Terminal:

```
Address origAddr = provider.getAddress ("5551111");
```

Now that the program has logical and physical endpoint objects for the originator, it creates a Call object with which they will be associated:

```
Call call = provider.createCall();
```

The program uses call to place an outgoing call from the originating address on the originating terminal to some other address. The call is made, two Connection objects are returned—one for the calling end and another for the called end—and the program is done:

```
Connection[] conns = call.connect (origTerminal, origAddr, "5552222");
```

12.2.4 JTAPI Event Listeners

The program in the previous section made an outgoing call and left it at that. Once the call is placed, the called end may answer or not, or it may be busy. But the program as written would never know. To do anything

more with a call, a JTAPI program has to be able to receive and res
to events that occur as the call proceeds. JTAPI uses the Java ever
tener mechanism that was described in Section 3.6 to intercept an
ward events to applications.[3] Event listeners do most of the wo
many JTAPI programs because telephony applications are inher
event driven. Like the factory-peer-provider pattern, the event li
pattern is one used in many other Java network APIs, for exa
Parlay and JAIN.

As shown in Fig. 12.12, JTAPI defines event listener interface
handle events generated by JTAPI objects of a particular sort. Th
get events from a `Terminal` object, one implements the `Terminal`
tener interface, creates an object of that type, and registers that
as a listener with the `Terminal` of interest. As that `Terminal` ger
events, they are forwarded to the `TerminalListener` object.

You could enhance the outgoing call program by implementir
JTAPI interface `ConnectionListener`. Let's call this class `MyCor`
tionListener. Because it's derived from `CallListener`, it can b
both `Call` and `Connection` events. To let the calling party know
going on at the called end, the program creates an instance of `M`
nectionListener and adds it to `call`, where it listens to
Connections:

```
Call call = provider.createCall();
Connection[] conns = call.connect (origTerminal, origAddr, "5552222");
MyConnectionListener listener = new MyConnectionListener();
call.addCallListener (listener);
```

The program reports any events that occur on the Connecti
listener, which responds to them as determined by the impler
tion of `MyConnectionListener`. For example, `MyConnectio`
tener would handle called-party ringing by implementing the
notification method `connectionAlerting` in `ConnectionList`
Answers would be handled by implementing `connectionConnec`

Event notification methods are passed event objects. Becau
event listener in this example implements `ConnectionLister`
gets `ConnectionEvents`. The code below uses the event object
play a status message when the called party starts ringing and to
up when they answer (the Automated Prank Call feature):

---

[3]The event listener model was introduced in JTAPI 1.3. JTAPI 1.2 and earlier release
variant known as the "event observer" model. Though JTAPI supports both th
observer and event listener models, new applications should use only the more
event listener model.

JTAPI

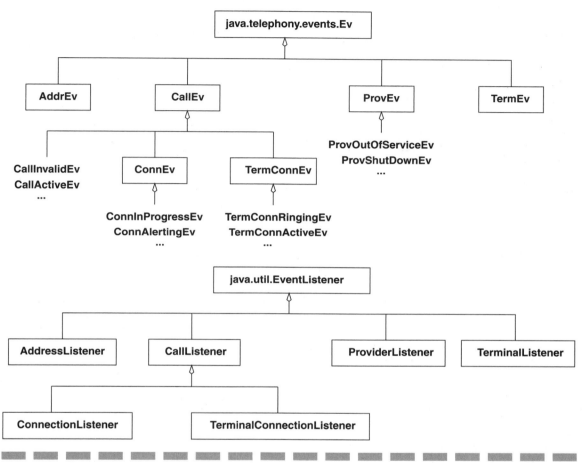

Figure 12.12
JTAPI event and listener interfaces.

```
void connectionAlerting (ConnectionEvent e) {
  Connection conn = e.getConnection();
  Address addr = conn.getAddress();
  System.out.println (addr.getName() + " is ringing");
}

void connectionConnected (ConnectionEvent e) {
  Connection conn = e.getConnection();
  conn.disconnect();
}
```

Besides `Connections`, JTAPI applications may attach event listeners to `Providers`, `Calls`, `Addresses`, `Terminals`, and `TerminalConnections`.

Each of these has its own listener interfaces and event classes simi[lar to] `ConnectionListener` and `ConnectionEvent`.

The program in Fig. 12.13 handles the called end of a call wi[th an] event listener. The object interactions of that program are shown i[n Fig.] 12.14. It first imports JTAPI definitions:

```
import javax.telephony.*;
```

Figure 12.13
A JTAPI incoming call application that uses event listeners.

```
import javax.telephony.*;
public class InCallListener
  implements TerminalConnectionListener {
  public void TerminalConnectionRinging
      (TerminalConnectionEvent e) {
    TerminalConnection tc = e.getTerminalConnection();
    Runnable r = new Runnable() {
      public void run () {
        tc.answer();
      }
    };
    Thread t = new Thread (r);
    t.start();
  }
}

public class InCall {
  public static void main (String [] args) {
    JtapiPeer peer = JtapiPeerFactory.getJtapiPeer (null);
    Provider prov = peer.getProvider (null);
    Terminal term = prov.getTerminal ("5552222");
    term.addCallListener (new InCallListener ());
  }
}
```

This program has two classes, one to implement the event listen[er, the] other to use it. The listener class implements `TerminalConnec`[tion]`Listener`:

```
public class InCallListener
    implements TerminalConnectionListener
```

To fully implement `TerminalConnectionListener`, you [would] need a lot more code than is shown here. All this listener does is w[ant] its `terminalConnectionRinging` method to be called when the [Ter]`minalConnection` starts ringing:

```
public void terminalConnectionRinging
    (TerminalConnectionEvent e)
```

JTAPI

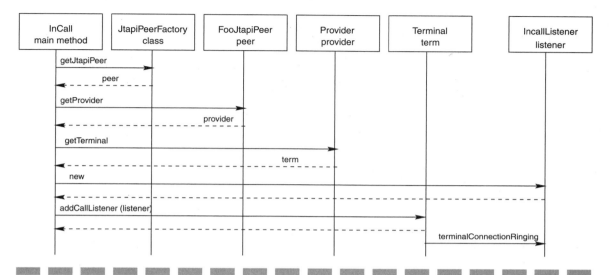

Figure 12.14
Object interactions for the `InCall` program.

The program extracts the `TerminalConnection` that's ringing from the event object passed to it:

```
TerminalConnection tc = e.getTerminalConnection();
```

Most of the code that follows is there to start a separate *thread*[4] that waits for an answer on the `TerminalConnection`. The only JTAPI code here is the call to `tc.answer()`. What happens when that method is called depends on how it's been implemented. Remember that JTAPI is mostly class interfaces, not class implementations, so it's up to you to supply most of the code:

```
Runnable r = new Runnable() {
  public void run () {
    tc.answer();
  }
};
Thread t = new Thread (r);
t.start();
```

---

[4]Threads are like processes, but they don't take as much work to start and stop as a full-fledged process. If you're interested in learning more, be forewarned that threads are a big topic of their own. There are many excellent references, such as [64]. Reference [68] provides a Java-centric view of threads and thread programming.

Now that there's a listener, there has to be something for it to list[en]. That's the purpose of the `InCall` class. Like `OutCall`, it has o[nly a] main method. That method first gets a `JtapiPeer`. Unlike `Out`[Call,] which named the peer it wanted, `InCall` gives null as the peer [name,] meaning it will accept whatever default peer the factory provides:

```
JtapiPeer peer = JtapiPeerFactory.getJtapiPeer (null);
```

From the peer, the program gets a provider, passing null to sho[w it will] accept the default provider:

```
Provider prov = peer.getProvider (null);
```

From the provider, the program gets a `Terminal` object that [repre]sents the phone it will be monitoring:

```
Terminal term = prov.getTerminal ("5552222");
```

Finally, the program creates an instance of `InCallListene`[r and] starts it listening to the `Terminal` of interest:

```
term.addCallListener (new InCallListener ());
```

When that `Terminal`'s phone starts ringing, it calls the `term`[inal]`ConnectionRinging` method of `InCallListener`.

12.3 JTAPI Extension Packages

The JTAPI extension packages provide capabilities that not every [JTAPI] application may need. Some of the more significant of the[m are] described in the following sections.

12.3.1 javax.telephony.callcenter

The `javax.telephony.callcenter` package has interfaces for f[eatures] used by call centers. This is a large package. Space—and the likely [inter]est of most readers—prohibits an exhaustive look at everything i[n it. To] give some sense of what it provides, I'll summarize two of it[s most] important features: call routing and ACD. Even so, the coverage [will be] cursory at best.

A JTAPI-enabled call center represents its switching element with an instance of `Provider`. When a call arrives for a particular `Address` (actually, for a particular `RouteAddress`, the call center package's extension of the core `Address` class), the switching element needs to know where to deliver that call. Its `Provider` notifies a call center application, which chooses among alternate destinations, guided by factors such as the calling number, the time of day, and agent availability. The application returns the destination it selects to its `Provider`, which tries to place the call. If the `Provider` can't place the call, it may ask for another route. The application needs to know that this is a continuation of the earlier request, not an entirely new request. The call center extension package provides a `RouteSession` interface for that purpose.

An application that handles routing for a particular `RouteAddress` implements the `RouteCallback` interface and registers it with the `RouteAddress` in question. When the `Provider` gets a new call for that `RouteAddress`, it creates a `RouteSession` for it. The `RouteSession` makes calls to methods in the `RouteAddress`'s associated `RouteCallback`. The `RouteCallback` methods that may be called by the `RouteSession` correspond to its states, as shown in Fig. 12.15:

ROUTE The `RouteSession` has been asked to route a call and has supplied a possible route. The `RouteSession` calls `routeEvent` in the `RouteCallback` associated with the `RouteAddress` it's routing.

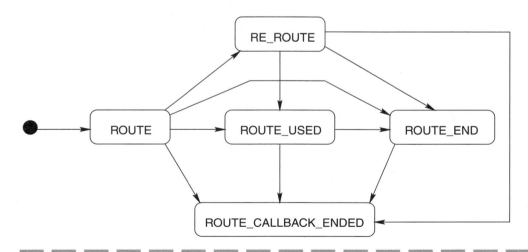

Figure 12.15
`RouteSession` states.

ROUTE_USED The `Provider` has informed the `RouteSession` [that] it's used a route supplied to it. The `RouteSession` calls r[oute] `UsedEvent` in the `RouteCallback`.

RE_ROUTE The `Provider` has informed the `RouteSession` [that the] route provided before didn't work, so the `RouteSession` has p[rovid]ed another. The `RouteSession` calls `reRouteEvent` in the R[oute] `Callback`.

ROUTE_END Call routing is finished and the call has been p[laced]. The `RouteSession` calls `routeEndEvent` in the `RouteCallba`[ck].

ROUTE_CALLBACK_ENDED All `RouteCallbacks` have [been] removed from this `RouteSession`. The `RouteSession` calls r[oute] `CallbackEndedEvent` in the `RouteCallback`. This is the fina[l event] for a `RouteSession`.

Another important feature of many call centers is *Automatic C*[all Dis]*tribution (ACD)*. An ACD defines an *ACD group* of zero or more [agent] *extensions*. Each agent extension can handle any call that's routed [to the] group. If no agent is available, calls to a group can be queued.

ACD support in the JTAPI call center extension package is bu[ilt on] the `ACDAddress` interface, which extends the `CallCenterAd`[dress] interface, which in turn extends the JTAPI core `Address` interface [. ACD] applications use these extensions to associate an `Address` with a [set of] available `AgentTerminals`, the ACD extension of the core `Ter`[minal] interface. These associations change as agents and calls come a[nd go]. JTAPI represents the current association between an `ACDAddres`[s and] an `AgentTerminal` by an instance of `Agent`. The state of an `A`[gent] `Terminal` changes as its associated `Agent` proceeds through the n[ormal] tasks of a work day. These states, shown in Fig. 12.16, determine w[hich] calls to its associated `ACDAddress` will be routed to it:

LOG_IN An `Agent` at this `AgentTerminal` has logged in[to the] `ACDAddress`.

LOG_OUT An `Agent` at this `AgentTerminal` has logged out [of the] `ACDAddress`.

NOT_READY The `Agent` at this `AgentTerminal` is doing work [other] than handling calls.

READY The `Agent` at this `AgentTerminal` is ready to handle c[alls].

WORK_NOT_READY The `AgentTerminal` is disconnected f[rom a] call while handling work associated with that call (for exampl[e, con]sulting with another agent).

JTAPI

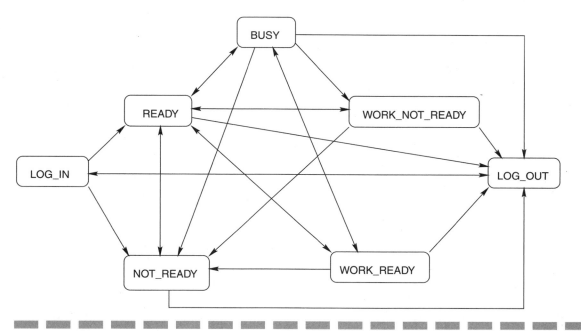

Figure 12.16
AgentTerminal states.

WORK_READY The AgentTerminal is ready to reconnect to and resume work on a call.

BUSY The AgentTerminal is working on a call and cannot accept any others.

12.3.2 javax.telephony.callcontrol

The javax.telephony.callcontrol package adds methods and states to objects in the JTAPI core. Among the features this package adds are:

- Conference calling
- Call transfer
- Call forwarding
- Call hold
- Message waiting

- Call park
- Call pickup

Among the more important interfaces and methods in this package

CallControlCall Adds new methods to the core `Call` interface most important are:

> **conference** Joins two calls together as a single call.
> **transfer** Moves the parties of one call to another call, after w one of the parties usually drops out.
> **drop** Provides a way to remove a party from a call that can be convenient to use in some situations than the core `disconne` method.
> **consult** Forks a consultation call from a call in progress. Onc consultation call has been forked, one or more of the parties original call may be transfered to it.

CallControlAddress Adds new methods to the core `Address` face. The most important are listed below. Keep in mind that n these methods apply to a call in progress; they give the `Pro` instructions on what to do with a specific `Address`

> **setForwarding** Tells the underlying telephony platform to fo calls to the affected `Address` using rules specified in an arra `CallControlForwarding` objects passed to it.
> **setDoNotDisturb** Tells the underlying telephony platform to block alerting for incoming calls to the affected `Address`.
> **setMessageWaiting** Tells the underlying telephony platform a message is waiting for the affected `Address`.

CallControlConnection Adds new methods to the core `Co` tion interface. The most important are:

> **accept** Accepts an incoming call at the `Address` associated w this `Connection`. Once accepted, the call starts alerting.
> **reject** Rejects an incoming call at the `Address` associated wi this `Connection`. Once rejected, the call never alerts and eve ually dies.
> **redirect** Sends an incoming call that has just arrived or is al alerting at the `Address` associated with this `Connection` to another `Address`. It's similar to `CallControlCall.transf` except that `redirect` is used on `Connections` that have yet become CONNECTED and are thus not participating in a ca
> **park** "Parks" a `Connection` at an `Address`. This moves the `Co` nection from its current `Address` and puts it into a waiting

CallControlTerminal Adds new methods to the core `Terminal` interface. The most important is:

pickup Picks up a parked `Connection`. This gives programs the ability to answer calls at an `Address` other than one associated with a `Terminal` for which they have responsibility.

CallControlTerminalConnection Adds new methods to the core `TerminalConnection` interface. The most important are:

hold Puts a `TerminalConnection` on hold in the call with which it is associated.

unhold Takes a `TerminalConnection` off hold and makes it active again.

leave Takes an active `TerminalConnection` and bridges it. A bridged `Terminal` is not an active participant of a call, but it can become one by joining the call. Any resources associated with the bridge are in use, even though the `Terminal` is not active in the call.

join Takes the `Terminal` of a bridged `TerminalConnection` and makes it active.

12.3.3 javax.telephony.media

The `javax.telephony.media` package has classes and interfaces for manipulating the media streams in calls. These allow an application to interact with:

- Media players
- Media recorders
- Signal generators
- Signal detectors

Like `javax.telephone.callcenter`, this is a large and complicated package, and I won't give much detail about it. Its foundation is the `BasicMediaService` class (unlike most of JTAPI, it's a class, not an interface), which has methods for playing and recording media streams, and for sending and receiving telephony signals. The media extension package provides event listener interfaces for recorders (`RecorderListener`), players (`PlayerListener`), and signal detectors (`SignalDetectorListener`). The methods in these interfaces let applications handle events like pause or resume on a player or recorder (`onPause` and

onResume), volume change on a player (onVolumeChanged), o
arrival of a signal (onSignalDetected).

12.3.4 javax.telephony.mobile

The `javax.telephony.mobile` package has interfaces that exten
JTAPI core to mobile networks. Many of the features in this pa
address roaming and the fact that, unlike wireline phones, wi
phones may be turned off from time to time. Among the interfa
provides are:

MobileProvider Extends the JTAPI core `Provider` interface. I
a new state, RESTRICTED_SERVICE, to the `Provider` state ma
that represent the situation in which services for a JTAPI appli
have been restricted. These restrictions occur because its user:
not subscribed to the `MobileProvider`'s network or were una
validate themselves to it because of an invalid or missing pas:
Associated with a `MobileProvider` is a set of `MobileNetwor`
means of which it provides service.

MobileNetwork Represents a cellular network. Applications
whether it's available or not by calling `isAvailable`. Other me
let an application get a network's name and network code, and
out if it's now restricted.

MobileTerminal Extends the JTAPI core `Terminal` interface.
methods for generating DTMF tones (which might otherw
tricky because mobile phones don't always do this themselve
getting information about a mobile phone (`getManufacturer`
`getSoftwareVersion`, and `getTerminalId`).

NetworkSelection Extends the `TerminalMobileNetwork` int
It has methods for selecting a network service provider when
than one is available and selection is allowed (`getPreferre`
works, `getSelectionModes`, and `setSelectionMode`).

12.3.5 javax.telephony.phone

The `javax.telephony.phone` package has interfaces that represe
physical characteristics of a terminal, such as a speaker, micropho
play, ringer, handset, or status indicator lamps (Fig. 12.17). It defines
interface, `PhoneTerminal`, that extends the core interface `Termi`

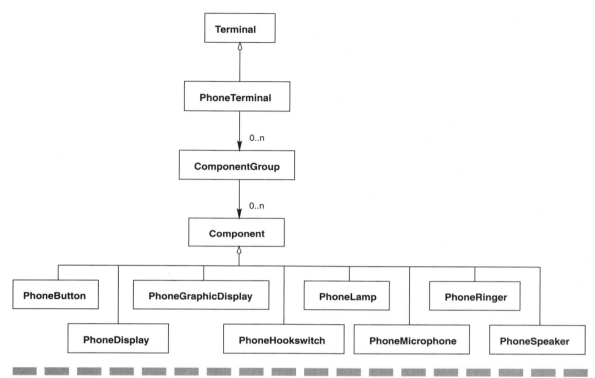

Figure 12.17
Interfaces in the `javax.telephony.phone` extension package.

`PhoneTerminal` has zero or more `ComponentGroups`, which have zero or more `Components`. `Component` is the parent class of a set of interfaces that represent elements of a phone:

PhoneButton Represents a button on a phone keypad. An application can "press" a button by calling `buttonPress`. Buttons have associated information, for example, their labels. Applications obtain and set this information by calling `getInfo` and `setInfo`. Buttons may have lamps. Objects representing these may be retrieved with `getAssociatedPhoneLamp`.

PhoneDisplay Represents the display (if any) on a phone. A display has some number of rows and columns, which an application can learn by calling `getDisplayColumns` and `getDisplayRows`. Applications obtain and set the text at row `r` and column `c` on the display by calling `getDisplay(r,c)` and `setDisplay(r,c)`.

PhoneGraphicDisplay Represents a pixel-oriented display (if a[…] a phone. Applications get the size of the display in pixels by c[…] `size`. Applications get an object they can use to write to the d[…] by calling `getGraphics`, which returns an instance of `Graph`[…] standard Java object for working with bit-mapped graphics.

PhoneHookswitch Represents the hookswitch on a phone. A […] switch can be in one of two states: `ON_HOOK` or `OFF_HOOK`. A[…] tions get its current state by calling `getHookSwitchState` and[…] by calling `setHookSwitch`.[5]

PhoneLamp Represents a lamp—for example, a message waiting l[…] on a phone. Lamps can be in a variety of states, which JTAP[…] *lamp modes*, including `OFF`, `STEADY`, `WINK`, and `FLUTTER`. Applic[…] get the current mode of a lamp by calling `getMode` and set it b[…] ing `setMode`.

PhoneMicrophone Represents a phone's audio input. Applicatio[…] and set input sensitivity by calling `getGain` and `setGain`. Th[…] three predefined gain levels: `FULL`, `MID`, and `MUTE`.

PhoneRinger Represents the audible ringer on a phone. Ringer[…] ring patterns, each with an associated numeric ID. Applicatio[…] get and set ring patterns by calling `getRingerPatter`[…] `setRingerPattern`. Applications can also get and set ringer v[…] by calling `getRingerVolume` and `setRingerVolume`. The[…] three predefined ringer volumes: `FULL`, `MIDDLE`, and `OFF`.[6]

PhoneSpeaker Represents a phone's audio output. Applicatio[…] and set the output volume by calling `getVolume` and `setV`[…] There are three predefined gain levels: `FULL`, `MID`, and `MUTE`.[7]

The reason a `PhoneTerminal` can have more than one `Compo`[…] `Group` is that an application may want to have two different views […] same physical endpoint. For example, sometimes you might want […] resent the physical endpoint of a call by the phone that sits on […] body's desk. Other times, you might want to represent the same p[…] endpoint with a GUI on their computer screen. However, even […]

---

[5] Why is it that no one in the software industry bothers to clean up silly incons[…] such as the one demonstrated here? The class is named `PhoneHookswitch` with […] case "s" on "switch," but that transmogrifies to upper case in the methods ge[…] `SwitchState` and `setHookSwitch`. Admittedly, it's a minor thing, but it causes […] of avoidable typing errors. And it would be so easy to get it right. Sorry. It's jus[…] those things that drives me batty because it's so egregiously confusing and I had to […]

[6] To continue the earlier rant, why not `FULL`, `MID`, and `MUTE`, as with `PhoneMicrop`[…]

[7] Whew, that was a close one. I fully expected these to be `HIGH`, `MEDIUM`, and `SILEN`[…]

PhoneTerminal has more than one ComponentGroup, only one of those may be "active" at a time. The phone extension package has methods for retrieving ComponentGroups and their Components:

PhoneTerminal.getComponentGroups Returns the array of ComponentGroups associated with a PhoneTerminal.

PhoneTerminal.getComponents Returns the array of Components associated with a ComponentGroup. Once it has these, an application can find out their type and invoke their methods:

```
Component components[] = groups[0].getComponents();
for (int i = 0; i < components.length; i++) {
  if (components[i] instanceof PhoneMicrophone) {
    PhoneMicrophone mic = (PhoneMicrophone) components[i];
    mic.setVolume (PhoneMicrophone.MUTE);
  }
}
```

ComponentGroup.activate Makes a ComponentGroup active on its PhoneTerminal.

ComponentGroup.inactivate Makes a ComponentGroup inactive on its PhoneTerminal.

12.3.6 javax.telephony.privatedata

The javax.telephony.privatedata package gives applications the ability to talk directly to switch hardware. This may be necessary when a switch provides functions that aren't otherwise represented within JTAPI.

12.4 For More Information

Sun's JTAPI Web site (http://java.sun.com/products/jtapi) is the most up-to-date and complete source of information. The JTAPI online documentation can be obtained there. There is currently only one book on the market devoted to JTAPI [83]. It's quite helpful when you're trying to make sense of the huge mass of information at the JTAPI Web site. Unfortunately, the book is somewhat out of date. One of the biggest differences between it and more recent versions of JTAPI is that it uses event observers—an earlier and now deprecated JTAPI event detection method—instead of event listeners.

CHAPTER 13

JAIN

A good listener is not only popular everywhere, but after a while he gets to know something.

—*Wilson Mizner*

13.1 Traditional Service Development

Until recently, phone companies depended on network equip[ment] providers to write service software. As we saw in Chap. 7, the goal [of] Intelligent Network (IN) was to let phone companies develop servi[ces] their own. By and large, IN succeeded, at least for those services th[at fell] within its scope.

IN was a significant accomplishment. But there was still roo[m for] improvement. With IN, phone companies could create services, bu[t only] within a limited domain consisting mostly of services that enh[anced] call routing. If a service concept fell outside that domain, it either [had to] be abandoned or have new features added in the network to supp[ort it.] New network features meant new software on network equipmen[t—soft]ware that had to be written by the makers of that equipment. An[d that] was the problem IN was supposed to solve in the first place.

IN also fell short of its promise to simplify service programmi[ng.] IN platform vendors provide their own *service creation environment*[s, or SCEs,] a tool for programming IN services. The output of an SCE is s[ervice] logic that generates and responds to IN messages. These messa[ges are] standardized. SCEs, on the other hand, vary widely from one ven[dor to] another. Consequently, IN services and the skills to develop them [can't] be moved from one platform to another. Every service must be [devel]oped separately using languages and tools unique to each vendor.

Furthermore, SCEs don't always fit the real needs of develop[ers. At] one time people thought IN SCEs would make development so [easy] that just about anybody would be able to write network services. [Rather] than using traditional programming languages, which were thou[ght to] be too difficult, many SCEs used *visual programming languages* [so that] programs could be "written" by drawing flowchart-like diagrams.

Visual programming is not unique to IN. It's common in man[y tools] for developing telephony applications for CTI. Several factors ha[ve con]tributed to this tendency:

- These applications often use a limited and well-defined set of functions that can easily be diagrammed.

- Many people thought "programming with pictures" would be easier than programming with traditional languages.

- SCEs and other telephony development tools emerged at a time when there was a surge of academic and industrial interest in t[he] largely untested field of visual programming.

The telecom industry's experience with visual SCEs provided a real-world test of the assumptions and expectations of the visual programming community. On the positive side:

- Non-programmers found it easy to understand a visual representation of a service.
- Services could be developed quickly as long as they were relatively simple.

On the other hand, IN service developers leveled some serious criticisms at visual programming in general and SCEs in particular. Let's consider these one by one.

1. *It's harder to build large, complex services with visual languages than with traditional programming languages.* Pictures and text are suited to different tasks. Diagrams are good at showing the high-level organization and behavior of a system. A diagram may not be as compact on the page as text, but it conveys a lot of information in the space it occupies. For simple programs, visual representations of both high-level and detailed designs are of roughly the same complexity. Going from a pictorial design to a pictorial implementation is straightforward. Design and programming can indeed be united when dealing with simple programs.

Unfortunately, many people assumed the same benefits would apply to larger systems as well. As it turned out, when used for large programs, visual programming environments were far more difficult to use than expected. Real IN services are much more complicated than simple demonstration services that leave out billing, logging, error handling, and a multitude of other things programmers deal with in the real world. Once a service graph grows to hundreds—or even thousands—of nodes, browsing and editing with graphical tools becomes unwieldy.

Pictures of programs usually take up more space than text versions of the same program. The more details you have in your system, the more pictures it takes to represent them. At some point, the sheer number of pictures, the space they occupy, and the constant need to move from one picture to another start to conceal, rather than reveal, what is going on. Text-based languages are better than pictures for implementing large systems. Ironically, the visual environments that were supposed to make development easier, actually made it harder to build complex services.

2. *Visual programming encourages poor programming styles.* A few decades ago, the programming world was consumed by the question of `goto` statements. Should they be used or not? One camp felt they were an

essential programming tool. Another camp believed they were da[ngerous]
ous because they led to indecipherable and unmaintainable "spa[ghetti]
code." The structured programming movement resolved this confl[ict by]
encouraging the use of a small set of language constructs that pres[erve]
a programmer's ability to alter a program's flow of execution—ju[st like]
a `goto`—while allowing them to do that in only a few well-defined

Flowcharts faded from use at about the same time as structure[d pro-]
gramming arose. In part, this was because flowcharts directly mo[deled]
`goto` statements and did not map well to the new structured pro[gram-]
ming constructs. Visual programming environments that follo[w the]
flowchart model encourage an outmoded, error-prone, unstruc[tured]
programming style.

3. *SCEs restrict the kinds of services that can be developed*. Many visual
achieve simplicity not only by presenting programs as "picture[s," but]
also by limiting the things developers can do. For example, servic[e pro-]
grams may not be able to invoke software outside the SCE or use o[perat-]
ing system facilities, such as file systems and connections to dat[a net-]
works. The designers of these environments decide what develope[rs can]
do. That leaves fewer opportunities for mistakes, but it also assum[es the]
SCE's designers can anticipate everything developers need or want [to do.]
If they leave something out, nothing can be done about it wi[thout]
changing the SCE itself.

4. *SCEs are inflexible and difficult to extend*. There are no standar[ds for]
visual programming or SCEs. Every SCE is a proprietary system. If [there are]
problems, users are at the mercy of its vendor. The lack of standar[ds also]
makes it difficult—or even impossible—to integrate third-part[y pro-]
gramming tools, which are usually text oriented, into the develo[pment]
process.

**5. *Because SCEs are outside the programming mainstream, it's harder [to find]
and train developers for them*.** The worldwide number of IN service [devel-]
opers is quite small. The dream of using nonprogrammers to wr[ite ser-]
vices has gone nowhere. Services are programs, whether developed [with]
visual or text-based tools, and programmers remain an indispe[nsable]
resource for writing service programs. That means service provide[rs need]
to tap the pool of general programming talent for their service de[velop-]
ers. There they find many programmers familiar with tools an[d lan-]
guages, such as C, C++, and Java. These tools and languages have [broad]
appeal, because many companies and systems use them. It's hard f[or ser-]
vice providers to attract these programmers, because they lack th[e spe-]
cialized skills needed to work with SCEs and, what's worse, they h

tle interest in acquiring them, because few companies look for SCE experience when hiring.

13.2 JAIN and Parlay to the Rescue?

In the late 1990s, Sun started working with a number of industry partners to define Java APIs for SS7 and IN, an effort it called *JAIN*, or *Java APIs for Intelligent Networks*. At first, JAIN's focus was on APIs that sat on top of the protocols beneath IN, such as SS7 TCAP, ISUP, and INAP. Over time, JAIN expanded to include call control. Soon, IP telephony—SIP, H.323, MGCP, Megaco, and the like—also came under JAIN's umbrella. Because JAIN was no longer just for IN, the meaning of its name changed to *Java APIs for Integrated Networks*. Java's "write once, run anywhere" mantra was expanded to "write once, run anywhere, on any network." JAIN promised to bring a new level of flexibility to service creation.

At around the same time, in March of 1998, British Telecom (BT), Microsoft, Nortel, Siemens, and Ulticom formed the Parlay Group. Together, they defined a set of language-independent object-oriented APIs for building network services. It's not hard to see why some people thought Parlay and JAIN were competing with each other. Seeing that cooperation was better for all concerned, the two efforts formally joined forces in 2000 (although they had never really been rivals), and JAIN is now mapping the Parlay APIs to the Java language.

Although the two groups are now collaborating, JAIN and Parlay are not identical:

- JAIN is a set of Java APIs. Parlay is specified in such a way that it can be used with any language for which a translation from the Parlay definition language has been defined.

- JAIN is a complete stack, with APIs that range from protocols all the way up to services. Parlay is an API, the sole purpose of which is services. Implementations of JAIN's protocol APIs might sit below a Parlay implementation, but they do not thus become a part of Parlay.

Until recently, JAIN and Parlay also favored slightly different developer communities. While both wanted to make it easier to develop network services, Parlay focused on the needs of developers working outside the networks on which their applications would depend, while JAIN focused on the needs of trusted developers working for network [operators] or for companies that provide software directly to network [operators]. As a consequence, JAIN's starting point was APIs for call c[ontrol] and network protocols, while Parlay started with what are c[alled] its *framework APIs*. The Parlay framework allows applications [by] untrusted third parties to obtain varying degrees of access to a ne[twork] and its capabilities (which Parlay refers to as *services*).

The trusted developer model is the one currently used in the [industry] for AIN. This has, not surprisingly, limited AIN's popularity [in the] developer community at large. Realizing this, it's been propose[d by] those who propose such things,[1] that network operators might do [better] if they emulated the model of operating systems such as UNIX o[r Win]dows. These provide a well-defined API that any developer can [use to] write new applications. That means there are many more applic[ations] available for these platforms than there would have been if AT&T [(pre-di]vestiture) or Microsoft had kept these APIs to themselves.

This works fine for desktop workstations and PCs where the [conse]quences of a bad program are likely to be fairly well contained. [When] you apply the same model to public networks, it soon becomes ap[parent] that the consequences of a bad program could extend far beyond [a sin]gle user. They could extend all the way to an entire network or e[ven to] all networks that connect to it. Obviously, this is not acceptable. A[nd that's] why, despite its attractiveness, the idea of open APIs for public ne[tworks] has been a hard one to realize in practice. The Parlay framewor[k pro]vides security and authentication features that allow an applicat[ion to] run outside a service provider's "trusted space" while using ne[twork] facilities inside that trusted space. One of the first outcomes of th[e part]nership between JAIN and Parlay was, in fact, the addition of the [equiv]alent of Parlay's framework to the JAIN architecture.

Although JAIN and Parlay are closely related, there's enough [differ]ence to warrant a separate chapter on each. It's important to reali[ze that] both are still works in progress. Some parts have been more com[pletely] worked out and will get more attention than others that are still [under] construction. Furthermore, some of this material is almost cer[tain to] change, as JAIN and Parlay continue to evolve.

---

[1]Perhaps there's a "Great Proposer" akin to the "Great Mentioner" who turns up ev[ery four] years around the time of the U.S. presidential elections, mentioning names t[hat then] appear in news stories as potential candidates.

13.3 The JAIN API Stack

Like JTAPI, JAIN defines interfaces, not implementations. The JAIN APIs fall into two groups:

- The *JAIN Protocol APIs* define Java interfaces for wireline, wireless and IP signaling protocols, including TCAP, ISUP, INAP, MAP, OAM, MGCP, and SIP.
- The *JAIN Application APIs* define interfaces for creating Java services that use network protocols. These APIs cover areas such as call control, secure network access, service creation, and service execution.

Everything in JAIN is defined using the JavaBeans naming conventions (see Sec. 3.7). That means implementations of the JAIN APIs may be imported into JavaBeans tools. Applications that use the JAIN APIs may also be JavaBeans, but it's not required.

Figure 13.1 shows the full JAIN API stack and how it relates to networks and protocols. Before we go on to look at each of the elements of this stack, one thing should be clarified. As shown, it might seem JCC needs implementations of one or more of the JAIN protocol APIs beneath it, and that JSLEE needs an implementation of JCC, and so on up the stack. While one could build a complete JAIN stack that way, with each element using implementations of the JAIN APIs directly beneath it, that's not necessary. Thus, one could build an implementation of JCC that interacts with a TCAP implementation that's not JAIN TCAP. To the user of such a JCC implementation, it makes no difference whether what lies beneath is JAIN or not, as long as the JCC behaves correctly. The same goes, of course, for every other element of the stack.

Of course, if a JAIN stack includes JAIN implementations at multiple layers, software at the higher layers is free to make calls to methods in JAIN objects at the lower layers. Thus, for example, a program that uses JCC could make a direct call to JAIN ISUP or JAIN SIP to do something JCC doesn't support. The advantage is flexibility. The disadvantage is loss of network portability since such a program will only work on a network that has the specific facilities such as ISUP or SIP that are used by the program.

At the very top of the stack is a *JAIN Service Creation Environment (JSCE)*. Service developers use a JSCE to write JAIN service programs.[2]

---

[2] Or, more accurately, "will use" since JSCE is still being defined. To keep things consistent, I've chosen to use the present tense since that will (hopefully) become correct in time.

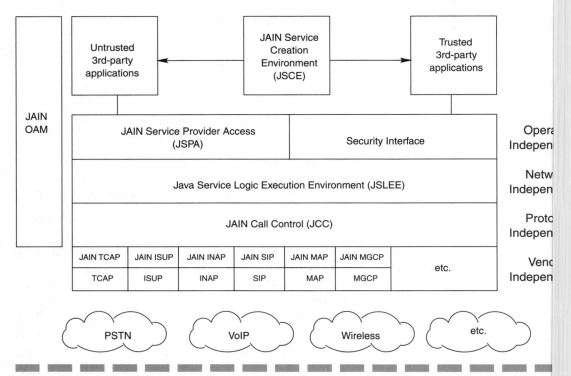

Figure 13.1
The JAIN API stack.

Like many AIN SCEs, a JSCE could be a visual programming e[n]viron]ment. It could also be a set of text-oriented tools that help a trad[itional] Java programmer write JAIN applications. Or it could combine [visual] programming with traditional programming, as is possible with [many] JavaBeans development tools. These are design decisions to be m[ade by] each JSCE vendor. However it's designed, the output of a JSCE is a [JAIN] application.

Below the JSCE are interfaces that allow JAIN applications to [run] in a live network while at the same time protecting that networ[k from] harm. Applications from trusted developers, who could be emplo[yees of] the network operator or third parties having a trust relationshi[p with] the network operator, interact directly with whatever security in[terface] sits in front of that network. This security interface is up to the n[etwork] operator and is not a part of JAIN. Security could be software bas[ed,] example, presenting credentials in the form of a digital signature. It could be based on secure physical access to a specific set of platforms that run JAIN applications, so that only trusted individuals can load new software on them.

Applications from untrusted third parties interact with the *JAIN Service Provider API (JAIN SPA)*, the JAIN implementation of Parlay's framework APIs, to gain the credentials they need for access to a particular network. Every network operator who supports JAIN services can provide a JAIN SPA interface. Because every operator provides the same interface (even though their implementations may differ), JAIN applications can run without change on any operator's JAIN SPA-enabled network—assuming the network provides everything the application needs. JAIN SPA is discussed at more length in the next chapter.

The next layer down is the *JAIN Service Logic and Execution Environment (JSLEE)*. This defines a standard environment within which JAIN services can be run. Like JAIN SPA, the interface to JSLEE is the same for every network: VoIP, PSTN, wireless, and so on. JSLEE is similar to the Enterprise JavaBeans (EJB) execution environment. Unlike regular JavaBeans, EJBs must run within something called a *container* that provides them with a consistent set of enterprise services, such as transaction processing and access management. JAIN applications have a similar need for a consistent set of services, so EJB containers are an obvious implementation option for JSLEE.

Below JSLEE is *JAIN Call Control (JCC)*, which, like JTAPI, provides a programmer's view of call processing. JCC includes everything needed for basic call processing and many features that enhance basic call processing. A JCC application can run without change on top of many telephony protocols. It thus offers protocol independence.

Until recently, JCC was paired with the *JAIN Coordination and Transactions (JCAT)* API. In fact, the two were frequently referred to jointly as *JCC/JCAT*. As work progressed on JCC, JCAT faded into the background as the working group decided to concentrate on call control APIs and leave the more complex issues of JCAT for later. At one point, in fact, the JCC working group's Web site said that references to JCC/JCAT should be interpreted as references to JCC alone. Early in 2001, when JCC 1.0 was completed, JCAT finally came into better focus as a new Java Specification Request (JSR) was approved for it. As specified in this JSR, JCAT will extend JCC to provide a higher degree of control over terminals and calls. Among the things it will add to JCC are AIN-like facilities, including the ability to invoke an application in the middle of a ca[...] use its results to direct subsequent call processing.[3]

The telephony protocols on which a JAIN application runs app[...] the lowest layer of the JAIN API stack. Among those now, or soon[...] supported are TCAP, ISUP, INAP, SIP, MGCP, MAP, H.323, SD[...] MEGACO. The protocol APIs give JAIN applications access to pr[...] stacks from many vendors. As shown in Figure 13.2, users of the pr[...]

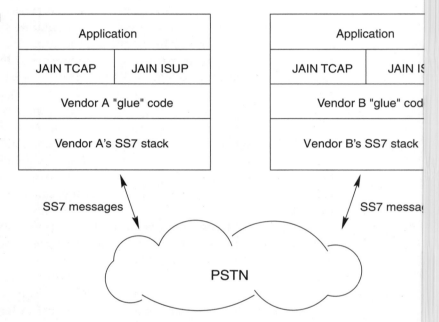

Figure 13.2
A JAIN application can use a protocol stack from any vendor who has built a JAIN implementation of that stack.

APIs see JAIN objects and methods, but the underlying imple[...] tions speak the language of the protocol they support.

There may be many JAIN implementations of a given prot[...] vendor that wants to put JAIN on top of a protocol they alrea[...] port must write "glue code" that connects their implementation [...] appropriate JAIN protocol API to their code for that protocol. W[...] vendor's protocol stack is written in a language other than Java, [...] done with the *Java Native Interface (JNI)*, a standard way of com[...] Java code with software written in other languages.

---

[3]It's also possible that, as stated in the new JSR, JCAT's name will be changed [...] describe what it's about.

Regardless of what's underneath, all JAIN implementations of a protocol present the same interface to users above, thus providing vendor independence. For example, if one vendor's JAIN TCAP implementation proves unsatisfactory, it should be possible to replace it with one from another vendor without having to rewrite any applications that sit on top of it.

Beneath the entire JAIN API stack are (of course) networks. To the extent two networks offer the same capabilities, these will be represented by the same classes and methods in the JAIN API stack. Thus, a JAIN application written for one network should run on another without change (assuming it doesn't use any specialized functions of either network).

Off to the side of all this is JAIN *Operations, Administration, and Management (OAM)*. JAIN OAM provides information and operations for provisioning and managing a JAIN protocol stack. Applications that use JAIN OAM can collect statistics and handle alarms. JAIN OAM 1.0 deals only with TCAP stacks. Future versions will cover all the JAIN protocol stacks as well as JCC. Because network operations is an entire subject to itself, and our focus is on software for developing services, JAIN OAM won't be covered in any more detail here.

13.4 JAIN's Development Process

The JAIN APIs are developed under Sun's *Java Community Process (JCP)*. The first draft of a particular API is written by an expert group drawn from companies active in that area. This draft goes through a *Community Review* by individuals and organizations that have signed the *Java Specification Participation Agreement (JSPA)*. Once that's finished, the draft is released for *Public Review*, after which it becomes a *Final Draft*. The last step in the process moves it to *Final Release* status. Vendors may then start to develop products based on the API (although many start before this, knowing full well that things may still change). APIs at the Final Release stage also have a reference implementation and a *Technology Compatibilty Kit (TCK)* that can be used to verify the correctness of an implementation of the API. As of this writing, there are more than 75 members of the JCP, working on over 20 new Java API specifications. This chapter concentrates on JAIN APIs (other than JAIN OAM) that had reached Public Review, Final Draft, or Final Release by early 2001.

13.5 JAIN Protocol APIs

This section covers a representative sample of JAIN's protocol
including JAIN TCAP, JAIN SIP, JAIN INAP, and JAIN MGCP. All h
common design framework. Once you've learned one, learning an
isn't too difficult. It's safe to assume that future protocol APIs wil
the same framework, especially with respect to their use of the fac
peer-provider and event-listener patterns.

13.5.1 JAIN TCAP

TCAP is an SS7 protocol that supports transactions in the PSTN
work elements and applications use TCAP to exchange inform
among themselves. It's also used when one network element v
another to execute an application on its behalf. JAIN TCAP puts a
API on top of TCAP. Because there's more than one version of
being used throughout the world, JAIN TCAP must offer a con
interface to them. A given implementation of JAIN TCAP takes c
these details for the version—or versions—it supports. Application
use the JAIN TCAP API should then be able to use different versio
TCAP without modification.

Most of JAIN TCAP is concerned with two activities:

- Getting TCAP messages from the network to a JAIN TCAP application, and vice versa.
- Giving a JAIN TCAP application access to the fields of a TCAP message.

A complete discussion of JAIN TCAP would require that we lo
every TCAP message and its corresponding classes and methods.
all follow essentially the same pattern, I'll show only a few repre
tive examples (Fig. 13.3). The interested reader may refer to the
TCAP specification (available at `http://java.sun.com/prodi
jain`) for more details.

SETTING UP A JAIN TCAP APPLICATION An applicatio
will process TCAP messages should implement the `JainTcapLis`
interface:

```
public class MyTcapListener implements JainTcapListener {
    private JainSS7Factory myFactory = null;
```

JAIN

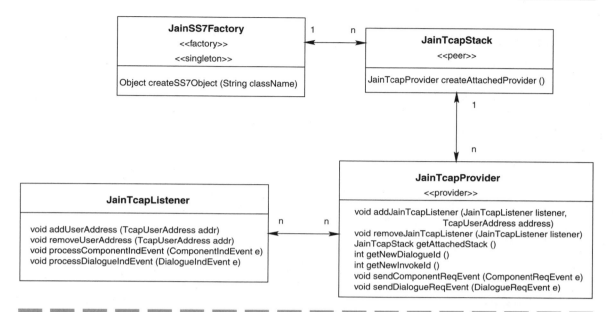

Figure 13.3
Core JAIN TCAP classes and some of their methods.

```
private JainTcapStack myStack = null;
private final byte mySignalingPointCode [3] = {255, 255, 254};
private final short mySubsystemNumber = 99;
private JainTcapProvider myProvider = null;
private TcapUserAddress myUserAddress = null;
```

`MyTcapListener` could be structured in many ways, but one of the easiest is to use a `main` method that creates an instance of `MyTcapListener`. That object then goes off and does its thing on behalf of the application:

```
public static void main(String[] args) {
  MyTcapListener myListener = new MyTcapListener();
}
```

So, the application has just created a `MyTcapListener` object by calling its constructor method. This constructor sets up the objects needed to exchange TCAP messages with the underlying SS7 stack. The structure of this code should already be familiar (see Fig. 13.4). It uses the same *factory-peer-provider* pattern as JTAPI, where a *peer* is a platform-specific implementation of a Java interface or API (JAIN TCAP here), a *factory*

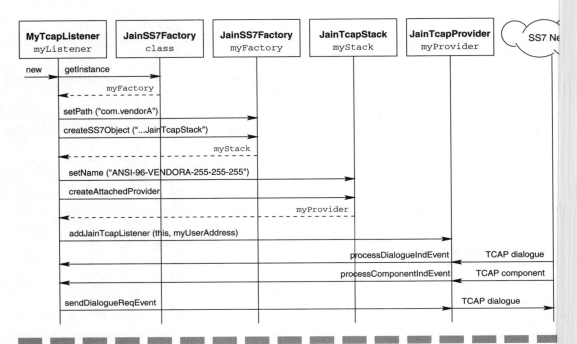

Figure 13.4
Object interactions in the example JAIN TCAP program.

creates instances of peers, and a *provider* supports a specific service
the platform-specific capabilities of the peer with which it is associ

```
public MyTcapListener () {
  myFactory = JainSS7Factory.getInstance();
  myFactory.setPath("com.vendorA");
  myStack = (JainTcapStack)myFactory.createSS7Object
              ("jain.protocol.ss7.tcap.JainTcapStack");
  myStack.setName("ANSI-96-VENDORA-255-255-255");
```

`JainSS7Factory` is a singleton class—there's only a single in
of it in the entire system—from which applications obtain JAIN
Here the application gets an instance of Vendor A's implementat
`JainTcapStack` from the `JainSS7Factory` and sets its name. T
ommended format for a stack's name is *specification-year-vendor-po*
where *pointcode* is formatted as *member-cluster-network*. Followin
recommendation, Vendor A's stack name is set to ANSI-96-VEN
255-255-255.

The application uses `myStack` to get an instance of `JainTcap-Provider` that's attached to `myStack`: (i.e., this code tells `myProvider` to use Vendor A's TCAP stack):

```
myProvider = myStack.createAttachedProvider();
```

As TCAP messages are sent and received, their dialogue and component primitives go through `myProvider`. The application registers as an event listener with `myProvider`, including its TCAP address (point code and subsystem number). When a TCAP message arrives, `myProvider` looks at the destination address. If it's the same as this application's TCAP address, `myProvider` creates event objects for each component and dialogue primitive in the message, and passes these to the application:

```
myUserAddress = new TcapUserAddress(mySignalingPointCode,
                                    mySubsystemNumber);
myProvider.addJainTcapListener(this, myUserAddress);
} // END constructor
```

RECEIVING TCAP MESSAGES As the SS7 software receives TCAP messages, it passes them up to Vendor A's implementation of JAIN TCAP. That software parses the messages into dialogue and component elements, which the provider objects send to their listeners as dialogue and component events. To send a dialogue event to the example program, `myProvider` calls the `processDialogueIndEvent` method in the application class that implements `JainTcapListener`:

```
public void processDialogueIndEvent(DialogueIndEvent event){
```

To process the event, this method has to figure out what kind of dialogue primitive it represents by calling its `getPrimitiveType` method. It then handles each dialogue primitive accordingly:

```
switch (event.getPrimitiveType()) {
```

If the event represents a BEGIN dialog primitive, the program converts[4] `event` to an object of type `BeginIndEvent`. Now it can get BEGIN information from, and do BEGIN things to the event object by calling methods in the `BeginIndEvent` interface. This code shows some of the things you can do, but it's not meant to be an example of what a real application would do. Other JAIN TCAP event objects have similar methods tailored to the TCAP primitive they represent:

---

[4]"Casts" in Java-speak.

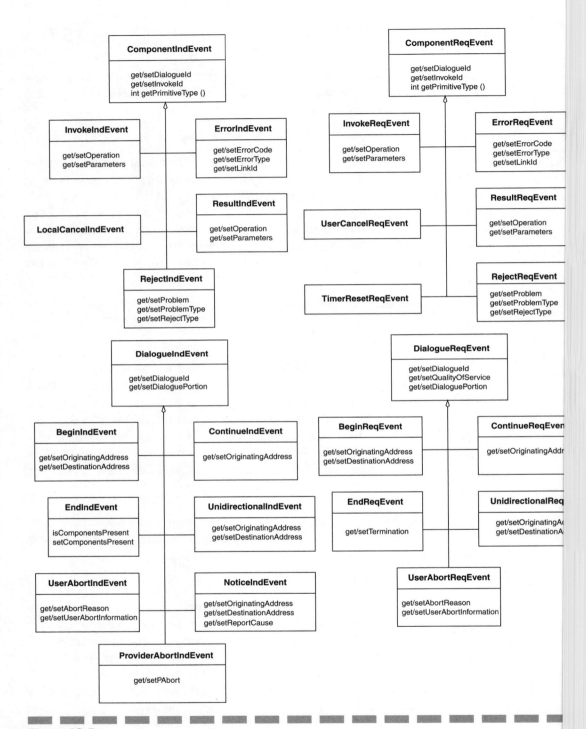

Figure 13.5
JAIN TCAP message classes and some of their methods.

TCAP Primitive	Received by Application	Sent by Application
BEGIN/Query	BeginIndEvent	BeginReqEvent
CONTINUE/Conversation	ContinueIndEvent	ContinueReqEvent
END/Response	EndIndEvent	EndReqEvent
ERROR	ErrorIndEvent	ErrorReqEvent
INVOKE	InvokeIndEvent	InvokeReqEvent
NOTICE	NoticeIndEvent	none
PROVIDER_ABORT	ProviderAbortIndEvent	none
REJECT	RejectIndEvent	RejectReqEvent
RESULT	ResultIndEvent	ResultReqEvent
TIMER_RESET	none	TimerResetReqEvent
UNIDIRECTIONAL	UnidirectionalIndEvent	UnidirectionalReqEvent
USER_ABORT	UserAbortIndEvent	UserAbortReqEvent
USER_CANCEL	none	UserCancelReqEvent

Figure 13.6
TCAP primitives and their corresponding JAIN TCAP event classes.

```
case TcapConstants.PRIMITIVE_BEGIN : {
  BeginIndEvent beginEvent = (BeginIndEvent)event;
  TcapUserAddress origination = beginEvent.getOriginatingAddress();
  TcapUserAddress destination = beginEvent.getDestinationAddress();
  if (beginEvent.isQualityOfServicePresent())
    byte qos = beginEvent.getQualityOfService();
  if (beginEvent.isComponentsPresent()) {
    // Do things to take account of the fact this dialog
    // primitive has associated components that will presumably
    // be handled when the provider calls processComponentIndEvent
    // in this application.
  }
  // Do other things
}
```

If the event represents another sort of dialog primitive, the program handles it similarly:

```
case TcapConstants.PRIMITIVE_CONTINUE : {
    // Cast to ContinueIndEvent and do stuff with it
  }
  case TcapConstants.PRIMITIVE_END : {
    // Cast to EndIndEvent and do stuff with it
  }
  case TcapConstants.PRIMITIVE_UNIDIRECTIONAL : {
    // Cast to UnidirectionalIndEvent and do stuff with it
  }
  default : {
    // Dialogue primitive either not recognized or we're
    // not prepared to handle it in this application.
  }
```

```
      } // END switch
    } // END processDialogueIndEvent
```

To send a component event to the application, `myProvider` ca[lls the] `processComponentIndEvent` method. Like `processDialogue`[Ind]`Event`, this method is defined in the application class that imple[ments] `JainTcapListener`. A real application would do far more th[an is] shown here. This example merely shows some of what's possible:

```
public void processComponentIndEvent(ComponentIndEvent event){
  switch (event.getPrimitiveType()) {
    case TcapConstants.PRIMITIVE_INVOKE : {
      InvokeIndEvent invokeEvent = (InvokeIndEvent)event;
      int dialogueId = invokeEvent.getDialogueId();
      int invokeId = invokeEvent.getInvokeId();
      Operation op = invokeEvent.getOperationType();
      if (op.getOperationType() == Operation.OPERATIONTYPE_LOC[AL]
        // this is a local/private operation
      }
      else {
        // this is a global/national operation
      }
      // Execute the Invoke primitive
    }
    case TcapConstants.PRIMITIVE_ERROR :
      {......};
    case TcapConstants.PRIMITIVE_REJECT :
      {......};
    case TcapConstants.PRIMITIVE_RESULT :
      {......};
    default : {
      // Component primitive either not recognized or we're
      // not prepared to handle it in this application.
    }
  } // END switch
} // END processComponentIndEvent
```

SENDING TCAP MESSAGES The program can send dialogu[e and] component primitives of its own via `myProvider`. It might do [this in] one of the `JainTcapListener` methods we've already seen, or it [might] do it elsewhere. Where it's done would depend on what the applica[tion is] trying to do. Regardless of where it's done, to send a TCAP reque[st, the] application creates event objects that represent the request's comp[onents] and sends these to `myProvider`. To start a dialogue, it first obtai[ns dia]logue (transaction) and invoke (component) IDs from `myProvider`[:]

```
int myDialogueId = myProvider.getDialogueId();
int myInvokeId = myProvider.getInvokeId(myDialogueId);
```

It then creates an instance of `Operation` and sets its operatio[n type,] family, and specifier:

```
Operation myOperation = new Operation();
myOperation.setOperationType
                (Operation.OPERATIONTYPE_GLOBAL);
myOperation.setOperationFamily
                (Operation.OPERATIONFAMILY_PARAMETER);
myOperation.setOperationSpecifier
                (OPERATION.PARAMETERSPECIFIER_PROVIDEVALUE);
```

It creates an instance of `InvokeReqEvent` using `myNewDialogueId` and `myOperation` and sets itself as the source of the event:

```
InvokeReqEvent myInvokeReqEvent = new InvokeReqEvent
                (this, myDialogueId, myOperation);
```

Finally, it sets values in the event and sends it to `myProvider`:

```
myInvokeReqEvent.setInvokeId(myInvokeId);
myInvokeReqEvent.setTimeout(5000);
myInvokeReqEvent.setLastInvoke(true);
byte parmval [1] = {MyApplicationConsts.AN_INTERESTING_PARAMETER};
Parameters parms = new Parameters
                (Parameters.PARAMETERTYPE_SINGLE,
                 parmval);
);
myInvokeReqEvent.setParameters(parms);
myProvider.sendComponentReqEvent(myInvokeReqEvent);
```

A similar process is used to create a dialogue request event and send it to `myProvider`:

```
TcapUserAddress destination = new TcapUserAddress(...);
BeginReqEvent myBeginRequestEvent =
    new BeginReqEvent(this, myDialogueId, myUserAddress, destination);
myProvider.sendDialogueReqEvent(myBeginRequestEvent);
```

13.5.2 JAIN SIP

SIP is an Internet protocol for establishing, modifying, and tearing down sessions. A phone call is just a particular kind of session, hence the importance of SIP for Internet telephony and convergent network services. JAIN SIP defines a Java API for SIP. Most of JAIN SIP is quite similar to JAIN TCAP. Its primary tasks include:

- Getting SIP messages from the network to a JAIN SIP program and vice versa.
- Giving a JAIN SIP program access to the fields of a SIP message.

Because the basic mechanisms of JAIN SIP and JAIN TCAP are so similar, I'll assume you're familiar by now with the factory-peer-provider and event listener patterns, and how they are used in JAIN TCA[P．An] application that uses JAIN SIP to process SIP messages implement[s the] `SipListener` interface. Notice how JAIN SIP has several different [kinds] of factory where JAIN TCAP has only one. The purpose of these factories will become clear shortly:

```
public class MySipListener implements SipListener
private SipFactory mySipFactory = null;
private AddressFactory myAddrFactory = null;
private HeaderFactory myHeaderFactory = null;
private MessageFactory myMsgFactory = null;
private SipStack mySipStack = null;
private SipProvider mySipProvider = null;
private ListeningPoint[] myListeningPts = null;
```

As in the JAIN TCAP example, this program has a `main` metho[d that] creates a `MySipListener` object named `myListener`. Then it [sends] a SIP invitation via the listener. Responses to that invitation w[ill be] reported back to `myListener`:

```
public static void main(String[] args) {
  MySipListener myListener = newMySipListener();
  myListener.sendInvitation();
}
```

The constructor for `MySipListener` sets up objects that exc[hange] SIP messages with the underlying SIP protocol software:

```
public MySipListener() {
```

Like `JainSS7Factory`, `SipFactory` is a singleton class from [which] applications obtain JAIN objects provided by a specific vendor. H[ere the] program gets an instance of Vendor B's implementation of `Sip[Stack`.] The program sets the stack's name, which for SIP can be any u[nique] string:

```
SipFactory mySipFactory = SipFactory.getInstance();
mySipFactory.setPathName("com.vendorB");
SipStack mySipStack = (SipStack)mySipFactory.createSipStack()
mySipStack.setStackName("Vendor B SIP stack");
```

A JAIN SIP stack has a set of `ListeningPoints` at which it ca[n listen] for SIP messages. Each `ListeningPoint` represents the combinat[ion of] a host, a port on that host, and a transport protocol, such as T[CP or] UDP, that's used on that port:

```
myListeningPts = mySipStack.getListeningPoints();
```

The program now gets an instance of `SipProvider`, telling it what `ListeningPoint` to monitor:

```
mySipProvider = mySipStack.createSipProvider
            ((ListeningPoint)myListeningPts.next());
```

Finally, the new `MySipListener` object adds itself to its provider. The provider will inform the listener when SIP events occur:

```
  mySipProvider.addSipListener(this);
} // END constructor
```

When a SIP request message arrives at a provider's listening point, the provider invokes the `processRequest` method of all its registered `SipListeners`. Application-specific code for handling requests goes in this method. It's passed a string containing a transaction ID and a `Request` object that represents the request message. Figure 13.7 shows some of the methods of a `Request` object. Since this is just an example, the program merely prints out some information about the request:

```
public void processRequest(String serverTransactionId,
                    Request request) {
  System.out.println("Request received with server transaction id"
                    + serverTransactionId + ":\n" + request);
} // END processRequest
```

The provider calls a listener's `processResponse` method when it gets a SIP response message. Here the listener sends an ACK if it gets a `200 OK` response to an invitation it sent earlier:

```
public void processResponse(String clientTransactionId,
                    Response response) {
  String method = response.getCSeqHeader().getMethod();
  int statusCode = response.getStatusCode();
  if((statusCode == Response.OK)
     && (method.equals(Request.INVITE))) {
    mySipProvider.sendAck(clientTransactionId);
  }
} // END processResponse
```

Sending a message is mostly a matter of filling in its fields. For example, the `sendInvitation` method below builds an invitation piece by piece, starting with the `From:` header. JAIN SIP makes dealing with all of SIP's different kinds of addresses fairly easy by putting the smarts needed to create them in an instance of `AddressFactory`. The program gets a SIP URL corresponding to the `From:` address by calling `createSipURL`, passing it the user ID and host of the originator, and then setting the address's port. The host and port are the same as those of this program's provider:

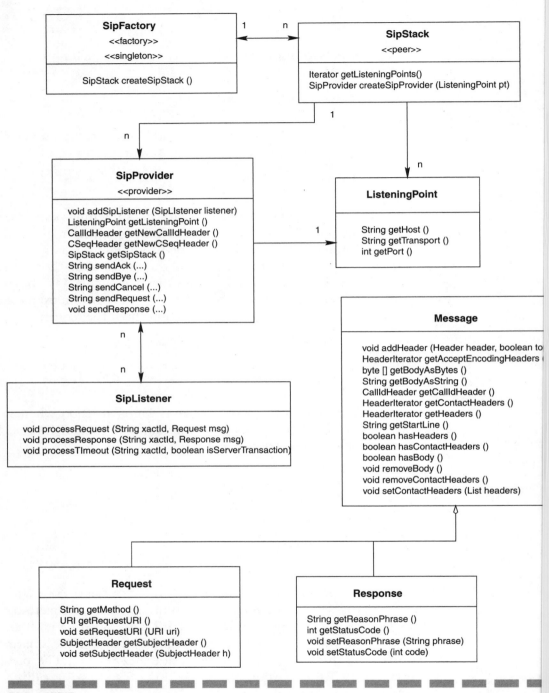

Figure 13.7
Core JAIN SIP classes and some of their methods.

JAIN

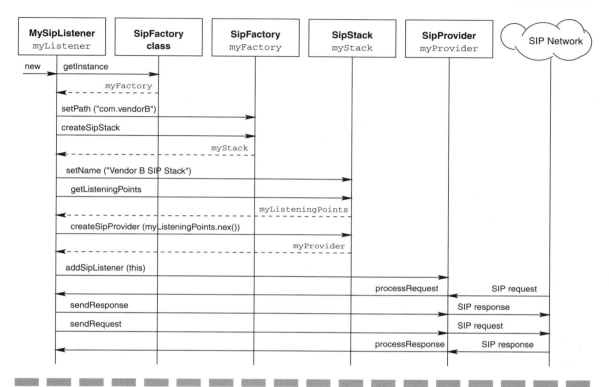

Figure 13.8
Object interactions in the example JAIN SIP program.

```
public void sendInvitation() {
  SipURL fromAddress = myAddrFactory.createSipURL("caller",
                    mySipProvider.getListeningPoint().getHost());
  fromAddress.setPort(mySipProvider.getListeningPoint().getPort());
```

Next, the program sets the calling party display name and address, again using the `AddressFactory`:

```
NameAddress fromNameAddress = myAddrFactory.createNameAddress
                                    ("Caller",fromAddress);
```

With the `From:` header set up, the program creates an object to represent it, this time using a `HeaderFactory` that knows all the peculiarities of `Header` objects:

```
FromHeader fromHeader =
          myHeaderFactory.createFromHeader(fromNameAddress);
```

Similarly, the program creates a `To:` header:

```
SipURL toAddress = myAddrFactory.createSipURL("callee",
                mySipProvider.getListeningPoint().getHost());
toAddress.setPort(mySipProvider.getListeningPoint().getPort());
NameAddress toNameAddress = myAddrFactory.createNameAddress
                                    ("Callee",fromAddress);
ToHeader toHeader = myHeaderFactory.createToHeader(fromNameAddress);
```

Followed by request URI, `CallId:`, `CSeq:`, `Via:`, and `Content Type:` headers:

```
SipURL requestURI = (SipURL)toAddress.clone();
requestURI.setTransport
            (mySipProvider.getListeningPoint().getTransport());
CallIdHeader callIdHeader = mySipProvider.getNewCallIdHeader();
CSeqHeader cSeqHeader = mySipProvider.getNewCSeqHeader(callIdHeader,
                                                fromHeader,
                                                toHeader,
                                                Request.INVITE);
ViaHeader viaHeader =
            myHeaderFactory.createViaHeader
               (mySipProvider.getListeningPoint().getHost(),
                mySipProvider.getListeningPoint().getPort(),
                mySipProvider.getListeningPoint().getTransport());
ArrayList viaHeaders = new ArrayList();
viaHeaders.add(viaHeader);
ContentTypeHeader contentTypeHeader =
     myHeaderFactory.createContentTypeHeader ("application", "sdp");
```

With all the invitation's pieces put together, the program u_ instance of `MessageFactory` to create the request message. Once done, the provider sends it out:

```
Request request = myMsgFactory.createRequest(requestURI,
                                Request.INVITE, callIdHe
                                cSeqHeader, fromHeader,
                                toHeader, viaHeaders,"bo
                                contentTypeHeader);
    String clientTransactionId = mySipProvider.sendRequest(requ
} // END sendInvitation
```

Perhaps the most important idea to be gleaned from this prog that, with minor exceptions, its overall structure is the same as t the JAIN TCAP example. That makes sense because both progra essentially the same thing: send and receive messages. Their bigge ferences are the message objects themselves. But the differences th mostly details of the underlying protocols. Message objects for bo tocols allow you to get and set message fields. JAIN TCAP m objects get and set TCAP message elements. JAIN SIP message get and set SIP message elements. A programmer who under

TCAP or SIP already should have little trouble learning JAIN TCAP or JAIN SIP. A programmer who doesn't know them will probably spend a lot more time learning the protocols themselves than learning their JAIN APIs.

13.5.3 JAIN MGCP

The structure of JAIN MGCP—and the rest of the JAIN protocol APIs—is so similar to that of JAIN TCAP and JAIN SIP, that I'll just go through highlights from here on. Most of the differences will be in the names in the factory-peer-provider and listener patterns, and the classes used to represent messages.

As with the other JAIN protocol APIs, a JAIN MGCP application implements a listener interface, `JainMgcpListener` in this case. The constructor for this listener gets a factory (`JainIPFactory`), a stack (`JainMgcpStack`), and a provider (`JainMgcpProvider`). The listener registers itself with the provider:

```
public class MyMgcpListener implements JainMgcpListener {
  JainIPFactory myFactory = null;
  JainMgcpStack myStack = null;
  JainMgcpProvider myProvider = null;
  public static void main(String[] args) {
    MyMgcpListener myListener = new MyMgcpListener();
    myListener.sendAuditMessage();
  }
  public MyMgcpListener () {
    myFactory = JainIPFactory.getInstance();
    myFactory.setPath("com.vendorA");
    myStack = (JainMgcpStack)myFactory.createIPObject
              ("jain.protocol.ip.mgcp.JainMgcpStack");
    myStack.setProtocolVersion("1.0");
    myProvider = myStack.createProvider();
    myProvider.addJainMgcpListener(this, myUserAddress);
  } // END constructor
```

The listener provides methods the provider should call when it receives messages be passed on to the listener. An MGCP listener provides separate methods for processing MGCP command and response messages:

```
void processMgcpCommandEvent (JainMgcpCommandEvent event) {
  // do stuff
} // END processMgcpCommandEvent
void processMgcpResponseEvent (JainMgcpResponseEvent event) {
  // do stuff
} // END processMgcpResponseEvent
```

A JAIN MGCP program can send messages. JAIN MGCP c
encapsulate all the various MGCP messages and provide useful con
for building those messages (Fig. 13.9). The code here builds an M
AuditConnection command and sends it via the provider:

```
void sendAuditMessage () {
   AuditConnection event [1];
   InfoCode info [2];
   EndpointIdentifier end = new EndpointIdentifier
                               ("hrd4/56", "gw23.example.net");
   info[0] = InfoCode.ConnectionMode;
   info[1] = InfoCode.ConnectionParameters;
   event[0] = new AuditConnection (this, end,
                              ConnectionIdentifier.AllConnections,
                              info);
   myProvider.sendMgcpEvents (events);
} // END sendMessages
```

13.5.4 JAIN INAP

As a member of the SS7 family of protocols, JAIN INAP shares
similarities with JAIN TCAP. As with all the other JAIN protoco
an application that uses JAIN INAP to send and receive INAP m
implements a listener interface:

```
public class MyInapListener implements JainInapListener {
```

Initializing a JAIN INAP application involves the usual factor
provider steps to obtain an INAP software vendor's factory, stac
provider. Once that's taken care of, the listener registers wi
provider:

```
JainSS7Factory myFactory;
JainINAPStack myStack;
JainInapProvider myProvider;
void MyInapListener () {

   JainSS7Factory myFactory = JainSS7Factory.getInstance();
   myFactory.setPathName("com.vendorC");
   myStack = (JainINAPStack)myFactory.createSS7Object(); }
   myProvider = myStack.createProvider();
   byte appcontext [128];
   myProvider.addJainInapListener(this, appcontext);
}
```

Sending INAP messages is easy (if you know INAP). You u
appropriate JAIN INAP classes to build up a message piece by pie
field by field. This example sends an Initiate Call Attempt messag

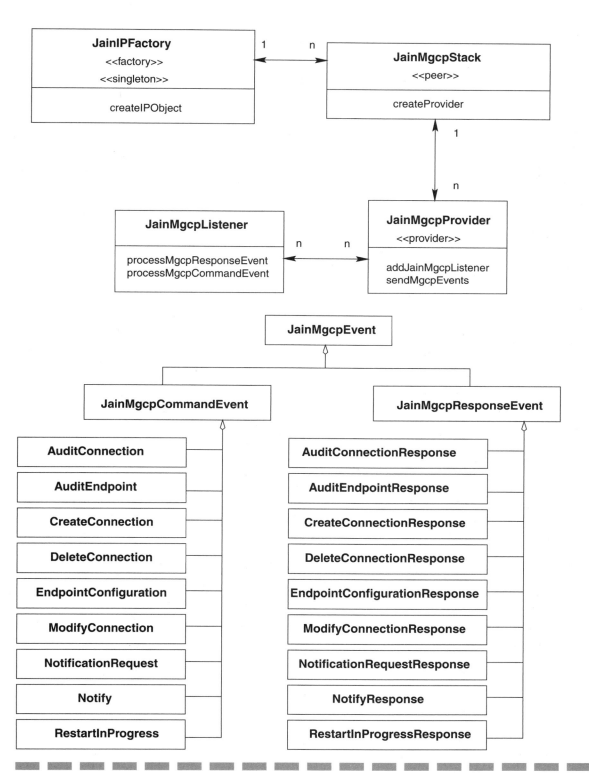

Figure 13.9
JAIN MGCP classes and some of their methods.

```
void sendInitiateCallAttempt (void) {
  int myNewCallID = myProvider.getCallId();
  CalledPartyNumber[] calledNumber = new CalledPartyNumber[1];
  int natureOfAddrIndicator = 1;
  int internalNwNumIndicator = 1;
  int numPlanIndicator = 1;
  String addrSignal = "5432156686666687";
  calledNumber[0]= new CalledPartyNumber(natureOfAddrIndicator,
                                  internalNwNumIndicator,
                                  numPlanIndicator, addrSignal)
  DestinationRoutingAddress destRteAddr =
                  new DestinationRoutingAddress(calledNumber);
  InitiateCallAttempt initCallAttempt =
                  new InitiateCallAttempt(destRteAddr);
  InvokeReqEvent invokeReqEvent =
                  new InvokeReqEvent(initCallAttempt);
  msg = new Vector;
  msg.add(invokeReqEvent);
  byte appcontext [128];
  InapSendEvent inapSendEvent =
                  new InapSendEvent(destinationAliasAddress,
                                  mynewCallID,
                                  appcontext,
                                  msgs);
  int invokeId = myProvider.sendINAPEvent(inapSendEvent);
}
```

The provider passes INAP messages it receives to registered listeners INAP events. Listeners implement corresponding methods of the `InapListener` interface to handle these. This next example handles response to the message sent by `sendInitiateCallAttempt`:

```
void processInapContinueEvent(InapContinueIndEvent event) {
  Vector msgs = new Vector();
  msgs = event.getMessageObjects();
  for ( i=0;i=msgs.size();i++) {
    ReceiveMessageObject msg = (ReceiveMessageObject)msgs.elementAt(i);
    switch(msg.getComponentPrimitiveType()) {
      case ComponentPrimitiveType.INVOKE : {
        InvokeIndEvent invokeIndEvent = (InvokeIndEvent)msg;
        Object opClass = invokeIndEvent.getMessage();
        switch((Continue)(opClass.getOperationCode())){
          case OperationCode.EVENT_REPORT_BCSM : {
            EventReportBCSM evrepBcsm = (EventReportBCSM)opClass;
            int evtypeBcsm = evrepBcsm.getEventTypeBCSM();
            // ...Now that we know what kind of message we have,
            // do application-specific stuff with it.
          }
          // ...Handle other opcodes
        } // END switch
      }
      // ... Handle other primitive types
    } // END switch
  } // END for
} // END
```

13.6 JAIN Call Control (JCC) API

JAIN Call Control (JCC) is the JAIN API for writing applications that manipulate phone calls. JCC supports both first-party and third-party call control. It's similar to JTAPI, after which it was modeled. But where JTAPI targets services for enterprise networks, JCC targets services for public networks. JCC's designers wanted to simplify and modernize the process of building IN services by moving it into the Java world. At the same time, they hoped that moving network software development to Java would expand the scope of IN services because so many more APIs and platforms are available there than with traditional IN. JCC is designed to work with any network or combination of networks and protocols in which sessions can be created, modified, or destroyed: circuit-switched (wireline and wireless) or packet. This is a particularly important way in which JCC differs from IN, which has little or no view of anything beyond traditional phone networks.

JCC is divided between two packages:

- *Java Call Processing (JCP)*. `java.application.services.jcp` supports the most elementary aspects of call control: making and answering calls.
- *Java Call Control (JCC)*. `java.application.services.jcc` adds facilities for observing and manipulating calls in progress, and for calling other applications during call processing.

The fact that JCC refers to both an API and a specific package within that API is unfortunate. From here on, I use "JCC" to refer to the package, and "JCC API" to refer to the entire API, consisting of both the JCC and the JCP packages.

13.6.1 Core JCP and JCC Classes

The core classes of the JCP and JCC packages are modeled after those of JTAPI. This can easily be seen by comparing Figs. 13.10 and 12.6. As with JTAPI, the JCP and JCC core classes have state machines that represent the state of a call. Also like JTAPI, JCP and JCC applications can register with providers as listeners to learn about and respond to events of interest during call processing. The biggest difference between JTAPI and JCC is that the JCC API has no classes that correspond to JTAPI's `Terminal` and `TerminalConnection`:

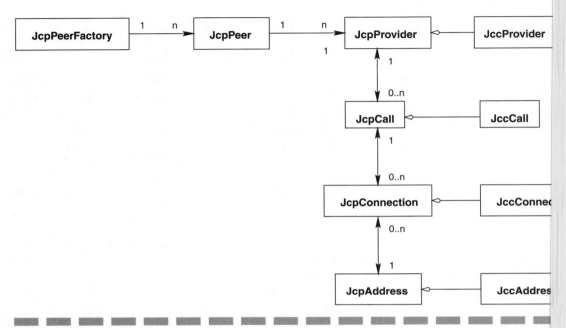

Figure 13.10
JCP and JCC core classes.

- A `JcpFactory` provides applications with instances of `JcpPeer` objects.
- A `JcpPeer` represents a specific implementation of JCP or JCC. Applications obtain `JcpProvider` and `JccProvider` objects from `JcpPeers`.
- A `JxxProvider`[5] represents call processing on behalf of an application. A `JcpProviderListener` registers with and listens for events from a `JxxProvider`. There is no `JccProviderListener`. A `JcpProviderListener` can listen to both `JcpProviders` and `JccProviders`.
- A `JxxCall` is a changing collection of call endpoints. A `JcpCallListener` registers with and listens for events from a `JcpCall`. A `JccCallListener` registers with and listens for events from a `JccCall`. (See Figs. 13.11 and 13.12.)

---

[5]Where `Jxx` can be either `Jcp` or `Jcc`.

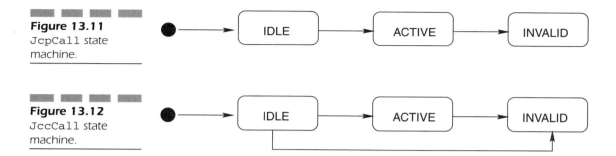

Figure 13.11 JcpCall state machine.

Figure 13.12 JccCall state machine.

- A `JxxAddress` is a logical endpoint of a call, for example, a phone number or IP address.
- A `JxxConnection` represents a dynamic relation between a `JxxAddress` and a `JxxCall` during call processing. If a `JxxAddress` is part of a `JxxCall`, a `JxxConnection` object will exist to connect the two for as long as that `JxxCall` exists and that `JxxAddress` is participating in it. A `JxxConnection` may belong to only one `JxxCall`. A `JcpConnectionListener` registers with and listens for events from a `JcpConnection`. A `JccConnectionListener` registers with and listens for events from a `JccConnection`. (See Fig. 13.13.)

13.6.2 JCP and JCC States

The JCC and JCP core classes have state machines similar to those of their corresponding JTAPI classes. The state machines in JCP are usually simpler than those in JCC. The `JcpCall` state machine is identical to that of JTAPI's `Call`. The state machine for `JccCall` is similar, but it adds a transition from IDLE to INVALID. JCC and JCP call states are similar to those in JTAPI:

IDLE When a `JxxCall` object is created, it enters the IDLE state. It has no `JxxConnections` associated with it.

ACTIVE Once a `JxxCall` has one or more `JxxConnections`, it enters the ACTIVE state and becomes the target of application activity.

INVALID This is the final state of a `JxxCall`. It has no `JxxConnections` because they've all moved to their DISCONNECTED state. It may no longer be the subject of any more actions by its application.

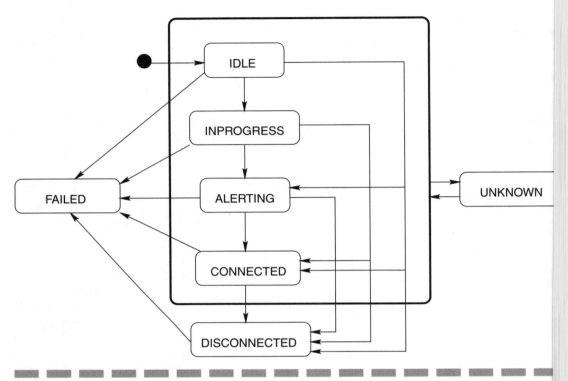

Figure 13.13
JcpConnection state machine.

The JcpConnection state machine in Fig. 13.13 is identical to t
JTAPI's Connection. The JCP call states are similar to those in JT

IDLE Newly created JcpConnections start in the IDLE stat
can't do much with an IDLE JcpConnection. It usually repre
party that has just joined a call, and it quickly moves to one
other JcpConnection states.

INPROGRESS A JcpConnection in this state has been contac
the calling party or is contacting the called party by exchangin
sages of the underlying protocol (SIP, SS7, and so forth). It's po
JcpConnection may not move beyond this state.

ALERTING The JcpAddress associated with a JcpConnect
this state is being alerted.

CONNECTED The JcpConnection and its associated JcpAd
are active participants in a call, i.e, parties are connected and tal

DISCONNECTED The `JcpConnection` used to be in a call but is no longer participating. However, its references to `JcpAddress` and `JcpCall` remain valid.

FAILED The `JcpConnection` has failed for some reason, perhaps because the party it represents was busy. It's still connected to the call.

UNKNOWN The implementation of the JCC API is unable to determine the current state of a `JcpConnection`. A `JcpConnection` may move into or out of this state at any time. Methods cannot usually be invoked on a `JcpConnection` while it is in the UNKNOWN state.

The state machine for `JccConnection` in Fig. 13.14 is quite different from that for `JcpConnection`. It shows how IN influenced the design of JCC. Beneath the extra detail, it is actually similar to the `JcpConnection` state machine. All that was done was to add some extra transitions and split certain `JcpConnection` states into more than one state. Because `JccConnection` extends `JcpConnection`, a `JccConnection` object has both a `JccConnection` and a `JcpConnection` state:

IDLE Same as for `JcpConnection`. The `JcpConnection` state when a `JccConnection` is in this state is also IDLE.

AUTHORIZE_CALL_ATTEMPT The calling or called party needs to be authorized before the call can be placed. As shown in Fig. 13.14, an originating `JccConnection` moves to ADDRESS_COLLECT or ADDRESS_ANALYZE. A terminating `JccConnection` moves to the CALL_DELIVERY or ALERTING state. The corresponding `JcpConnection` state when a `JccConnection` is in this state is INPROGRESS.

ADDRESS_COLLECT The calling party has been authorized, and at least some called party address information has been collected. As information is collected, it's examined to see if any more is needed. The calling party leaves this state and goes to ADDRESS_ANALYZE once all information has been collected or collection has failed because of timeout or other errors. The corresponding `JcpConnection` state is INPROGRESS.

ADDRESS_ANALYZE The calling party has entered enough address information for the program to determine a route for the call. The calling party leaves this state once it finds a usable route for the call, when it moves to the CALL_DELIVERY state. It also leaves this state when it finds it cannot route the call because there are no available routes, the address information had errors, the calling party has

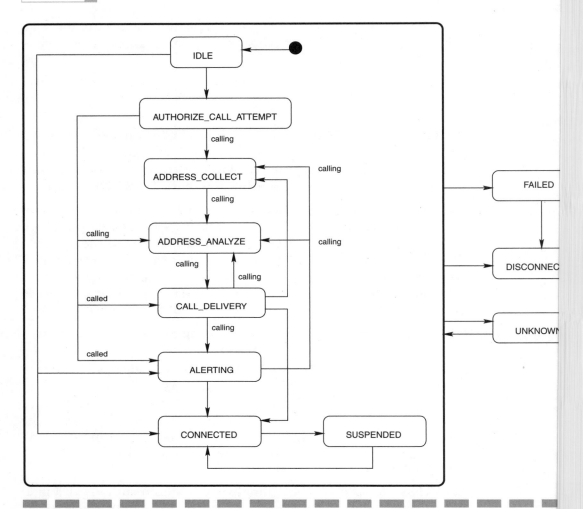

Figure 13.14
JccConnection state machine.

abandoned the call, and so forth. The corresponding JcpConne
state is INPROGRESS.

CALL_DELIVERY On the calling side, a routing address and ca
are available. Call setup messages are sent to the called end. O
called-party side, a call to that address has been authorized, a
checked to see if it's busy or idle. The corresponding JcpConne
state is INPROGRESS.

ALERTING The called party is being alerted. The corresponding `JcpConnection` state is ALERTING.

CONNECTED The called party has answered. The corresponding `JcpConnection` state is CONNECTED.

DISCONNECTED Same as for `JcpConnections`. The corresponding `JcpConnection` state is DISCONNECTED.

SUSPENDED The `JccConnection` object is suspended from the call, for example, when the party it represents flashes the hookswitch. The associated `JccCall` and `JccAddress` remain valid. The corresponding `JcpConnection` state is CONNECTED.

UNKNOWN Same as for `JcpConnections`. The corresponding `JcpConnection` state is UNKNOWN.

13.6.3 JCC Call Processing

OUTGOING CALL The first example of JCC is a program that applies first-party call control on the originating side of a connection. To be able to receive events once it launches the call, it has to be a listener of some sort. It implements the `JccConnectionListener` interface so it can hear about both `JccCall` and `JccConnection` events:

```
public class MyOutgoingListener implements JccConnectionListener
```

The `main` routine gets the peer and provider, creates a listener, and registers it with the provider. The `EventFilter` passed to `addConnectionListener` is something that must be supplied whenever a JCC listener registers with an event generator. When an event occurs, the event generator uses that filter to see if the event should be sent to the listener. JCC implementation provides several standard filters. To keep things simple, the code here merely hints at all of that. Once the listener has registered, the program creates a call with no connections and routes it from the calling to the called address. The null arguments in the call to `routeCall` are optional. If they're not null, they contain the first called address for this call and the last address from which it was redirected. The biggest difference between this and the corresponding JTAPI code is that JTAPI makes a call from a specific address on an originating terminal to a terminating address. JCC makes calls from one address to another and has no notion of terminals:[6]

---

[6]Exception handling try-catch blocks have been omitted in this and all other examples.

```
public static void main (String [] args) {
  MyJcpPeer peer =
      (MyJcpPeer)JcpPeerFactory.getJcpPeer ("my.jcp.MyJcpPeer
  Provider provider = peer.getProvider ("MyJccProvider");
  MyOutgoingListener listener = new MyOutgoingListener ();
  EventFilter ef = provider.createEventFilter...(...);
  provider.addConnectionListener (listener, ef);
  JccCall call = provider.createCall ();
  String calledAddr = "5125551234";
  String callingAddr = "5125554321";
  call.routeCall (calledAddr, callingAddr, null, null);
}
```

Because `MyJccConnectionListener` implements `JccConnect Listener`, it must provide methods to handle all the events a `ConnectionListener` might receive. Only a few of these are ac shown here. The first of these, `connectionCallDelivery`, is when the originating `JccConnection` moves to the CALL_DELI state:

```
public void connectionCallDelivery (JccConnectionEvent event)
  System.out.println("Call has been deliver\n");
}
```

The second, `connectionConnected`, is called when the origi `JccConnection` moves to the CONNECTED state. This mear called end has answered and the two parties may now talk. Just to a bit more being done with the call, this method forces the calling off after 60 seconds:

```
public void connectionConnected (JccConnectionEvent event) {
  System.out.println ("You may now talk for one minute\n");
  wait (60000);
  JccConnection conn = (JccConnection)event.getConnection();
  conn.release();
}
```

INCOMING CALL The next example shows first-party call on the terminating side. As before, this program implements the `ConnectionListener` interface:

```
public class MyIncomingListener implements JccConnectionListe
```

The `main` routine does pretty much what it did for the outgoi example, up to the point where it registers a listener with the p After that, it doesn't do anything more and just waits for an i ing call:

```
public static void main (String [] args) {
  MyJcpPeer peer =
```

```
            (MyJcpPeer)JcpPeerFactory.getJcpPeer ("my.jcp.MyJcpPeer");
        Provider provider = peer.getProvider ("MyJccProvider");
        MyIncomingListener listener = new MyIncomingListener ();
        EventFilter ef = provider.createEventFilter...(...);
        provider.addConnectionListener (listener, ef);
    }
```

When a call arrives, the called party is alerted and the provider calls the listener's `connectionAlerting` method, which displays a message that tells the called party there is a call and lets them answer it:

```
public void connectionAlerting (JccConnectionEvent event) {
  System.out.println("Call from "
                + event.getConnection().getAddress().getName()
                + ".Press any key to answer.\n");
  System.in.read ();
  event.getConnection().answer();
}
```

THIRD-PARTY CALL CONTROL The next program uses third-party call control to make two outgoing calls and join them together. As in all the other examples, this program implements the `JccConnectionListener` interface:

```
public class MyJoinCallListener implements JccConnectionListener {
    String party1 = "5125551234";
    String party2 = "5125554321";
    JccConnection conn1 = null;
    JccConnection conn2 = null;
```

The `main` routine gets a peer and a provider, creates a listener, registers it with the provider, creates a call object with no connections, and uses that object to place a call to the first party. By using the same address in the calling and called address arguments to `routeCall`, the program does a third-party call setup to that address alone:

```
    public static void main (String [] args) {
        MyJcpPeer peer =
            (MyJcpPeer)JcpPeerFactory.getJcpPeer ("my.jcp.MyJcpPeer");
        Provider provider = peer.getProvider ("MyJccProvider");
        MyJoinCallListener listener = new MyJoinCallListener ();
        EventFilter ef = provider.createEventFilter...(...);
        provider.addConnectionListener (listener, ef);
        JccCall call = provider.createCall ();
        conn1 = call.routeCall (party1, party1, null, null);
    }
```

The provider calls `connectionConnected` when either party answers. If `conn2` is null, this is an answer from the first party. That means the program may now place a call to the second party. When answered, the second party is connected to the first party (that'[s wh]y the program includes both `party2` and `party1` in the argume[nts to] `routeCall`). If `conn2` is not null, this is an answer by the second [party,] and there's nothing more to be done, because `routeCall` has a[lready] connected the two parties:

```
public void connectionConnected (JccConnectionEvent event) {
  if (conn2 == null) {
    JccCall call = (JccCall)event.getCall();
    conn2 = call.routeCall (party2, party1, null, null);
  }
}
```

PRIVATE DIALING PLAN The final JCC example in this c[hapter] shows how a private dialing plan—for example, four-digit dialing [in an] office building—could be implemented with JCC:

```
public class MyDialingPlanListener implements JccConnectionLis[tener]
```

The first few steps of the `main` routine should be familiar by n[ow. It] gets a peer and a provider and creates a listener:

```
public static void main (String [] args) {
  MyJcpPeer peer =
      (MyJcpPeer)JcpPeerFactory.getJcpPeer ("my.jcp.MyJcpPeer[");]
  Provider provider = peer.getProvider ("MyJccProvider");
  MyJoinCallListener listener = new MyDialingPlanListener ();
```

So far, I've done a lot of handwaving where event filters wer[e con]cerned. That won't do here, since event filters are what makes th[e pro]gram work. The next few lines of code call the provider's [cre]ateEventFilterEventSet method to get one of the standard [event] filters. The two arrays passed to that method specify events a reg[istered] listener wants to hear about. When the provider sees an ev[ent in] `blockEvents`, it informs the listener and waits for the listener to [finish] handling the event before continuing with call processing. Wh[en the] provider sees an event in `notifyEvents`, it informs the listen[er but] returns immediately to call processing without waiting for the l[istener] to handle the event. When the provider sees an event that's in nei[ther of] these arrays, it does nothing and just continues with call processin[g.]

```
    int blockEvents[] = {JccConnectionEvent.CONNECTION_ADDRESS_ANALYZE};
    int notifyEvents[] = {JcpConnectionEvent.CONNECTION_ALERTING,
                          JcpConnectionEvent.CONNECTION_FAILED};
EventFilter ef1 = provider.createEventFilterEventSet(blockEvents,
                                                     notifyEvents);
```

Next the program creates a filter that checks the endpoints of events before sending them to the listener. Any event with an endpoint address in the range from `lowAddr` to `highAddr` will be handled as indicated by `match`. Any event with an address outside that range will be handled as indicated by `nomatch`. As written below, the program provides four-digit dialing for any phone in the range 512-555-4000 to 512-555-9999. Any event with an address in that range causes the provider to inform the listener and pause call processing. Any other event is discarded and call processing continues as usual:

```
String lowAddr = "5125554000";
String highAddr = "5125559999";
int match = EventFilter.EVENT_BLOCK;
int nomatch = EventFilter.EVENT_DISCARD;
EventFilter ef2 = provider.createEventFilterAddressRange(lowAddr,
                                                         highAddr,
                                                         match,
                                                         nomatch);
```

The last bit of event filter hocus-pocus combines `ef1` and `ef2` into a single event filter. Consequently, only events that pass through both filters make it to the program:

```
EventFilter filters[] = {ef1,ef2};
int nomatch = EventFilter.EVENT_DISCARD;
EventFilter ef = provider.createEventFilterAnd(filters, nomatch);
```

The program registers the listener and filter, creates a call with no connections, and makes a call from one member of the private dialing plan to another:

```
    provider.addConnectionListener (listener, ef);
    JccCall call = provider.createCall ();
    String calling = "5125555413";
    String called = "9876";
    call.routeCall (called, calling, null, null);
} // END main
```

The provider processes the call. Behind the scenes, all sorts of things happen up to the point in call processing where the originating connection's state machine enters the ADDRESS_ANALYZE state. This triggers the event filter's blocking event condition. Because the originating address of 512-555-5413 is in the filter's range, the event passes that condition also. The provider calls the listener's `connectionAddressAnalyze` method. That method translates the dialed four-digit address to an actual network address. Using the `selectRoute` method, it puts the actual address into the originating connection in place of the address. Once that's done, call processing continues as usual:

```
private String translate (String dialed) {
  // Code to translate dialed digits to an actual address
  // goes here.
}

public void connectionAddressAnalyze (JccConnectionEvent event
  JccConnection origConn = (JccConnection)event.getConnection(
  String dialedDigits = origConn.getDestinationAddress();
  String actualAddr = translate (dialedDigits);
  origConn.selectRoute (actualAddr);
}
```

13.7 For More Information

The official source for JAIN information is Sun's Web http://java.sun.com/products/jain. JAIN specifications n obtained here as they become public. The IEEE has published s papers on JAIN, including Refs. [3], [4], [13], and [63]. The Java Con ity Process and activities being carried out under it are descril more length on the JCP Web site, http://www.jcp.org.

CHAPTER 14

Parlay

Could you parlay a bit quicker? This guy here's gonna make me a goner!

(A Dungeons and Dragons Web page)

Hinky, dinky, parley-voo.

(World War I soldier's song)

14.1 Overview

The Parlay Group was formed in March 1998, when British Te(BT), Microsoft, Nortel, Siemens, and Ulticom (formerly DGM&S) sworking together to define a set of language-independent, ooriented APIs for network service applications. When the Parlaywas first organized, membership was by invitation only. It hasmoved to open membership and grown to more than 40 full andate members. It has also established significant liaisons with JAIN,ETSI, and the OMG. The first version of the Parlay APIs, Parlay 1published in December 1998. Since then, the Parlay Group has cued to expand and revise the APIs, releasing Parlay 2.0 in 2000, anlay 3.0 in 2001. At this writing, work on Parlay 4 has just begun.

The Parlay APIs are divided into two groups, *framework APIs* a*vice APIs*.

- *Framework APIs*. The framework APIs provide a secure environment—often referred to simply as the *framework*—withiwhich Parlay applications can be run. Parlay's Phase 2 framewoffers the following capabilities:[1]
    - The Parlay framework gives applications and networks a way authenticate themselves to each other. It's not hard to see why network would want to check the identity of an application t wants to use its facilities. It may not be obvious why an applic would want to authenticate a network it's going to use until y remember that an application may need to hand over persona information about its users to a network. That means Parlay authentication has to be a two way street.
    - Applications can use the framework to discover network capa (usually referred to in Parlay as *services*).
    - Applications can subscribe to services they have discovered.
    - The framework can notify applications when network events
    - The framework manages network and application integrity.
- *Service APIs*. The service APIs give applications access to the resources and capabilities (i.e., services) of a Parlay-enabled netw

    Call Control Services provide call processing for generic, multipar conference calls. Generic calls have only two parties that com cate through a single medium. Multiparty calls have many p

---

[1] Phase 1 is versions 1.0, 1.1, and so forth. Phase 2 is versions 2.0, 2.1, and so forth.

communicating over many media. Conference calls are multiparty calls that add features such as subconferences, reservations, chair selection, and speaker selection.

User Interaction Services provide tools to get information to and from a caller.

Messaging Services support messages, mailboxes, and mailbox folders.

Mobility Services provide information about user location and status (for example, reachable, unreachable, busy).

Parlay 3 and 4 introduce new framework capabilities and service APIs, for example, policy management and presence and availability management services. Given the size and scope of the Parlay API, describing all of Parlay 2—much less Parlay 3 and 4—would take an entire book in itself. Consequently, we can look at only a small subset of Parlay's many capabilities in this chapter.

As we've already seen, JAIN and Parlay are closely related. Both provide object-oriented APIs for developing convergent network services, and JAIN supports a Java implementation of Parlay. However, it's worth pointing out (again) how they differ. These differences are more matters of approach and emphasis than of basic philosophy. The intent of both Parlay and JAIN is to open network service programming to a wider body of developers and to encourage applications that span networks:

- Unlike JAIN, Parlay has no APIs for talking directly to network protocols such as TCAP and SIP. It assumes protocols like these are present but, like JCC, it has no explicit view of them. An implementation of Parlay connects the APIs that programmers see to the real world of hardware and protocols. Such an implementation might use the JAIN protocol APIs. It might use something else. It might use a mix of JAIN and other protocol stacks. It's all the same to Parlay, because Parlay's focus is entirely on services.

- While both JAIN and Parlay can be used to create network services, a major aim of Parlay is to provide an environment in which third parties can write applications that run outside a service provider's "trusted space," even while using facilities inside that trusted space. Security, authentication, and policy management are thus important elements of Parlay, elements that were introduced into JAIN via Parlay.

- Parlay defines a language-independent API. Implementations can be written for many languages, such as C++, C, and Java. JAIN is a Java-only API.

- Because JAIN is a Java API, application portability comes for free. Parlay's language independence means that Parlay applications m[ay] or may not be portable from one environment to another, depending on the portability of the language for which it has be[en] implemented.
- Because JAIN is built on a single language, Java, reference implementations are available in that language. Parlay's language-independence gives it a degree of freedom, but makes i[t] more difficult to offer reference implementations.

Standard Java mappings for Parlay are now being developed w[ith] JAIN. The *JAIN Service Provider API (JAIN SPA)*, a mapping of the [Parlay] framework APIs, was released in mid-2001. JAIN Call Control (JCC) w[as] mapped to the Parlay *Multiparty Call Control Service (MPCCS)*. Other [JAIN] to Parlay mappings will include *JAIN SPA Mobility, JAIN SPA GUI (Gra[phical] User Interface), and JAIN SPA PAM (Presence and Availability Management)*.

14.2 Parlay's Interface Definition Language

Because Parlay is supposed to be independent of any particula[r lan]guage or technology, it uses UML, the Universal Modeling Lan[guage,] and a language-independent interface definition language (IDL) based on the OMG's CORBA IDL. To implement Parlay, one must [trans]late from Parlay's IDL to the target language. To that end, standard [map]pings to Java and CORBA IDL, for example, are being defined. B[ecause] Parlay is language-independent, there is nothing to prevent addi[tional] mappings from being created, for example, to the Simple Object [Access] Protocol (SOAP). This section introduces the Parlay IDL and its tr[ansla]tion into Java via JAIN SPA.

14.2.1 Naming Conventions

The Parlay APIs use several naming conventions:[2]

- Type names begin with `Tp`: `TpBoolean` and `TpAddress`.

---

[2]In this chapter, Parlay IDL "code" uses *italic typewriter* font. Java, C++, and [...] uses regular typewriter font.

- The names of interfaces, the methods of which are invoked by client applications, begin with `Ip`: `IpCall` and `IpAuthentication`. Parlay implementations provide classes that implement these interfaces.
- The names of interfaces that must be implemented by client applications begin with `IpApp`: `IpAppAccess` and `IpAppAuthentication`.
- The names of interfaces in the Parlay framework, the methods of which are invoked from within a Parlay implementation, begin with `IpFw`: `IpFwServiceRegistration`. Parlay implementations provide classes that implement these interfaces.
- The names of Parlay service interfaces, the methods of which are invoked from within a Parlay implementation, begin with `IpSvc`: `IpSvcFactory`. Parlay implementations provide classes that implement these interfaces.

14.2.2 Basic Data Types

Parlay's basic data types are shown in Fig. 14.1. These types are used throughout the Parlay APIs.

Figure 14.1 Parlay's basic data types.

Parlay Type	Purpose	Java Equivalent
`TpBoolean`	true or false	`boolean`
`TpInt32`	32-bit signed integer	`int`
`TpFloat`	Single-precision real number	`float`
`TpString`	Character strings	`String`
`TpLongString`	Really long character strings	`String`

14.2.3 Arrays

Parlay calls a one-dimensional array a *Numbered List of Data Elements*. Java uses its normal array syntax to represent these. Thus, a Parlay *Numbered List of Data Elements* of type `TpInt32` would be declared as an `int[]` in Java.

14.2.4 Constructed Types

Parlay has "constructed types" analogous to the structure, enumer
and union types found in many programming languages. CORBA
supports these directly. Java doesn't, so Java implementations can u
standard CORBA IDL-to-Java translations for these. Rather than
the classes that result from those translations (see [69] for instructio
how to do that), examples will be given of how instances of these
might be used in a Java program.

SEQUENCES A Parlay type may be declared as a *Sequence of*
Elements, which is the same as a C++ struct. Consider a l
sequence named *TpSomeStruct*:[3]

```
Sequence of Data Elements TpSomeStruct {
  field1 TpInt32;
  field2 TpString;
};
```

This Java code uses the equivalent TpSomeStruct class:

```
TpSomeStruct s1 = new TpSomeStruct();
s1.setField1(10);
s1.setField2(20);
TpSomeStruct s2 = new TpSomeStruct (100, 200);
s1.setField1(s2.getField2());
```

ENUMERATIONS A Parlay enumeration type is a set of name
associated values. The names and values of an enumeration n
TpSomeEnum are as follows (the P_ that starts each name is a Parlay
ing convention for enumeration elements):

Name	Value
P_FIRST_VALUE	0
P_SECOND_VALUE	1

---

[3] At their current level of maturity, the Parlay specification documents, in places, la
tional consistency. For example, the names of sequence fields are usually capitali
not always. There is also no standard notation for the definitions of constructed ty
notation for a sequence used here occurs in several places throughout the Parlay s
tion. It is not the most common, but it is conveniently compact, which is why I
use it.

In a Java program, the equivalent class could be used as follows. Note that `TpSomeEnum.P_FIRST_VALUE` and `TpSomeEnum.P_SECOND_VALUE` are instances of `TpSomeEnum` that behave like constants. These constants are created as part of the translation from IDL to Java:

```
TpSomeEnum e1 = TpSomeEnum.P_FIRST_VALUE;
TpSomeEnum e2 = TpSomeEnum.P_SECOND_VALUE;
if (e1 == TpSomeEnum.P_FIRST_VALUE)
   System.out.println ("e1 is P_FIRST_VALUE");
if (e1 == e2)
   System.out.println ("e1 and e2 have the same enumeration value");
e1 = e2;
e2 = TpSomeEnum.P_FIRST_VALUE;
```

TAGGED CHOICES A Parlay type may be declared as a *Tagged Choice of Data Elements,* which corresponds to a C++ tagged union. A tagged union lets you uses the same memory to hold different kinds of data. Another data item is set aside as a *tag.* The tag's value indicates what sort of data is currently stored in the union. Here's a Parlay tagged choice named `TpSomeTaggedChoice`—the tag type is `TpSomeEnum`, as defined above—with the following tag values and their corresponding choice types and names:

Tag Value	Choice Type	Choice Name
P_FIRST_VALUE	TpInt32	theInt
P_SECOND_VALUE	TpString	theString

If the tag for an instance of `TpSomeTaggedChoice` named `tc` is `P_FIRST_VALUE`, then `tc` holds a `TpInt32` that can be accessed as `tc.theInt`. If the tag is `P_SECOND_VALUE`, then `tc` hold a `TpString` that can be accessed as `tc.theString`. The equivalent Java class could be used like this (note the method for getting at the tag is named `discriminator`):[4]

```
TpSomeTaggedChoice tc = new TpSomeTaggedChoice ();
// This sets tc's tag to TpSomeEnum.P_SECOND_VALUE. It sets
// the aString value of tc to "Abcdef".
tc.setTheString ("Abcdef");
try {
```

---

[4] Just to pull back the covers a bit, a Java class can't use the same memory for the different choice fields. But it can make sure the type of value stored matches the tag.

```
        // This sets s to the value "Abcdef"
        String s = tc.getTheString ();
        // This throws a InvalidUnionAccessorException because tc's
        // is TpSomeEnum.P_SECOND_VALUE and this tries to get
        // an integer out of tc.
        int i = tc.getTheInt ();
    }
    catch (InvalidUnionAccessorException e) { ... }
    // There's no problem switching tc to its other mode.
    // Now its tag is TpSomeEnum.P_FIRST_VALUE. The value of
    // theString is undefined and theInt is 123.
    tc.setTheInt (123);
    // Get the tag from tc.
    TpSomeEnum tag = tc.discriminator();
```

14.2.5 Method Definitions

Parlay method definitions are of the form:

interface-name method-name (arguments) : TpResult

All Parlay methods return a value of type `TpResult` tha[t] whether the method succeeded or failed and, if it failed, why. I[n] programs, failure results become exceptions. The actual exception[s] must be thrown vary from method to method and may be deter[mined] by referring to the Parlay specifications to see what sorts of `TpRe`[sult] can be returned from a given method.

Arguments, separated from each other by commas, are of the fo[rm]

argument-name : in-out argument-type

The *in-out* part is either the keyword `in` or the keyword `out`. Argu[ments] marked `in` correspond to the input arguments of a Java method. [Parlay] methods may have at most one `out` argument, corresponding [to the] value returned from a Java function.

In the Parlay specification, many arguments passed to and [from] methods have a type with a name that ends with `Ref` (for `in` argu[ments]) or `RefRef` (for `out` arguments), for example, `TpInt32R`[ef or] `TpInt32RefRef`. This highlights the fact that Parlay uses pa[ss-by-] reference rather than pass-by-value for getting values to and from [meth]ods. Since Java does this anyway, these suffixes are left off in Java [imple]mentations of Parlay. Thus, in Java, an int passed to a method is [just] an `int`, not an `intRef`. An int returned from a method is also [just] an `int`, not an `intRefRef`. In this chapter, the `Ref` and `RefRef` s[uffixes] appear wherever the text of a Parlay method definition is quoted [direct]ly. Everywhere else, the suffixes are left off.

The following Parlay method definition:

```
IpAccess selectService
(serviceID : in TpServiceID,
 serviceProperties : in TpServicePropertyList,
 serviceToken : out TpServiceTokenRef)
: TpResult
```

corresponds to the following Java method declaration in an implementation of `IpAccess`:

```
public String selectService
   (String serviceID,
    ServiceProperty[] serviceProperties)
   throws ServiceAccessDeniedException,
          InvalidServiceIDException,
          InvalidPropertyException,
          InvalidServiceTokenException
```

14.3 Setting Up a Parlay Application

As usual, we'll explore the Parlay APIs through sample Java code. The code that interacts with the Parlay framework will use the recently defined JAIN SPA mapping of Parlay's framework APIs to the Java language. As with the JTAPI and JAIN examples, it's important to keep in mind that none of this code will work without a Parlay implementation against which to compile and run it. To keep things from getting bogged down in too many details, the example code in this chapter uses only a small portion of Parlay's framework and call processing APIs. In addition, because many of the arguments for Parlay methods are complex data structures, rather than put complete descriptions of these in line with the code walkthrough, they've been moved to Sec. 14.5 at the end of this chapter.

From Parlay's perspective, an application is a client of the services a Parlay implementation provides. Figure 14.2 shows the structure of a Parlay client and server, and their relationship to each other. At the upper left of the diagram is a client application written in Java. This application uses the JAIN SPA, JAIN SPA Call Control, and other Java APIs, and executes inside a Java Virtual Machine (JVM). The client talks to a server using an agreed-upon wire protocol. The application doesn't care what this is since the details are hidden under the Parlay APIs with

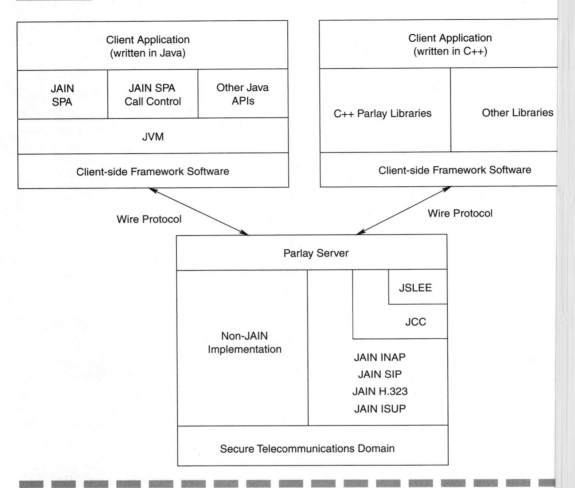

Figure 14.2
Parlay's client-server architecture.

which it deals. The wire protocol could be based on CORBA, a
choice because the client application will be making method c
objects that sit on the server. It could be based on Java RMI, also
cal choice if this is to be an all-Java implementation of Parlay. It
be based on DCOM. It could be a handcrafted custom pr
unique to this implementation of Parlay. At the upper right of F
is another client, this one written in C++. Like the Java client, i
acts with the server using a wire protocol, the details of which a
den beneath the Parlay APIs.

The server at the bottom of Figure 14.2 shows both a Java and a non-Java implementation. At the top is the software that deals with the wire protocol, whatever it is, and passes method requests and results back and forth to a Parlay implementation that sits in the middle of the server architecture. Beneath the Parlay implementation is a secure telecommunications domain, within which the Parlay services come to life. The non-Java Parlay implementation in Fig. 14.2 is shown as a monolithic block, since there is no standard way to decompose this software in other languages as there is with Java. JAIN, because it provides the standard mapping of Parlay to Java, provides a framework within which to build a Java implementation of Parlay. There are three ways in which this can be done:

- The Parlay server software at the top can talk directly to JAIN protocol APIs that talk to the telecommunications infrastructure.
- The Parlay server software can talk to JAIN Call Control (JCC) software. That software in turn interacts with the protocol APIs.
- The Parlay server software can talk to JAIN Service Logic Execution Environment (JSLEE) software, which interact with JCC.

The class that contains the example application is named `ClientApp`. Besides logic for the application itself, `ClientApp` also implements the `JainSpaAppAuthentication`, `JainSpaAppAccess`, and `IpAppCallControlManager` interfaces. The first two of these provide methods Parlay expects to find on the client side when it and the client are authenticating each other and negotiating a service agreement. The third provides methods that Parlay can use while processing generic (two-party, single medium) calls. All these interfaces could have been implemented in separate classes, but this is a convenient way to handle them:

```
public class ClientApp
   implements JainSpaAppAuthentication,
         JainSpaAppAccess,
         IpAppCallControlManager {
  JainSpaFactory myFactory = null;
  JainSpaInitial myFwInitial = null;
  JainSpaAuthentication myFwAuthentication = null;
  JainSpaAccess myFwAccess = null;
  JainSpaServiceDiscovery myFwDiscovery = null;
  IpCallControlManager myCallControlMgr = null;
  IpCall myCall = null;
```

The top-level code for the application is found in the class's main routine, which creates an instance of the `ClientApp` class. The program authenticates itself to the Parlay framework (see Fig. 14.4), discovers what

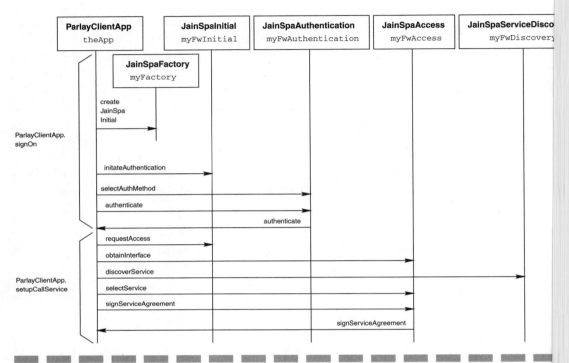

Figure 14.3
Setup phase for a Parlay client application.

services the framework supports, obtains an instance of a call m_
service from the framework (see Fig. 14.5), and uses that service t_
port a private dialing plan, similar to the one shown with JAIN _
previous chapter:

```
public static void main (String [] args) {
  ClientApp theApp = new ClientApp();
  theApp.signOn();
  myCallControlManager = theApp.setupCallService();
  theApp.registerForCallEvents();
  qqq
}
```

14.3.1 Authentication

CLIENT OBTAINS INSTANCE OF IpInitial The first thing _
Parlay application must do is authenticate itself with its Parlay _
provider. That's taken care of by the `signOn` method in `ClientA`_

Figure 14.4 signOn method of ClientApp.

```
public void signOn () {
  myFactory = JainSpaFactory.getInstance();
  myFactory.setPathName("com.jainvendor");
  myFwInitial = myFactory.createJainSpaInitial("MyParlayInitial");
  JainSpaDomainID theDomainId = new JainSpaDomainID ();
  theDomainId.setClientAppId ("MyId");
  JainSpaAuthDomain appDomain = new JainSpaAuthDomain (theDomainId,
                                                        this);
  JainSpaAuthDomain fwDomain =
    myFwInitial.initiateAuthentication (appDomain,
                                        "JAIN_SPA_AUTHENTICATION");
  myFwAuthentication =
    (JainSpaAuthentication)fwDomain.getAuthInterface();
  String prescribedMethod =
    myFwAuthentication.selectAuthMethod ("P_DES_56,P_DES_128");
  String unencryptedChallenge = "This is a challenge string";
  Cipher cipher = Cipher.getInstance("RSA");
  PublicKey networkPublicKey = getNetworkPublicKey();
  cipher.init(Cipher.ENCRYPT_MODE, networkPublicKey);
  byte[] byteChallenge =
    cipher.doFinal(unencryptedChallenge.getBytes());
  String challenge = new String(byteChallenge);
  String response =
    myFwAuthentication.authenticate (prescribedMethod, challenge);
  if (! unencryptedChallenge.equals(response)) {
    System.err.println("Failed authenticating server, exiting");
    System.exit(1);
  }
}
```

start authentication, the application obtains a reference to an implementation of the Parlay `IpInitial` interface (in JAIN SPA, this interface is called `JainSpaInitial`). There's no set way to do this. The Parlay specification says it "may be gained through a URL, an Application Support Broker, a stringified object reference, etc." The JAIN SPA solution is to use the factory pattern we've already seen with JTAPI and JAIN:

```
myFactory = JainSpaFactory.getInstance();
myFactory.setPathName("com.jainvendor");
myFwInitial = myFactory.createJainSpaInitial
              http://www.jainvendor.com/("MyParlayInitial");
```

CLIENT INITIATES AUTHENTICATION The program next invokes `initiateAuthentication` on its instance of `IpInitial` (or `JainSpaInitial` for a program that uses JAIN SPA) to obtain references to framework objects it will use during authentication. Parlay defines this method as follows:

```
private IpCallControlManager setupCallService () {
  myFwAccess = (JainSpaAccess)
                myFwInitial.requestAccess ("JAIN_SPA_ACCESS",
                                           this);
  myFwDiscovery = (JainSpaServiceDiscovery)
                   myFwAccess.obtainInterface ("P_DISCOVERY");
  ServiceProperty[] svcProps = new ServiceProperty[1];
  svcProps[0] = new ServiceProperty();
  svcProps[0].setServicePropertyName("PARLAY_SERVICE_NAME");
  String[] props1 = {"PARLAY_CALL_CONTROL"};
  svcProps[0].setServicePropertyValueList(props1);
  String serviceTypeName = "P_CALL_CONTROL";
  Service[] discoveredSvcs = myFwDiscovery.discoverService
                                (serviceTypeName, svcProps, (short)5);
  String serviceID = discoveredSvcs[0].getServiceID();
  String serviceToken = myFwAccess.selectService(serviceID, svcProps);
  JainSpaSignatureAndServiceMgr ssMgr = null;
  String agreement = "The agreed upon text";
  String signingAlgorithm = "P_MD5_RSA_1024";
  ssMgr = myAccess.signServiceAgreement(serviceToken,
                                        agreement,
                                        signingAlgorithm);
  Signature dsa = Signature.getInstance("MD5withRSA");
  PublicKey networkPublicKey = getNetworkPublicKey();
  dsa.initVerify(networkPublicKey);
  dsa.update(agreement.getBytes());
  if (dsa.verify(result.getDigitalSignature().getBytes())) {
    return null;
  }
  return (IpCallControlManager)ssm.getServiceMgrInterface();
}
```

Figure 14.5
setupCallService method of ClientApp.

IpInitial initiateAuthentication

(appDomain : in TpAuthDomain,
 authType : in TpAuthType,
 fwDomain : out TpAuthDomainRef)
: TparlayResult

where:

appDomain identifies the application's domain to the framework and provides the framework with an object it can use to authenticate itself within that domain (see Sec. 14.5 for explanations of this and many other Parlay types referred to throughout this discussion).

authType is the authentication method the client wants to use. Its TpAuthType, is equivalent to a TpString. If the string is either empty or "P_AUTHENTICATION", the client is requesting Parlay's own authentication method. If some other operator-specific method is to be used, it's specified by supplying a string that starts with the characters "OP_".

fwDomain is an instance of *TpAuthDomain* that contains a reference to an instance of the framework's authentication interface. Its type matches whatever was supplied by the client in the *authType* argument.

JAIN SPA defines classes and methods for working with these types. Using those, the code for initiating authentication with Parlay is as follows:

```
JainSpaDomainID theDomainId = new JainSpaDomainID ();
theDomainId.setClientAppId ("MyId");
JainSpaAuthDomain appDomain = new JainSpaAuthDomain
                                    (theDomainId, this);
JainSpaAuthDomain fwDomain = myFwInitial.initiateAuthentication
                                    (appDomain,
                                    "JAIN_SPA_AUTHENTICATION");
```

Because the client specified "JAIN_SPA_AUTHENTICATION", the AuthInterface returned to it inside fwDomain is an instance of JainSpaAuthentication:

```
myFwAuthentication = (JainSpaAuthentication)
                        fwDomain.getAuthInterface();
```

CLIENT SELECTS AN AUTHENTICATION METHOD The client invokes selectAuthMethod on its instance of IpAuthentication (JainSpaAuthentication) to tell the framework what authentication methods it can use, for example, P_DES_128 (128 bit DES encryption) or P_RSA_1024 (1024 bit RSA encryption). The framework selects one of these and returns it to the client:

IpAuthentication selectAuthMethod

```
(authCaps: in TpAuthCapabilityList,
 prescribedMethod : out TpAuthCapabilityRef)
: TpResult
```

where:

authCaps is a *TpString*[5] of values separated by commas, each naming an authentication capability the client supports.

---

[5] Like many Parlay types, *TpAuthCapabilityList* is defined to be the same as a *TpString*. Since this is so common, I won't state it explicitly any more. Instead, I'll just say the item in question is a *TpString*.

prescribedMethod is the authentication method preferred by the framework. If the client cannot use this method, it must stop exeing. To keep things simple, the example code assumes the client c support whatever the framework returns.

Translating into Java:

```
String prescribedMethod =
    myFwAuthentication.selectAuthMethod ("P_RSA_512,P_RSA_1024")
```

CLIENT AND FRAMEWORK AUTHENTICATE EACH OT

The client may now authenticate itself using the authenticate m of IpAuthentication (JainSpaAuthentication):

```
IpAuthentication authenticate

  (prescribedMethod : in TpAuthCapability,
   challenge : in TpString,
   response : out TpStringRef)
  : TpResult
```

The first argument should be the same as the *TpAuthCapabi* value returned by selectAuthMethod. The second argument is a lenge string encrypted according to *prescribedMethod*. The f work decrypts this string and, using techniques described in [32], putes a value that's encrypted and returned in *response*. If resp matches what the client expects, the framework has authenticated to the client. Depending on the value of *prescribedMethod*, aut cation may require several calls to authenticate. Assuming onl call is needed, and putting this into Java, with some calls to classe take care of the cryptography and challenge-response computation code is as follows:

```
String unencryptedChallenge = "This is a challenge string";
Cipher cipher = Cipher.getInstance("RSA");
PublicKey networkPublicKey = getNetworkPublicKey();
cipher.init(Cipher.ENCRYPT_MODE, networkPublicKey);
byte[] byteChallenge =
    cipher.doFinal(unencryptedChallenge.getBytes());
String challenge = new String(byteChallenge);
String response =
    myFwAuthentication.authenticate (prescribedMethod, challenge
if (! unencryptedChallenge.equals(response)) {
    System.err.println("Failed while authenticating server");
    System.exit(1);
}
```

For this to work, a `getNetworkPublicKey` method must be provided by `ClientApp`. Its details aren't important for our purposes:

```
PublicKey getNetworkPublicKey(){
  // obtain network public key from somewhere
  return ...;
}
```

To authenticate the client, the framework goes through a similar process, invoking the `authenticate` method on the instance of `IpApp-Authentication` (`JainSpaAppAuthentication`) that was passed to it when the client called `initiateAuthentication`. The framework calls this method with a challenge string. From this, the client must compute the expected response to authenticate itself with the framework:

```
public String authenticate
   (String prescribedMethod, String challenge)
   throws InvalidAuthCapabilityException {
  String response = null;
  // Challenge must be decrypted using my private key
  Cipher cipher = Cipher.getInstance("RSA");
  PrivateKey myPrivateKey = getPrivateKey();
  cipher.init(Cipher.DECRYPT_MODE, myPrivateKey);
  // Encrypt my unencrypted challenge
  byte[] byteResponse = null;
  byteResponse = cipher.doFinal(challenge.getBytes());
  response = new String(byteResponse);
  return response;
}
```

The method `getPrivateKey` is provided by `ClientApp`. The details are irrelevant:

```
PrivateKey getPrivateKey(){
  // obtain private key from somewhere
  return ...;
}
```

14.3.2 Requesting Access to Parlay

Once the framework and client are satisfied they're not dealing with an impostor, the client program can start using Parlay's services. To do this, a client must find out what services the Parlay implementation supports

and subscribe to the services it will need. This is taken care of b
setupCallService method of ClientApp. The first thing it d
call requestAccess on its instance of IpInitial (JainSpaIni
The client uses this method to exchange objects with the frame
that each will use as a point of contact for setting up a mutual s
agreement:

IpInitial requestAccess

```
(accessType : in TpAccessType,
 appAccessInterface : in IparlayInterfaceRef,
 fwAccessInterface : out IparlayInterfaceRefRef)
 : TpResult
```

where:

accessType specifies what kind of access object the client is passi
the framework and what kind of access object it expects to get b
It's a *TpString* that takes one of the following values (other acce
types specific to a network operator may be indicated with strin
preceded by the characters "OP_"):

empty string Requesting default access.

P_PARLAY_ACCESS The client is passing an instance of a class t
implements IpAppAccess and expects to get back an instanc
IpAccess. This is Parlay's default access.

JAIN_SPA_ACCESS The client is passing an instance of a class t
implements JainSpaAppAccess and expects to get back an
instance of JainSpaAccess.

appAccessInterface is an instance of a class the framework can
to access the client, as indicated by *accessType*.

fwAccessInterface is an instance of a class the client can use to
the framework, as indicated by *accessType*.

The JAIN SPA code is very simple. Because ClientApp imple
the JainSpaAppAccess interface, it supplies itself as the object thr
which the framework will access the client:

```
myFwAccess = (JainSpaAccess) myFwInitial.requestAccess
                                    ("JAIN_SPA_ACCESS",
                                     this);
```

14.3.3 Discovering Parlay Services

The client may now gain access to services supported by the Parlay implementation through a process called *discovery*. It first obtains an instance of the framework's discovery interface by calling `obtainInterface` on its instance of `IpAccess` (`JainSpaAccess`):

```
IpAccess obtainInterface
   (interfaceName : in TpInterfaceName,
    fwInterface : out IparlayInterfaceRefRef)
  : TpResult
```

where

interfaceName is a `TpString` that names the interface to be obtained. Parlay interface names, of which there are about 16, start with "P_". Operator-specific interfaces have names that start with "OP_". Since the example application wants the discover interface, it uses "P_DISCOVERY" for *interfaceName*.

fwInterface is a framework object that supports the requested interface.

Here's the Java code:

```
myFwDiscovery = (JainSpaServiceDiscovery)
                    myFwAccess.obtainInterface ("P_DISCOVERY");
```

Once it has a discovery interface, the client invokes `discoverService` to find out if the Parlay implementation with which it is dealing supports the services it needs:

`IpServiceDiscovery discoverService`

```
(serviceTypeName : in TpServiceTypeName,
 desiredPropertyList : in TpServicePropertyList,
 max : in TpInt32,
 serviceList : out TpServiceListRef)
: TpResult
```

where

serviceTypeName is the name of the desired service.

desiredPropertyList is an array of `TpServiceProperty`. These identify properties the client wants the discovered services to have. See Ref. [75] for more information on the `TpServiceProperty` data structure.

max is the maximum number of services to be returned.

serviceList is an array of *TpService* objects that match the su[pplied] criteria.

If the client doesn't know what services are supported, it can fi[nd out using] *listServiceTypes* on the discovery interface. To keep things s[imple,] we'll assume this application already knows the generic call contr[ol ser]vice ("PARLAY_CALL_CONTROL") is supported and can go straight [to dis]covery of that service:

```
ServiceProperty[] svcProps = new ServiceProperty[1];
svcProps[0] = new ServiceProperty();
svcProps[0].setServicePropertyName("PARLAY_SERVICE_NAME");
String[] props1 = {"PARLAY_CALL_CONTROL"};
svcProps[0].setServicePropertyValueList(props1);
String serviceTypeName = "P_CALL_CONTROL";
Service[] discoveredSvcs = myFwDiscovery.discoverService
                        (serviceTypeName, svcProps, (sho
```

14.3.4 Subscribing to a Parlay Service

The client now knows what services are supported. To subscribe [to one] of these, it calls *selectService* on its instance of IpA[ccess] (JainSpaAccess):

```
IpAccess selectService

  (serviceID : in TpServiceID,
   serviceProperties : in TpServicePropertyList,
   serviceToken: out TpServiceTokenRef)
  : TpResult
```

where

serviceID is the ID returned from *discoverService* of the se[rvice the client] wants.

serviceProperties is an array of *TpServiceProperty* the service must support. See Ref. [75] for more information on the *TpServiceProperty* data structure.

serviceToken is a string that uniquely identifies the client's inst[ance] of the selected service. This token has a limited lifetime. Once i[t] expires, any method that takes that token as an argument fails [with an] invalid service token result. Clients and frameworks can force a [token] to expire by calling *endAccess* on each other's access interface (IpAccess or IpAppAccess).

With real world sanity checks omitted (such as whether any services were returned at all from `discoverService`), the Java code looks like this:

```
String serviceID = discoveredSvcs[0].getServiceID();
String serviceToken = myFwAccess.selectService(serviceID, svcProps);
```

Finally, the client invokes `signServiceAgreement` on its instance of `IpAccess` (`JainSpaAccess`). This asks the framework to sign an agreement that will give the application access to the desired service. If the framework agrees, both parties sign the agreement:

IpAccess signServiceAgreement

(serviceToken : in TpServiceToken,
agreementText : in TpString,
signingAlgorithm : in TpSigningAlgorithm,
signatureAndServiceMgr : out TpSignatureAndServiceMgrRef)
: TpResult

where

serviceToken is the token returned from *selectService*.

agreementText is the text to be signed by the framework using its private key. Establishing an agreement is a business transaction, so the text that goes here is whatever the parties agree to. They could establish this through offline negotiations, apart from the Parlay framework, or they could establish them online, using the Parlay subscription framework.

signingAlgorithm is the algorithm that will be used to compute the digital signature with which the agreement is signed. The following algorithms have been defined (operator-specific algorithms may be designated with names starting with "OP_"):

empty string   No signing algorithm required.

P_MD5_RSA_512   Use the MD5 algorithm to convert an arbitrary length input string to a 128-bit message digest, which is then encrypted with 512-bit private key using the RSA algorithm.

P_MD5_RSA_1024   Same as above, using 1024-bit private key.

signatureAndServiceMgr holds the framework's digital signature of the service agreement and a reference to the service's service manager interface:

Sequence of Data Elements TpSignatureAndServiceMgr {
  digitalSignature : TpString;

```
    serviceMgrInterface : IparlayInterfaceRef;
};
```

The Java code uses an unrealistic (but mercifully brief) service ment string. Once the agreement has been signed, the last line o extracts and returns an instance of the Parlay call control ma which gets the program (at last) to the point where it can start to processing:

```
JainSpaSignatureAndServiceMgr sigSvcMgr = null;
String agreement = "The agreed upon text";
String signingAlgorithm = "P_MD5_RSA_1024";
sigSvcMgr = myAccess.signServiceAgreement(serviceToken,
                                         agreement,
                                         signingAlgorithm);
Signature dsa = Signature.getInstance("MD5withRSA");
PublicKey networkPublicKey = getNetworkPublicKey();
dsa.initVerify(networkPublicKey);
dsa.update(agreement.getBytes());
if (dsa.verify(result.getDigitalSignature().getBytes())) {
  return null;
}
return (IpCallControlManager)ssm.getServiceMgrInterface();
```

The last thing to take care of for this part of the application client's own `signServiceAgreement` method. The framewor this, just as the client calls the corresponding method in the fran to ensure both sides of the transaction leave a record of their agr to provide and use the subscribed service. `ClientApp` provid method itself because it implements the `JainSpaAppAuthenti` interface:

```
public String signServiceAgreement(String serviceToken,
                                   String agreementText,
                                   String signingAlgorithm)
throws InvalidServiceTokenException, InvalidAgreementTextExc
InvalidSigningAlgorithmException
{
  Signature dsa = Signature.getInstance("MD5withRSA");
  PrivateKey myPrivateKey = getPrivateKey();
  dsa.initSign(myPrivateKey);
  dsa.update((agreementText + serviceToken).getBytes());
  byte[] byteSignature = dsa.sign();
  return new String(byteSignature);
}
```

14.4 Parlay Call Processing

14.4.1 Registering for Call Events

Now that Parlay and the client have successfully negotiated access to Parlay services between themselves, the client can start using those services (see Fig. 14.6). Because the application supports a private dialing plan, it needs to know whenever a subscriber dials an address in that plan so it can route the call to its actual address. The `registerForCallEvents` method of `ClientApp` takes care of this (see Fig. 14.7). In it, the client invokes the Parlay call control manager's `enableCallNotification` method to tell Parlay it wants to be notified when particular call events occur (in the event listener pattern, this makes the client an "event listener"):

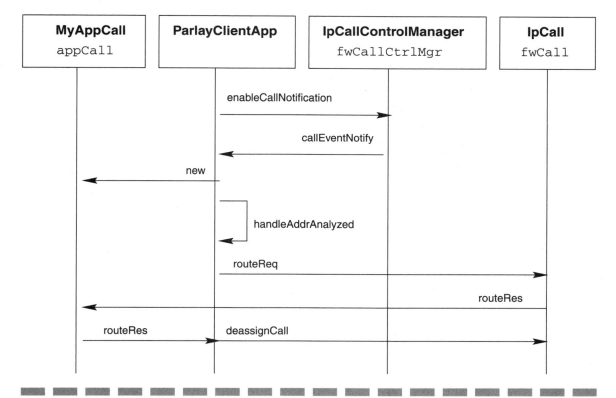

Figure 14.6
Call processing phase of Parlay client application.

```
private void registerForCallEvents () {
  TpCallEventCriteria events;
  events.setCallNotificationType (TpCallNotificationType.P_ORIGINATING);
  events.setCallEventName
          (TpCallEventName.P_EVENT_GCCS_ADDRESS_ANALYZED_EVENT);
  events.setOrigAddrPlan (TpAddressPlan.P_ADDRESS_PLAN_E164);
  events.setOrigAddrAddrString ("15125556???");
  events.setOrigAddrName ("");
  events.setOrigAddrPresentation
          (TpAddressPresentation.P_ADDRESS_PRESENTATION_UNDEFINED);
  events.setOrigAddr.SubAddressString ("");
  events.setDestAddrPlan (TpAddressPlan.P_ADDRESS_PLAN_E164);
  events.setDestAddrAddrString ("????");
  events.setDestAddrName ("");
  events.setDestAddrPresentation
          (TpAddressPresentation.P_ADDRESS_PRESENTATION_UNDEFINED);
  events.setDestAddrSubAddressString ("");
  events.setMonitorMode (TpCallMonitorMode.P_CALL_MONITOR_MODE_INTERRUPT);
  int assignmentId = myCallCtrlMgr.enableCallNotification (this,
                                                           events);
}
```

Figure 14.7
registerForCallEvents method of ClientApp.

IpCallControlManager enableCallNotification

```
(appInterface    : in  IpAppCallControlmanagerRef,
 eventCriteria   : in  TpCallEventCriteria,
 assignmentID    : out TpAssignmentIDRef)
: TpResult
```

where `appInterface` is an instance of a class that implements I CallControlManager, `eventCriteria` identifies events the a tion wants to hear about, and `assignmentID` is a unique inte assigned by Parlay call control to this event notification request. cations that issue more than one call to `enableCallNotificati` this value to correlate events with the criteria through which the passed.

The `registerForCallEvents` method first creates an obje describes the events in which it is interested. The client specifi it wants to hear about events that pertain to the originating the call:[6]

---

[6]In the previous section, I used JAIN SPA as the basis for the Java code. As mentio ier, a comparable standard mapping of Parlay's Generic Call Control Service to Jav this writing, still being developed. Consequently, the code in this section demonst one way Parlay call control might be programmed in Java.

```
TpCallEventCriteria events;
events.setCallNotificationType (TpCallNotificationType.P_ORIGINATING);
```

The client indicates the specific event in which it is interested. The event here, P_EVENT_GCCS_ADDRESS_ANALYZED_EVENT, occurs when the network has collected and analyzed an address from the calling party, an address that is, at that time, known to be valid and complete:

```
events.setCallEventName
        (TpCallEventName.P_EVENT_GCCS_ADDRESS_ANALYZED_EVENT);
```

The client doesn't want to know about this event every time it occurs. It should be notified only when the originator's phone number starts with the digits 15125556 (the last three digits can be anything). It doesn't care about any other aspects of the originator:

```
events.setOrigAddrPlan (TpAddressPlan.P_ADDRESS_PLAN_E164);
events.setOrigAddrAddrString ("15125556???");
events.setOrigAddrName ("");
events.setOrigAddrPresentation
        (TpAddressPresentation.P_ADDRESS_PRESENTATION_UNDEFINED);
events.setOrigAddr.SubAddressString("");
```

Furthermore, it only wants to know when these originators dial numbers that are exactly four digits long. It doesn't care about any other aspects of the destination address:

```
events.setDestAddrPlan (TpAddressPlan.P_ADDRESS_PLAN_E164);
events.setDestAddrAddrString ("????");
events.setDestAddrName ("");
events.setDestAddrPresentation
        (TpAddressPresentation.P_ADDRESS_PRESENTATION_UNDEFINED);
events.setDestAddrSubAddressString ("");
```

Finally, the client indicates that Parlay call control should wait for further instructions after sending a notification of the event:

```
events.setMonitorMode (TpCallMonitorMode.P_CALL_MONITOR_MODE_INTERRUPT);
```

With the events of interest fully described, the client invokes enableCallNotification to register itself as a listener with Parlay call control. Note that it supplies itself in the arguments as the object that implements the IpAppCallControlManager interface:

```
int assignmentId = myCallCtrlMgr.enableCallNotification (this,
                                                         events);
```

14.4.2 Handling Call Events

To notify a client application of call events, the Parlay call control ager calls the `callEventNotify` method of a class that implemen `IpAppCallControlManager` interface. Parlay defines this meth follows

```
IpAppCallControlManager callEventNotify

    (callReference : in TpCallIdentifier,
     eventInfo : TpCallEventInfo,
     assignmentID : in TpAssignmentID,
     appInterface : out IpAppCallRefRef)
    : TpResult
```

where

callReference is a *Sequence of Data Elements* containing:
- A reference to the `IpCall` object that produced the event.
- An integer ID assigned to the call.

eventInfo is a *Sequence of Data Elements* with information about the reported event.

assignmentID should be the same as the one of the assignment I returned from a call to `enableCallNotification`. Since this application calls `enableCallNotification` only once, there's n much to be done with it (other than perhaps doing a sanity chec that it's the same as the value returned by `enableCallNotific` tion).

appInterface is an instance of a class that implements the `IpApp` interface.

When the `callEventNotify` method of `ClientApp` is called network has collected and analyzed a valid and complete four- address from a calling party. The method creates an instance of `My` `Call`, an application class that implements the `IpAppCall` inte passing it a reference to the client application object. After it's done it calls an application-specific method—`handleAddrAnalyzed`— does everything else necessary to handle the event. Finally, it return application call object it just created to the Parlay call control mana

```
public IpAppCall callEventNotify (TpCallIdentifier callReference,
                    TpCallEventInfo eventInfo,
                    TpAssignmentID assignmentId) {
  TpCallEventName eventName = eventInfo.getCallEventName();
```

```
switch (eventName) {
  case TpCallEventName.P_EVENT_GCCS_ADDRESS_ANALYZED_EVENT :
    // Network has collected and analyzed an address
    // from the calling party. The address is known
    // to be valid and complete.
    MyAppCall appCall = new MyAppCall (this);
    handleAddrAnalyzed (callReference,
                       eventInfo,
                       assignmentId);
    return appCall;
  default :
    System.err.println
           ("Received notification of unexpected event, exiting");
    System.exit (1);
  }
}
```

The method `handleAddrAnalyzed` extracts the calling and called addresses from the reported event and translates them into a new destination address using another application-specific method—translate—the details of which we needn't explore. (You may, if you wish, write a translation method of your own if it really bugs you not to have the code at hand.) It then invokes `routeReq` on the Parlay call object to route the call to its new destination. Parlay defines `routeReq` as follows:

```
IpCall routeReq

    (callSessionID : in TpSessionID,
     responseRequested : in TpCallReportRequestSet,
     targetAddress : in TpAddress,
     originatingAddress : in TpAddress,
     originalDestinationAddress : in TpAddress,
     redirectingAddress : in TpAddress,
     appInfo : in TpCallAppInfoSet,
     callLegSessionID : out TpSessionIDRef)
    : TpResult
```

where

`callSessionID` is the session ID for the call.

`responseRequested` specifies events that should cause Parlay to call `routeRes` in the application call object.

`targetAddress` is the destination to which the call should be routed.

`originatingAddress` is the address of the calling party.

`originalDestinationAddress` is the original destination address of the call (here, it's the four-digit private dialing plan number).

`redirectingAddress` is the address from which the call was last redirected (which this program ignores).

appInfo is application-related information concerning the call, information this program doesn't use.

callLegSessionID is the session ID of the call leg that's created b[y] routing the call to its destination. This ID is provided as an argu[ment] whenever Parlay calls `routeRes` (or `routeErr`) on the client side. It does this so the application can match a request with its corresponding result (or error). It's useful only when `routeReq` is called more than once on the same call (for example, for each participant in a multiparty call).

The method `handleAddrAnalyzed` translates the four dialed [digits] to a complete network address and then reroutes the call, telling P[arlay] call processing to notify it of any progress in routing the call on its [way] (P_CALL_REPORT_PROGRESS). This time, Parlay can just send the n[otifi]cation and continue processing the call without waiting for instruct[ions] from the client (P_CALL_MONITOR_MODE_NOTIFY):

```
private void handleAddrAnalyzed (TpCallIdentifier callReference,
                    TpCallEventInfo eventInfo,
                    TpAssignmentID assignmentId) {
 myCall = callReference.getCallReference();
 TpAddress orig = eventInfo.getOrigAddr();
 TpAddress dest = eventInfo.getDestAddr();
 TpAddress newDest = translate (orig, dest);
 TpCallReportRequest[] responseRequested =
                    new TpCallReportRequest [1];
 responseRequested[0].setCallReportType =
            (TpCallReportType.P_CALL_REPORT_PROGRESS);
 responseRequested[0].setMonitorMode
            (TpCallMonitorMode.P_CALL_MONITOR_MODE_NOTIFY);
 responseRequested[0].setAdditionalReportCriteria (null);
 TpSessionID callLegSessionID =
        myCall.routeReq (callReference.getCallSessionID(),
                    responseRequested,
                    newDest,
                    orig,
                    dest,
                    eventInfo.getOriginalDestAddr(),
                    eventInfo.getRedirectingAddr(),
                    null);
}
```

14.4.3 Handle Results from a Request

Every Parlay request method—such as `routeReq`— has a match[ing] result method provided by the client. Parlay calls these to report resu[lts] of a request back to the client. The result method for `routeReq` [is] `IpAppCall.routeRes`. Parlay calls it when a call routed by `routeR`[eq] has been successfully routed to its destination. (The client should also provide an error reporting method—`routeErr` in this case—but that won't be covered here.):

```
IpAppCall routeRes

    (callSessionID : in TpSessionID,
     eventReport : in TpCallReport,
     callLegSessionID : in TpSessionID)
    : TpResult
```

where

callSessionID is the session ID of the call.

eventReport contains information about what happened in the call (for example, called party is busy or has answered).

callLegSessionID is the same as the session ID that was returned from `routeReq`.

Here's how the method might be implemented in Java for the `My-AppCall` class. It does nothing but pass the buck to a method in the client object:

```
public void routeRes (TpSessionID callSessionId,
                      TpCallReport eventReport,
                      TpSessionID callLegSessionId) {
  myClientApp.routeRes (callSessionId,
                        eventReport,
                        callLegSessionId);
}
```

`ClientApp.routeRes` could be implemented as shown below. The most interesting part here is the call to `deassignCall`. This is a method of `IpCall` the application uses to tell Parlay it's no longer interested in that call. The call will continue to be processed as usual in the network, but Parlay will not interact any more with the application regarding it:

```
public void routeRes (TpSessionID callSessionId,
                      TpCallReport eventReport,
                      TpSessionID callLegSessionId) {
  System.out.println ("Routing successful at "
                      + eventReport.getCallEventTime());
  switch (eventReport.getCallReportType()) {
    case TpCallReportType.P_CALL_REPORT_ALERTING : {
      System.out.println ("Called party being alerted);
      break;
    }
    case TpCallReportType.P_CALL_REPORT_ANSWER : {
```

```
        System.out.println ("Called party answered");
        break;
      }
      case TpCallReportType.P_CALL_REPORT_BUSY : {
        System.out.println ("Called party busy);
        break;
      }
      case TpCallReportType.P_CALL_REPORT_NO_ANSWER : {
        System.out.println ("Call did not answer");
        break;
      }
      default : {
        System.out.println ("Other result");
        break;
      }
    }
    myCall.deassignCall();
}
```

14.5 Parlay Types Used in This Chapter

TpAddress A *TpAddress* represents a network address, for example phone number:

```
Sequence of Data Elements TpAddress {
  Plan : TpAddressPlan;
  AddrString : TpString;
  Name : TpString;
  Presentation : TpAddressPresentation;
  Screening : TpAddressScreening;
  SubAddressString : TpString;
};
```

Plan indicates the type of address or numbering plan that's used.

AddrString contains the address as a string of characters. Its format depends on the value of *Plan*. For example, if *Plan* is P_ADDRESS_PLAN_IP, *AddrString* will be four numbers separated by periods, plus an optional port number set off by a colon: "127.0.0.1:122". if *Plan* is P_ADDRESS_PLAN_E164, *AddrString* will be an international number: "15125559876".

Name contains a (possibly empty) name string that's associated with the address.

Presentation specifies whether information in this *TpAddress* can be presented to an end user.

Screening specifies whether this `TpAddress` was provided by a user or network. If provided by a user, it says whether the address has been verified and, if so, whether it passed.

SubAddressString can be used to provide a subaddress that falls under an address, when the numbering plan supports these.

TpAddressPlan `TpAddressPlan` is an enumeration of network address plans:

P_ADDRESS_PLAN_NOT_PRESENT (0) No address present

P_ADDRESS_PLAN_UNDEFINED (1) Undefined

P_ADDRESS_PLAN_IP (2) IP

P_ADDRESS_PLAN_MULTICAST (3) Multicast

P_ADDRESS_PLAN_UNICAST (4) Unicast

P_ADDRESS_PLAN_E164 (5) E.164

P_ADDRESS_PLAN_AESA (6) AESA

P_ADDRESS_PLAN_URL (7) URL

P_ADDRESS_PLAN_NSAP (8) NSAP

P_ADDRESS_PLAN_SMTP (9) SMTP

P_ADDRESS_PLAN_MSMAIL (10) Microsoft Mail

P_ADDRESS_PLAN_X400 (11) X.400

TpAddressPresentation `TpAddressPresentation` is an enumeration of dispositions for address presentation:

P_ADDRESS_PRESENTATION_UNDEFINED (0) Undefined

P_ADDRESS_PRESENTATION_ALLOWED (1) Presentation allowed

P_ADDRESS_PRESENTATION_RESTRICTED (2) Presentation restricted

P_ADDRESS_PRESENTATION_ADDRESS_NOT_AVAILABLE (3) Address not available for presentation

TpAddressRange A `TpAddressRange` is a `TpAddress`, with an `AddrString` field that can use the wildcard characters *, which matches zero or more characters, and ?, which matches exactly one character. Wildcards may appear only at the beginning or the end of `AddrString`. The following are thus illegal:

1?3 has a wildcard in the middle.

1*3 has a wildcard in the middle.

*123? has wildcards at both beginning and end.

The following are legal:

123? matches any string of four characters starting with 123.

123?? matches any string of five characters starting with 123.

123* matches any string of three or more characters starting with 1̲

123??* matches any string of five or more characters starting with 1

TpAddressScreening *TpAddressScreening* is an enumeratior results from address screening:

P_ADDRESS_SCREENING_UNDEFINED (0) Undefined

P_ADDRESS_SCREENING_USER_VERIFIED_PASSED (1) User ̲ vided address verified and passed

P_ADDRESS_SCREENING_USER_NOT_VERIFIED (2) User ̲ vided address not verified

P_ADDRESS_SCREENING_USER_VERIFIED_FAILED (3) User ̲ vided address verified and failed

P_ADDRESS_SCREENING_NETWORK (4) Network provid address

TpAuthCapabilityList A *TpAuthCapabilityList* is a *TpStri̲* of values separated from each other by commas and lists either su ported or requested encryption capabilities. Each value is one of the f̲ lowing strings (other capabilities specific to a network operator may ̲ indicated with strings preceded by the characters "OP_"):

empty string No capabilities supported.

P_DES_56 Security between client and framework provided ḇ encrypting using DES algorithm with 56-bit shared secret key.

P_DES_128 Same as previous, but using 128-bit shared secret key.

P_RSA_512 Security provided by encrypting using public ke cryptography with 512-bit keys.

P_RSA_1024 Same as previous, but using 1024-bit keys.

TpAuthDomain The framework passes information a client needs tc authenticate itself in an instance of *TpAuthDomain*:

```
Sequence of Data Elements TpAuthDomain {
  DomainID : TpDomainID;
  AuthInterface : IparlayInterfaceRef;
};
```

DomainID is a tagged union that can be used to identify either the framework itself or some entity, such as a client application, that wants to access the framework. The specific IDs recognized by a given framework are an implementation detail.

AuthInterface is a reference to an object with which the framework can authenticate itself to the application. Its type, *IparlayInterfaceRef*, and the related *IparlayInterface*, are interfaces from which many other Parlay interfaces derived, including *IpAppAccess*, the object type to be used in this example.

TpCallEventCriteria A client application passes information about events for which it wants notifications in an instance of *TpCallEventCriteria*:

```
Sequence of Data Elements TpCallEventCriteria {
  DestinationAddress : TpAddressRange;
  OriginatingAddress : TpAddressRange;
  CallEventName : TpCallEventName;
  CallNotificationType : TpCallNotificationType;
  MonitorMode : TpCallMonitorMode;
};
```

DestinationAddress is a range of called-party addresses that should appear in any events passed on to the application.

OriginatingAddress is a range of calling-party addresses that should appear in any events passed on to the application.

CallEventName identifies the event or events to be passed on to the application.

CallNotificationType specifies whether the events to be passed on to the application relate to the originating or terminating end of a call.

MonitorMode specifies how the call control service should handle event notifications (notify and wait, or notify and continue).

TpCallEventInfo An instance of *TpCallEventInfo* holds information about an event being reported from Parlay to a client application. Many of the fields are the same as those in *TpCallEventCriteria*. There they are used as a filter on the events to be passed to the application. Here they provide information on a specific event that has

occurred. Other fields contain additional information about the
that may be of interest to some applications, but not to the one i
chapter.

```
Sequence of Data Elements TpCallEventInfo {
    DestinationAddress : TpAddress;
    OriginatingAddress : TpAddress;
    OriginalDestinationAddress : TpAddress;
    RedirectingAddress : TpAddress;
    CallAppInfo : TpCallAppInfoSet;
    CallEventName : TpCallEventName;
    CallNotificationType : TpCallNotificationType;
    MonitorMode : TpCallMonitorMode;
};
```

TpCallEventName *TpCallEventName* is an enumeration of
that occur during call processing. The event values can be com
with the logical OR operator when requesting notification of more
one type of event (that's why the values go up by powers of 2):

P_EVENT_UNDEFINED (0)   Undefined event.

P_EVENT_GCCS_OFFHOOK_EVENT (1)   Offhook event i
Generic Call Control Service.

P_EVENT_GCCS_ADDRESS_COLLECTED_EVENT (2)   Networ
collected address from calling party, but has yet to analyze it
address may be incomplete.

P_EVENT_GCCS_ADDRESS_ANALYZED_EVENT (4)   Networ
collected and analyzed address from calling party. The address i
known to be valid and complete.

P_EVENT_GCCS_CALLED_PARTY_BUSY (8)   Called party is bu

P_EVENT_CALLED_PARTY_UNREACHABLE (16)   Called pa
unreachable, for example, when using a mobile phone that has
turned off.

P_EVENT_GCCS_NO_ANSWER_FROM_CALLED_PARTY (32
answer from called party.

P_EVENT_GCCS_ROUTE_SELECT_FAILURE (64)   Network
not route call.

P_EVENT_GCCS_ANSWER_FROM_CALLED_PARTY (128)
party has answered call.

TpCallIdentifier   When Parlay notifies a client that a call eve
occurred, it passes along an instance of *TpCallIdentifier* to l
client know what call generated the event:

```
Sequence of Data Elements TpCallIdentifier {
  CallReference : IpCallRef;
  CallSessionID : TpSessionID;
};
```

`CallReference` is a reference to a call object associated with an event that has been reported from Parlay to a client application.

`CallSessionID` is an integer that uniquely identifies that call.

TpCallNotificationType `TpCallNotificationType` is an enumeration that specifies whether an event relates to the terminating or to the originating end of a call:

P_TERMINATING (0)   Event relates to terminating end of call.

P_ORIGINATING (1)   Event relates to originating end of call.

TpCallReport   Results of a call request are packaged in an instance of `TpCallReport`:

```
Sequence of Data Elements TpCallReport {
  MonitorMode : TpCallMonitorMode;
  CallEventTime : TpDateAndTime;
  CallReportType : TpCallReportType;
  AdditionalReportInfo : TpCallAdditionalReportInfo;
};
```

`MonitorMode` specifies how this result notification is being handled (notify and wait, or notify and continue).

`CallEventTime` is a time at which the event occurred.

`CallReportType` is the type of event being reported.

`AdditionalReportInfo` contains additional information about the event (which won't be covered here).

TpCallReportRequest   When a client makes an asynchronous request of Parlay (i.e., calls a method, the name of which ends with *Req*, it needs to tell Parlay under what conditions to notify it of results (i.e., call a client method, the name of which ends in *Res*). It puts that information in an instance of *TpCallReportRequest*.

```
Sequence of Data Elements TpCallReportRequest {
  MonitorMode : TpCallMonitorMode;
  CallReportType : TpCallReportType;
  AdditionalReportCriteria : TpCallAdditionalReportCriteria;
};
```

`MonitorMode` specifies how result notifications should be handled (notify and wait, or notify and continue).

`CallReportType` is the type of event the client wants to hear abo

`AdditionalReportCriteria` contains additional criteria that m[...] used to filter report notifications (which won't be covered here).

TpCallReportRequestSet A `TpCallReportRequestSet` is an [...] of `TpCallReportRequest` instances.

TpCallReportType `TpCallReportType` is an enumerati[...] events that may occur after a client has called a request (`Req`) meth[...] Parlay. Any one or all of them may cause Parlay to call a results [...] method in the client to inform it of the event:

P_CALL_REPORT_UNDEFINED (0) Undefined results.

P_CALL_REPORT_PROGRESS (1) Progress has been made in ro[...] the call to its destination.

P_CALL_REPORT_ALERTING (2) Called party is being alerted.

P_CALL_REPORT_ANSWER (3) Called party has answered.

P_CALL_REPORT_BUSY (4) Called party is busy.

P_CALL_REPORT_NO_ANSWER (5) Called party has not answe[...]

P_CALL_REPORT_DISCONNECT (6) Party has disconnected.

P_CALL_REPORT_REDIRECTED (7) Call has been redirected [...] new address.

P_CALL_REPORT_SERVICE_CODE (8) Mid-call service code rece[...]

P_CALL_REPORT_ROUTING_FAILURE (9) Call has failed rou[...] May be possible to reroute.

TpDateAndTime A `TpDateAndTime` is a `TpString` that specif[...] date and time as a string of characters of the form:

```
YYY-MM-DD HH:MM:SS.mmm
```

A `z` may be appended to indicate Universal Time (UTC).

TpMonitorMode `TpMonitorMode` is an enumeration of values [...] specify how event monitoring is to be handled:

P_CALL_MONITOR_MODE_INTERRUPT (0) Call control ser[...] should halt call processing when event occurs, pass the event o[...] the application, and wait for further instructions or events be[...] continuing.

P_CALL_MONITOR_MODE_NOTIFY (1) Call control service should pass event to application when it occurs and immediately resume execution.

P_CALL_MONITOR_MODE_DO_NOT_MONITOR (2) Call control service should not monitor for event.

TpService

```
Sequence of Data Elements TpService {
  ServiceID : TpServiceID;
  ServicePropertyList : TpServicePropertyList;
};
```

`ServicePropertyList` is an array of `TpServiceProperty` that describe properties of the service. The example is this chapter didn't use service properties, so that's all there is to say about that. See Ref. [75] for more information on the `TpServiceProperty` data structure.

`ServiceID` is a string that uniquely identifies an instance of a service interface. The framework generates a service ID for each service it instantiates. A service ID, for example, "34857234/P_CALL_CONTROL/P_MULTIPARTY" for the Multiparty Call Control Service, has three parts—a `TpUniqueServiceNumber`, a `TpServiceNameString`, and a `TpServiceSpecString`—separated by forward slashes (/):

TpServiceNameString This is a string that uniquely identifies a service interface, chosen from the following set (operator-specific interfaces may be created with names starting with "OP_"):

empty string No service

`P_CALL_CONTROL` Generic Call Control Service

`P_MESSAGING` Generic Messaging Service

`P_USER_INTERACTION` Generic User Interaction Service

`P_USER_LOCATION_EMERGENCY` Mobility Emergency User Location Service

`P_USER_STATUS` User Status Service

`P_CONNECTIVITY_MANAGER` Connectivity Manager Service

`P_USER_LOCATION` Mobility User Location Service

`P_USER_LOCATION_CAMEL` Mobility User Location Camel Service

TpServiceSpecString This is a string that uniquely identifies vice specification interface, chosen from the following set (op specific interfaces may be created with names starting with "OP_"

empty string No service specialization

P_MULTI_PARTY   The multiparty specialization of the Generi Control Service

P_MULTI_MEDIA   The multimedia specialization of the Multi Call Control Service

P_CONFERENCE   The conference specialization of the Multimedi Control Service

P_VMAIL   The voice mail specialization of the Generic Mess Service

P_EMAIL   The email specialization of the Generic Messaging Servi

P_CALL   The call specialization of the Generic User Interaction Se

P_TRIGGERED_USER_LOCATION   The Triggered (i.e. by location ch specialization of the Mobility User Location Service

TpSessionID This is an integer that uniquely identifies a call. U when a single call object may manage many different calls, eac which is considered a separate session.

TpUniqueServiceNumber This is a unique string of digits.

For More Information

All Parlay specifications that have been released to the public may found at http://www.parlay.org. Data structure used through Parlay are defined in Ref. [73]. The documents that describe the Pa framework are Refs. [74] through [78]. Generic call processing described in Refs. [79], [71], [80], and [72]. The sample code in this chap that interacts with the Parlay framework is based on examples from t JAIN SPA online documentation, which can be obtained from Su Web site, http://java.sun.com/products/jain.

APPENDIX A

ACRONYMS

3GPP 3rd Generation Partnership Project
ACD Automatic Call Distributor
AD ADjunct (IN)
ADPCM Adaptive DPCM
AIN Advanced Intelligent Network
ANSI American National Standards Institute
API Application Programming Interface
AS Autonomous System
ASN.1 Abstract Syntax Notation 1
ASR Automated Speech Recognition
ATIS Alliance for Telecommunications Industry Solutions
ATM Asynchronous Transfer Mode
BCP Best Current Practice (IETF)
BCSM Basic Call State Model (IN)
BER Basic Encoding Rules
BGP Border Gateway Protocol
BIND Berkeley Internet Name Domain
BOA Basic Object Adapter (CORBA)
BOC Bell Operating Company
BRI Basic Rate Interface (ISDN)
CCIS Common Channel Interoffice Signaling
CCITT International Telephone and Telegraph Consultative Committee (French acronym)
CCSSO Common Channel Signaling Switching Office (SS7)
CCS Common Channel Signaling
CFBL Call Forward Busy Line
CGI Common Gateway Interface
CHAP Challenge Handshake Authentication Protocol
CLEC Competitive Local Exchange Carrier

CO Central Office
COM Component Object Model (Microsoft)
CORBA Common Object Request Broker Architecture
CPE Customer Premises Equipment
CPL Call Processing Language
CS-1 Capability Set 1 (IN)
CSS Cascading Stylesheets (XML)
CSTA Computer-Supported Telecommunications Applications
CTI Computer Telephony Integration
DAS Directory Assistance Service (RFC 1202)
DCOM Distributed Component Object Model (Microsoft)
DEN Directory Enabled Network
DES Data Encryption Standard
DII Dynamic Invocation Interface (CORBA)
DIXIE Directory Interface to X.500 Implemented Efficie (RFC 1249)
DN Distinguished Name (X.500 and LDAP) *or* Directory Number
DNS Domain Name System/Service
DOM Document Object Model (XML)
DP Detection Point (IN)
DPC Destination Point Code (SS7)
DPCM Differential PCM
DTD Document Type Definition (XML)
DSSSL Document Style Semantics and Specification Language (SGN
ECTF Enterprise Computer Telephony Forum
EDGE Enhanced Data rates for GSM Evolution (3GPP)
EDP Event Detection Point (IN)
EJB Enterprise JavaBeans
EO End Office
ETSI European Telecommunications Standards Institute
FAQ Frequently Asked Questions
FTP File Transfer Protocol
GIOP General Inter-ORB Protocol (CORBA)

Acronyms

GMPLS Generalized MPLS
GPRS General Packet Radio Service (3GPP)
GSTN Global Switched Telephone Network
GTT Global Title Translation (SS7)
HTML HyperText Markup Language
HTTP HyperText Transfer Protocol
IANA Internet Assigned Numbers Authority
ICW Internet Call Waiting
IDL Interface Definition Language
IEC Interexchange Carrier
IEEE Institute of Electrical and Electronics Engineers
IETF Internet Engineering Task Force
IIOP Internet Inter-ORB Protocol (CORBA)
ILEC Incumbent Local Exchange Carrier
IN Intelligent Network
INAP Intelligent Network Application Part
IP Internet Protocol *or* Intelligent Peripheral (IN)
ISDN Integrated Services Digital Network
ISO International Standards Organization
ISP Internet Service Provider
ISUP ISDN User Part (SS7)
ITU-T International Telecommunications Union—Telecommunications Sector
IVR Interactive Voice Response
IXC Interexchange Carrier
JAIN Java APIs for Integrated (or Intelligent) Networks
JAIN SPA JAIN Service Provider API
JCAT JAIN Coordination and Transactions
JCC JAIN Call Control
JCP Java Community Process *or* Java Call Processing (JAIN)
JDK Java Development Kit
JSCE JAIN Service Creation Environment
JSLEE JAIN Service Logic and Execution Environment

JSPA Java Specification Participation Agreement
JSR Java Specification Request
JTAPI Java Telephony API
JVM Java Virtual Machine
LAN Local Area Network
LAP Link Access Protocol
LATA Local Access and Transport Areas
LDAP Lightweight Directory Access Protocol (LDAPv2 in RFC LDAPv3 in RFC 2251)
LEC Local Exchange Carrier
MAP Mobile Application Part (SS7)
MDTP Multiprotocol Datagram Transport Protocol
MEGACO MEdia GAteway COntrol (RFC 3015)
MG Media Gateway
MGC Media Gateway Control
MGCP Media Gateway Control Protocol (RFC 2705)
MIDL Microsoft IDL (DCOM)
MPLS Multi-Protocol Label Switching
MSC Message Sequence Chart
NANP North American Numbering Plan
NDS Novell Directory Services
NPA Numbering Plan Area
NTP Network Time Protocol (RFC 1305)
OA&M Operations, Administration, and Maintenance
OASIS Organization for the Advancement of Structured Informati Standards
O-BCSM Originating Basic Call State Model (IN)
OLE Object Linking and Embedding (Microsoft)
OMAP Operations, Maintenance and Administration Part (SS7)
OMG Object Management Group
OOAD Object-Oriented Analysis and Design
OOP Object-Oriented Programming
OPC Origination Point Code (SS7)

Acronyms

ORB   Object Request Broker (CORBA)
OSGi   Open Systems Gateway Initiative
OSI   Open Systems Interconnect
OSS   Operations Support System *or* Operator Services System (SS7)
PAM   Presence and Availability Management
PBX   Private Branch Exchange
PCM   Pulse Code Modulation
PER   Packed Encoding Rules
PIC   Point In Call (IN) *or* Primary InterLATA Carrier (PSTN)
PINT   PSTN/Internet Interworking (RFC 2848)
POA   Portable Object Adapter (CORBA)
POP   Point Of Presence
POTS   Plain Old Telephone Service
PRI   Primary Rate Interface (ISDN)
PSTN   Public Switched Telephone Network
QoS   Quality of Service
RAS   Registration, Admission, and Status (H.323)
RBOC   Regional Bell Operating Company
RDN   Relative Distinguished Name (X.500 and LDAP)
RFC   Request For Comments (IETF)
RMI   Remote Method Invocation (Java)
RPC   Remote Procedure Call
RSA   Rivest Shamir Adelman (algorithm)
RTCP   Real-Time Control Protocol
RTP   Real-Time Transport Protocol
RUDP   Real-time UDP
SASL   Simple Authentication and Security Layer
SAX   Simple API for XML
SCCP   Signaling Connection Control Part (SS7)
SCE   Service Creation Environment (IN)
SCN   Switched Circuit Network
SCP   Service Control Point (SS7/IN)
SCTP   Simple Control Transport Protocol

SDK   Software Development Kit

SDP   Session Description Protocol (RFC 2327) *or* Service Data Poi[nt]

SEP   Signaling End Point (SS7)

SG   Signaling Gateway

SGCP   Simple Gateway Control Protocol

SGML   Standard Generalized Markup Language

SIP   Session Initiation Protocol (RFC 2543)

SMS   Service Management System (IN)

SMTP   Simple Mail Transfer Protocol

SN   Service Node (IN)

SOAP   Simple Object Access Protocol

SP   Signaling Point (SS7)

SPC   Stored Program Control

SPI   Service Provider Interface

SPIRITS   Service in the PSTN/IN Requesting Internet Service

SS6   Signaling System 6

SS7   Signaling System 7

SSP   Service Switching Point (SS7/IN)

STP   Signaling Transfer Point (SS7/IN)

TAPI   Telephony API (Microsoft)

T-BCSM   Terminating Basic Call State Model (IN)

TCAP   Transaction Capabilities Application Part (SS7)

TDM   Time Division Multiplexing

TDP   Trigger Detection Point (IN)

TIA   Telecommunication Industry Association

TIPHON   Telecommunications and Internet Protocol Harmoniz[ation] Over Networks (ETSI)

TLS   Transport Layer Security

TN   Telephone Number

TRIP   Telephony Routing Information Protocol

TSAPI   Telephony Services API

TTS   Text To Speech

URI   Universal Resource Identifier

Acronyms

URL Universal Resource Locator
VoIP Voice over IP
W3C World Wide Web Consortium
XML Extensible Markup Language
XSL Extensible Stylesheet Language (XML)

APPENDIX B

WEB REFERENCES

3GPP http://www.3gpp.org
ATM Forum http://www.atmforum.org
CORBA http://www.corba.org
DCOM http://www.microsoft.com
ECTF http://www.ectf.org
ETSI http://www.etsi.org
IEEE http://www.ieee.org
IETF http://www.ietf.org
ITU-T http://www.itu.int
JAIN http://java.sun.com/products/jain
Java http://java.sun.com
JTAPI http://java.sun.com/products/jtapi
PAMforum http://www.pamforum.org
Parlay http://www.parlay.org
RMI http://java.sun.com
SOAP http://www.w3c.org
T1 http://www.t1.org
TIA http://www.tiaonline.org
VoiceXML http://www.voicexml.org
W3C http://www.w3c.org
XML http://www.xml.org

BIBLIOGRAPHY

1. Paul Albitz and Cricket Liu. *DNS and BIND, Third Edition.* O'Reilly and Associates, 1998.
2. Stephen Asbury and Scott Weiner. *Developing Java Enterprise Applications.* Wiley, 1999.
3. Simon Beddus, Gary Bruce, and Steve Davis. "Opening up networks with JAIN Parlay." *IEEE Communications Magazine,* April 2000.
4. Ravi Raj Bhat and Rajeev Gupta. "JAIN Protocol APIs." *IEEE Communications Magazine,* January 2000.
5. Uyless Black. *Voice Over IP.* Prentice Hall, 2000.
6. David Chadwick. *Understanding X.500 The Directory.* Chapman and Hall, 1994.
7. Patrick Chan. *Java Developer's Almanac.* Addison-Wesley, 1999, 2000, etc.
8. David Chappell. "Abstract Syntax Notation One (ASN.1)." *Journal of Data and Computer Communications,* Spring 1989.
9. Marion Cole. *Introduction to Telecommunications: Voice, Data, and The Internet.* Prentice Hall, 1999.
10. George Coulouris, Jeen Dollimore, and Tim Kindberg. *Distributed Systems: Concepts and Design,* Second Edition. Addison-Wesley, 1996.
11. R. Cox and P. Kroon. "Low bit-rate speech coders for multimedia communication." *IEEE Communications Magazine,* December 1996.
12. Data Connection Ltd. *Directory Services: The Role of LDAP and X.500,* 2001. http://www.dataconnection.com.
13. John de Keijzer, Douglas Tait, and Rob Goedman. "JAIN: A new approach to services in communication networks." *IEEE Communications Magazine,* January 2000.
14. Bill Douskalis. *IP Telephony: The Integration of Robust VoIP Services.* Prentice Hall, 2000.
15. Jim Farley. *Java Distributed Computing.* O'Reilly and Associates, 1998.
16. Igor Faynberg, Lawrence R. Gabuzda, Marc P. Kaplan, and Nitin J. Shah. *The Intelligent Network Standards: Their Application to Services.* McGraw-Hill, 1997.
17. Igor Faynberg, Lawrence R. Gabuzda, and Hui-Lan Lu. *Converged Networks and Services: Internetworking IP and the PSTN.* Wiley, 2000.

18. Martin Fowler and Kendall Scott. *UML Distilled.* Addison-Wesley, 1997.
19. James Gosling, Bill Joy, and Guy Steele. *The Java Language Specification.* Addison-Wesley, 1996.
20. Richard Grimes. *Professional DCOM Programming.* Wrox Press, 1997.
21. Tim Howes and Mark Smith. *LDAP: Programming Directory-Enabled Applications with Lightweight Directory Access Protocol.* Macmillan, 1997.
22. Tim Howes, Mark Smith, and Gordon Good. *Understanding and Deploying LDAP Directory Services.* Macmillan, 1999.
23. IETF. *RFC 1202: Directory Assistance Service,* 1991.
24. IETF. *RFC 1249: DIXIE Protocol Specification,* 1991.
25. IETF. *RFC 1305: Network Time Protocol (Version 3) Specification, Implementation and Analysis,* 1992.
26. IETF. *RFC 1487: X.500 Lightweight Directory Access Protocol,* 1993.
27. IETF. *RFC 1771: A Border Gateway Protocol 4 (BGP-4),* 1995.
28. IETF. *RFC 1777: Lightweight Directory Access Protocol,* 1995.
29. IETF. *RFC 1823: The LDAP Application Program Interface,* 1995.
30. IETF. *RFC 1889: RTP: A Transport Protocol for Real-Time Applications,* 1996.
31. IETF. *RFC 1890: RTP Profile for Audio and Video Conferences with Minimal Control,* 1996.
32. IETF. *RFC 1994: PPP Challenge Handshake Authentication Protocol (CHAP),* 1996.
33. IETF. *RFC 2045: Multipurpose Internet Mail Extensions (MIME) Part One: Format of Internet Message Bodies,* 1996.
34. IETF. *RFC 2046: Multipurpose Internet Mail Extensions (MIME) Part Two: Media Types,* 1996.
35. IETF. *RFC 2205: Resource ReSerVation Protocol (RSVP)—Version 1 Functional Specification,* 1997.
36. IETF. *RFC 2222: Simple Authentication and Security Layer,* 1997.
37. IETF. *RFC 2251: Lightweight Directory Access Protocol (v3),* 1997.
38. IETF. *RFC 2327: SDP: Session Description Protocol,* 1998.
39. IETF. *Internet Draft: The SIP Servlet API,* September 1999.
40. IETF. *RFC 2246: The TLS Protocol, Version 1.0,* 1999.
41. IETF. *RFC 2543: SIP: Session Initiation Protocol,* 1999.

Bibliography

42. IETF. *RFC 2705: Media Gateway Control Protocol (MGCP) Version 1.0*, 1999.
43. IETF. *Internet Draft: Common Gateway Interface for SIP*, June 2000.
44. IETF. *Internet Draft: CPL: A Language for User Control of Internet Telephony Services*, November 2000.
45. IETF. *Internet Draft: Guidelines for Authors of SIP Extensions*, November 2000.
46. IETF. *Internet Draft: SIP Telephony Call Flow Examples*, November 2000.
47. IETF. *Internet Draft: SIP Telephony Service Examples*, November 2000.
48. IETF. *Internet Draft: SPIRITS Protocol Requirements*, August 2000.
49. IETF. *Internet Draft: The SPIRITS Architecture*, August 2000.
50. IETF. *RFC 2806: URLs for telephone calls*, 2000.
51. IETF. *RFC 2848: The PINT Service Protocol: Extensions to SIP and SDP for IP Access to Telephone Call Services*, 2000.
52. IETF. *RFC 2916: E.164 number and DNS*, 2000.
53. IETF. *RFC 2976: The SIP INFO Method*, 2000.
54. IETF. *RFC 2995: Pre-SPIRITS Implementations of PSTN-initiated Services*, 2000.
55. IETF. *RFC 3015: Megaco Protocol Version 1.0*, 2000.
56. IETF. *Internet Draft: Event Notification in SIP*, February 2001.
57. IETF. *Internet Draft: SIP Call Control—Transfer*, February 2001.
58. ITU-T. *Recommendation Q.931, ISDN User-Network Interface Layer 3 Specification for Basic Call Control*, 1993.
59. ITU-T. *Recommendation X.500, The Directory: Overview of concepts, models and services*, 1997.
60. ITU-T. *Recommendation H.225.0, Call Signaling Protocols and Media Stream Packetization for Packet-Based Multimedia Communications Systems*, 1998.
61. ITU-T. *Recommendation H.245, Control Protocol for Multimedia Communication*, 1998.
62. ITU-T. *Recommendation H.323, Packet-Based Multimedia Communications Systems*, 1998.
63. Ravi Jain, Farooq Anjum, Paolo Missier, and Subramanya Shastry. "Java Call Control, Coordination, and Transactions." *IEEE Communications Magazine*, January 2000.

64. Bil Lewis and Daniel Berg. *Threads Primer: A Guide to Multithreaded Programming.* Prentice Hall, 1996.
65. Neil Matthew and Richard Stones. *Beginning Linux Programming.* Wrox Press, 1996.
66. David G. Messerschmitt. "The convergence of telecommunications and computing: What are the implications today?" *IEEE Proceedings,* August 1996.
67. Pat Niemeyer and Jonathan Knudsen. *Learning Java.* O'Reilly and Associates, 2000.
68. Scott Oaks and Henry Wong. *Java Threads,* Second edition. O'Reilly and Associates, 1999.
69. OMG. *IDL to Java Language Mapping Specification,* 1999.
70. Robert Orfali and Dan Harkey. *Client/Server Programming with Java and CORBA,* Second Edition. Wiley, 1998.
71. Parlay Group. *Parlay APIs 2.1: Call Processing Class Diagrams,* 2000.
72. Parlay Group. *Parlay APIs 2.1: Call Processing Sequence Diagrams,* 2000.
73. Parlay Group. *Parlay APIs 2.1: Common Data Definitions,* 2000.
74. Parlay Group. *Parlay APIs 2.1: Framework Class Diagrams,* 2000.
75. Parlay Group. *Parlay APIs 2.1: Framework Data Definitions,* 2000.
76. Parlay Group. *Parlay APIs 2.1: Framework Interfaces, Client Application View,* 2000.
77. Parlay Group. *Parlay APIs 2.1: Framework Interfaces, Parlay Service View,* 2000.
78. Parlay Group. *Parlay APIs 2.1: Framework Sequence Diagrams,* 2000.
79. Parlay Group. *Parlay APIs 2.1: Generic Call Control Service Data Definitions,* 2000.
80. Parlay Group. *Parlay APIs 2.1: Generic Call Control Service Interfaces,* 2000.
81. Greg M. Perry. *The Absolute Beginner's Guide to Programming.* Sams, 1993.
82. Larry L. Peterson. *Computer Networks: A Systems Approach.* Morgan Kaufman, 1996.
83. Spencer Roberts. *Essential JTAPI.* Prentice Hall, 1999.
84. Jonathan Rosenberg, Jonathan Lennox, and Henning Schulzrinne. "Programming Internet Telephony Services." *IEEE Internet Computing Magazine,* May/June 1999.
85. Travis Russell. *Signaling System #7,* Second Edition. McGraw-Hill, 1998.

Bibliography

86. Walter Savitch and Richard Johnsonbaugh. *Java: An Introduction to Computer Science and Programming.* Prentice Hall, 1998.
87. Kennard Scribner and Mark Stiver. *Understanding SOAP.* Sams, 2000.
88. Clemens Szyperski. *Component Software: Beyond Object-Oriented Programming.* Addison-Wesley, 1998.
89. Andrew S. Tanenbaum. *Computer Networks,* Third Edition. Prentice Hall, 1996.
90. Thuuan L. Thai. *Learning DCOM.* O'Reilly and Associates, 1999.
91. VoiceXML Forum. *Voice eXtensible Markup Language: VoiceXML 1.0,* 2000.

INDEX

A

Abstract Syntax Notation 1 (ASN.1), 37—39
ACM (Address Complete Message), 177
Action lines, 296—297
ActiveX, 114
Adaptive DPCM (ADPCM), 194
ACD (Automatic Call Distributor), 313
Address Complete Message (ACM), 177
ADPCM (adaptive DPCM), 194
Advanced Intelligent Network (AIN), 182—183
Agent monitoring, 314
AIN (see Advanced Intelligent Network)
Alliance for Telecommunications Industry Solutions (ATIS), 16
American National Standards Institute (ANSI), 16
Analog switches, 160—161
ANM (Answer Message), 177
ANSI (American National Standards Institute), 16
Answer Message (ANM), 177
APIs (see Application programming interfaces)
Applets, 55
Application layer, 42, 44
Application programming interfaces (APIs), 31—34
arpa (domain), 139
Arrays (Parlay), 387
AS (Autonomous System), 215
ASCII, 36
ASN.1 (see Abstract Syntax Notation 1)
ASR (Automated Speech Recognition), 303
Asynchronous Transfer Mode (ATM), 16
ATIS (Alliance for Telecommunications Industry Solutions), 16
ATM (Asynchronous Transfer Mode), 16
ATM Forum, 16
Attributes, 75, 146
Audible ringing, 164
Authentication (Parlay), 394—399
Authority, 140
Automated Speech Recognition (ASR), 303
Automatic Call Distributor (ACD), 313
Autonomous System (AS), 215

B

Basic Call State Model (BCSM), 183—185
Basic Encoding Rules (BER), 38, 39
Basic Object Adapter (BOA), 109
Basic Rate Interface (BRI), 166
BCPs (see Best Current Practices)
BCSM (see Basic Call State Model)
Beanboxes, 61

Beans, 61—62
Bell System, 8n2
Bellcore, 182
BER (see Basic Encoding Rules)
Berkeley Internet Name Domain (BIND), 143
Best Current Practices (BCPs), 15—16
BGP (Border Gateway Protocol), 215
Binary protocols, 34—39
BIND (Berkeley Internet Name Domain), 143
Bitways, 3
BOA (Basic Object Adapter), 109
Border Gateway Protocol (BGP), 215
Bound properties, 62
BRI (Basic Rate Interface), 166
Broadcast message, 44
Built-in type, 82
Byte code, 51
Byte Code Verifier, 55, 56

C

C++, 53
C language, 53
Caching, 142n3
Call agents, 204
Call control, 314
Call Forward Busy Line (CFBL), 277—279
Call processing, 163—165, 314
  JCC, 377—382
  Parlay, 405—412
Call Processing Language (CPL), 77, 287—293
  design objectives of, 288—289
  features of, 288
  simple scripts in, 290—293
Call reference number, 167
Capability Set 1 (CS-1), 183
Cascading Style Sheets (CSS), 84—86
CCIR (International Radio Consultative Committee), 13
CCIS (Common Channel Interoffice Signaling), 171
CCITT (International Telephone and Telegraph Consultative Committee), 13
CCSSO (Common Channel Signaling Switching Office), 172
CDL (Component Description Language), 119
CDR (Common Data Representation), 103
Cell centers, 312—314
Central offices (COs), 160
CFBL (see Call Forward Busy Line)
Chemical Markup Language (CML), 77
Child nodes, 137

Circuit-associated signaling, 170
Circuit switching, 9—10
Class 4 offices, 160
CLECs (Competitive LECs), 160
Client-server architecture, 4, 5
Client stubs, 99
CML (Chemical Markup Language), 77
CORBA (see Common Object Request Broker Architecture)
Coders, 194—195
com (domain), 139
Common Channel Interoffice Signaling (CCIS), 171
Common channel signaling, 170—171
Common Channel Signaling Switching Office (CCSSO), 172
Common Data Representation (CDR), 103
Common Gateway Interface (CGI), 56—58
Common Object Request Broker Architecture (CORBA), 17, 103—109
Competitive LECs (CLECs), 160
Complex type definitions, 81
Component-based software, 29—31
Component Description Language (CDL), 119
Components, 147
Computer-Supported Telecommunications Applications (CSTA), 314—315
Computer Telephony Integration (CTI) applications, 312—316
Connections, 207
Constrained properties, 62
Constructed types (Parlay), 388—390
Content, 68
Control, 11—12
Convergence, network (see Network convergence)
Cordboards, 160
Core INAP, 183
COs (central offices), 160
Country code, 165
CPL (see Call Processing Language)
Crossbar switches, 160
CS-1 (Capability Set 1), 183
CSS (see Cascading Style Sheets)
CSTA (see Computer-Supported Telecommunications Applications)
CTI applications (see Computer Telephony Integration applications)

D

DAP (see Directory Access Protocol)
DAS (Directory Assistance Service), 149
Data Encryption Standard (DES), 42
Data models, 134
Database access systems, 93
Datagram sockets, 94—95
Datagrams, 44, 94
Datalink layer, 40
DCOM (see Distributed Component Object Model)
Decoders, 194—195
Deferred applications, 4
DEN (see Directory Enabled Network)
DES (Data Encryption Standard), 42
Detection points (DPs), 183—185
Dialogue, 178
DIB (see Directory Information Base)
Differential PCM (DPCM), 194
Digit translation, 164
Digital Subscriber Line (DSL), 167
Digital switches, 161
DII (see Dynamic Invocation Interface)
Directories, 134—136
Directory Access Protocol (DAP), 146, 149, 150
Directory Assistance Service (DAS), 149
Directory Enabled Network (DEN), 154—155
Directory Information Base (DIB), 145, 146
Directory Information Shadowing Protocol (DISP), 146
Directory Information Tree (DIT), 147
Directory Interface to X.500 Implemented Efficiently (DIXIE), 149
Directory Service Protocol (DSP), 146
Directory services, 134
Directory System Agents (DSAs), 145
Discovery, 401—402
DISP (Directory Information Shadowing Protocol), 146
Distinguished names (DNs), 147
Distributed Component Object Model (DCOM), 114—117
Distributed computing, 92—131
  CORBA standard for, 103—109
  with DCOM, 114—117
  definition of, 92
  with Java RMI, 109—114
  middleware for, 92—102
  with SOAP, 117—129
  technologies for, 129—130
Distributed objects, 93, 101—102
DIT (Directory Information Tree), 147
DIXIE (Directory Interface to X.500 Implemented Efficiently), 149
DNS (see Domain Name System)
DNs (Distinguished names), 147
DNS (Domain Name Service), 134
Document Object Model (DOM), 89—90
Document Style Semantics and Specification Language (DSSSL), 86
Document type declaration, 76
Document Type Definition (DTD), 73—78
DOM (see Document Object Model)
Domain Name System (DNS), 134, 136—144, 216
  namespace of, 137—139
  nameservers of, 140—142
  resource records in, 142—144
Domain names, 138—139
Domains, 138
DPCM (Differential PCM), 194
DPs (see Detection points)
Draft standards, 15
DSAs (Directory System Agents), 145

Index

DSL (Digital Subscriber Line), 167
DSP (Directory Service Protocol), 146
DSSSL (Document Style Semantics and Specification Language), 86
DTD (*see* Document Type Definition)
Dynamic call model, 320—321
Dynamic Invocation Interface (DII), 107—108
Dynamic object activation, 113

E

E.164 numbers, 165
ECMA (European Computer Manufacturer's Association), 315
ECTF (*see* Enterprise Computer Telephony Forum)
EDGE (Enhanced Data Rates for GSM Evolution), 16
EDPs (Event detection points), 184
edu (domain), 139
800 numbers, 175, 182
EJB container, 62
EJB (Enterprise JavaBeans), 62
Elements, 74
EmbeddedJava, 49—50
Encoding, speech, 194—195
End offices (EOs), 160
Endpoints, 207
Enhanced Data Rates for GSM Evolution (EDGE), 16
Enterprise Computer Telephony Forum (ECTF), 17, 316
Enterprise Java, 50
Enterprise JavaBeans (EJB), 62
Entity beans, 62
Entries, 146
EOs (end offices), 160
ETSI (European Telecommunications Standards Institute), 16
European Computer Manufacturer's Association (ECMA), 315
European Telecommunications Standards Institute (ETSI), 16
Event detection points (EDPs), 184
Events (Java), 58—61
Executive systems, 265
Experimental RFCs, 15
Extensible Markup Language (XML), 66—90
 displaying, 84—88
 and HTML, 66, 68—73
 parsing, 88—90
 schemas in, 78—84
 and SOAP, 119—124
 tag definition in, 73—77
 VoiceXML, 303—309
Extensible Stylesheet Language (XSL), 84—88

F

Features, 261
File Transfer Protocol (FTP), 42
First-party call control, 314
Flow control, 41

Forked invitations, 249—253
Forking, 225
Formatting tags, 68, 69
Framework APIs, 348, 384
FTP (File Transfer Protocol), 42

G

Garbage collection, 52
Gateways, 12
General Inter-ORB Protocol (GIOP), 103
General Packet Radio Service (GPRS), 16
Generalized MPLS (GMPLS), 197
GET, 126—127
GIOP (General Inter-ORB Protocol), 103
Global/General Switched Telephone Network (GSTN), 158
Global System for Mobile Communication (GSM), 16
GMPLS (Generalized MPLS), 197
gov (domain), 139
GPRS (General Packet Radio Service), 16
Group-based triggers, 185
GSM (Global System for Mobile Communication), 16
GSTN (Global/General Switched Telephone Network), 158

H

H.323, 197—202, 220
 architecture of, 197—200
 call processing by, 201—202
 and SIP, 253—255
H.GCP, 211
Horizontally integrated applications, 5—6
HTML (*see* HyperText Markup Language)
HTTP (*see* HyperText Transfer Protocol)
Hush-A-Phone, 8n2
HyperText Markup Language (HTML), 66, 68—74, 77, 84—88
HyperText Transfer Protocol (HTTP), 118, 121, 124—129

I

IAM (*see* Initial Address Message)
IANA (Internet Assigned Numbers Authority), 212
ICW (*see* Internet Call Waiting)
IDL (*see* Interface Definition Language)
IEEE (Institute of Electrical and Electronics Engineers), 16
IETF (*see* Internet Engineering Task Force)
IFRB (*see* International Frequency Registration Board)
IIOP (*see* Internet Inter-ORB Protocol)
ILECs (Incumbent LECs), 160
Immediate applications, 4
IN (*see* Intelligent Network)
In-band signals, 170
INAP (*see* Intelligent Network Application Part)
Incumbent LECs (ILECs), 160
Informational RFCs, 15
Initial Address Message (IAM), 176—177

Index

Institute of Electrical and Electronics Engineers (IEEE), 16
int (domain), 139
Integrated Services Digital Network (ISDN), 166—169
Intelligent Network Application Part (INAP), 183, 368, 370
Intelligent Network (IN), 181—188, 344
  architecture of, 183—186
  example, 186—188
Interactive Voice Response (IVR), 303
Interexchange Carriers (IXCs), 160
Interface Definition Language (IDL), 101, 104—106
Interfaces, 105, 114
International Frequency Registration Board (IFRB), 13
International phone numbers, 166
International Press Telecommunications Council (IPTC), 77
International Radio Consultative Committee (CCIR), 13
International Telecommunications Union—Telecommunications Sector (ITU-T), 13—14
International Telecommunications Union (ITU), 36, 39, 165
International Telephone and Telegraph Consultative Committee (CCITT), 13
Internet, 7—11, 39, 43
Internet address, 94
Internet Assigned Numbers Authority (IANA), 212
Internet Call Waiting (ICW), 279—281
Internet Drafts, 15
Internet Engineering Task Force (IETF), 12, 14—16, 36, 39, 207, 215—216, 220—222, 263, 264
Internet Inter-ORB Protocol (IIOP), 103, 104
Internet Protocol (IP), 7
Internet Protocol (IP) layer, 43
Internetworks, 43
InterNIC, 139
Interpreters, 50
Intranets, 7
INVOKE components, 178
IP addresses, 134
IP Device Control (IPDC) protocol, 207
IP (Internet Protocol), 7
IP networks, 7, 8
IP telephony, 190—217
  definition of, 190
  features of, 190—191
  and gateways, 202—214
  and H.323, 197—202
  QoS for, 193—197
  routing and translation for, 214—217
  scope of, 190
IPDC (IP Device Control) protocol, 207
IPTC (International Press Telecommunications Council), 77
ISDN (see Integrated Services Digital Network)
ISDN User Part (ISUP), 176—177
ISUP (see ISDN User Part)
ITU (see International Telecommunications Union)
ITU-D (Development) Sector, 13
ITU-R (Radiocommunication) Sector, 13
ITU-T (see International Telecommunications Union—Telecommunications Sector)
ITU-T draft standards, 14
ITU-T Recommendations, 14
IVR (Interactive Voice Response), 303
IXCs (Interexchange Carriers), 160

J

JAIN (see Java APIs for Intelligent Networks)
JAIN Call Control (JCC), 351, 371—382
  call processing by, 377—382
  core classes of, 371—373
  state machines in, 373—377
JAIN Coordination and Transactions (JCAT), 351
JAIN Service Creation Environment (JSCE), 349, 350
JAIN Service Logic and Execution Environment (JSLEE), 351
JAIN Service Provider API (JAIN SPA), 351
JAIN SPA (JAIN Service Provider API), 351
JAIN TCAP, 354—361
Java, 48—63
  applets in, 55
  class libraries in, 53—55
  development of, 48
  event model of, 58—61
  and security, 55—56
  servlets in, 56—58
Java APIs for Intelligent Networks (JAIN), 50, 347—382
  API stack of, 349—353
  development process for, 353
  INAP, JAIN, 368, 370
  and JCC, 371—382
  MGCP, JAIN, 367—369
  Parlay vs., 347—348
  SIP, JAIN, 361—367
  TCAP, JAIN, 354—361
Java Applet Class Loader, 56
Java Call Processing (JCP), 371—375
Java Card, 50
Java Community Process (JCP), 353
Java Native Interface (JNI), 352
Java Remote Method Invocation (Java RMI), 109—114
Java RMI (see Java Remote Method Invocation)
Java Security Manager, 56
Java Specification Participation Agreement (JSPA), 353
Java Speech API Grammar Format (JSGF), 306
Java Telephony API (JTAPI), 50, 312—341
  call model of, 319—322
  core classes of, 318—319
  and core object states, 322—325
  and CTI technology, 312—316
  and event listeners, 327—332
  extension packages for, 332—341
  sample application, 325—327
Java Virtual Machine (JVM), 50—52, 318
JavaBeans, 61—63
`javax.telephony.callcenter` package, 332—335

Index

`javax.telephony.callcontrol` package, 335—337
`javax.telephony.media` package, 337—338
`javax.telephony.mobile` package, 338
`javax.telephony.phone` package, 338—341
`javax.telephony.privatedata` package, 341
JCAT (JAIN Coordination and Transactions), 351
JCC (*see* JAIN Call Control)
JCP (*see* Java Call Processing)
JCP (Java Community Process), 353
Jini, 50
JNI (Java Native Interface), 352
JSCE (*see* JAIN Service Creation Environment)
JSGF (Java Speech API Grammar Format), 306
JSLEE (JAIN Service Logic and Execution Environment), 351
JSPA (Java Specification Participation Agreement), 353
JTAPI (*see* Java Telephony API)
JVM (*see* Java Virtual Machine)

L

Labels, 137
Ladder diagrams, 33—35
LAP (Link Access Protocol), 40
LATAs (Local Access and Transport Areas), 160
LDAP (*see* Lightweight Directory Access Protocol)
LECs (Local Exchange Carriers), 160
Lightweight Directory Access Protocol (LDAP), 134, 149—154
Linksets, 172
Local Access and Transport Areas (LATAs), 160
Local Exchange Carriers (LECs), 160
Local exchanges, 160
Location nodes, 290—291
Location transparency, 101
Logical call destination, 238
Lucent, 211

M

Mail exchangers, 144
MAP (Mobile Application Part), 177
Mapping, 101
Markup languages, 66
Marshaling, 99
Math Markup Language (MML), 77
MDCP (Media Device Control Protocol), 211
MDTP (*see* Multiprotocol Datagram Transport Protocol)
Media Device Control Protocol (MDCP), 211
Media Gateway (MG), 203, 204
Media Gateway Controller (MGC), 204—213
Media Gateway Controller Protocol (MGCP), 207—208, 256, 367—369
Media streams, 201, 231—235
MEGACO, 211—213, 256
Member states, 13
Memory leaks, 52

Message Sequence Charts (MSCs), 33
Message Transfer Part (MTP), 173, 174
Messerschmitt, David, 3
Metalanguage, 68
Method definitions (Parlay), 390—391
MG (*see* Media Gateway)
MGC (*see* Media Gateway Controller)
MGCP (*see* Media Gateway Controller Protocol)
Microsoft IDL (MIDL), 115
Middleware, 42, 92—93
MIDL (Microsoft IDL), 115
mil (domain), 139
MML (Math Markup Language), 77
Mobile Application Part (MAP), 177
Monikers, 115
MSCs (message sequence charts), 33
MTP (*see* Message Transfer Part)
Multicast message, 44
Multiprotocol Datagram Transport Protocol (MDTP), 213—214
Multivendor Interaction Forum, 182—183
Music on hold, 273—277
MX records, 143—144

N

Nameservers, 140—142
Namespaces, 79, 137—139
Naming Service, 105
NANP (North American Numbering Plan), 166
National number, 165
National numbering plan, 165—166
net (domain), 139
Network convergence, 2—17
    approaches to, 11—13
    and the Internet, 7—11
    organizations involved in, 13—17
    and phone networks, 6—7
Network layer, 40, 41, 43
Networks, 2
NewsML, 77
Nodes, 137
Nonremote procedure calls, 96—99
North American Numbering Plan (NANP), 166
NPAs (Numbering Plan Areas), 166
Numbering Plan Areas (NPAs), 166

O

OAM (Operations, Administration, and Management), 353
Object classes, 146
Object Linking and Embedding (OLE), 114
Object Management Group (OMG), 17
Object managers, 93, 102
Object-oriented programming, 22—29
Object Request Brokers (ORBs), 103
Object serialization, 109
Object skeletons, 102*n*5

Offhook signal, 163
OLE (Object Linking and Embedding), 114
OMG (Object Management Group), 17
Open interfaces, 5—6
Open Systems Gateway Initiative (OSGi), 17
Open Systems Interconnection (OSI) Reference Model, 39—43
Operations, Administration, and Management (OAM), 353
Operations Support Systems (OSSes), 185
Operator Services System (OSS), 173
Option names, 261
OPTIONS requests, 262—263
ORBs (Object Request Brokers), 103
org (domain), 139
Origination, 163
OSGi (Open Systems Gateway Initiative), 17
OSI Model (*see* Open Systems Interconnection Reference Model)
OSS (Operator Services System), 173
OSSes (Operations Support Systems), 185
Out-of-band signals, 170

P

Packed Encoding Rules (PER), 38, 39
Packet switching, 9, 10
PAM (Presence and Availability Management), 17
PAMforum, 17
Parents, 137
Parlay, 50, 384—420
   arrays in, 387
   authentication by, 394—399
   basic data types of, 387
   call processing by, 405—412
   constructed types in, 388—390
   discovery of services, 401—402
   interface definition language of, 386—391
   JAIN vs., 347—348
   method definitions in, 390—391
   naming conventions of, 386—387
   requesting access to, 399—400
   subscribing to services, 402—405
   types, Parlay, 412—420
Parlay Group, 16, 384
PBX (Private Branch Exchanges), 159
PCM (*see* Pulse code modulation)
Peer terminals, 4
Peer-to-peer architecture, 4, 5
PER (*see* Packed Encoding Rules)
PersonalJava, 49
Phone Markup Language (PML), 303
Phone networks, 6—7
Physical layer, 40
PICs (points in call), 183
Ping-pong diagrams, 33
PINT (*see* PSTN/Internet Internetworking)
Plain Old Telephone Service (POTS), 158
Plumbing, 92

PML (Phone Markup Language), 303
POA (Portable Object Adapter), 109
Points in call (PICs), 183
Portable Object Adapter (POA), 109
POST, 126—127
POTS (Plain Old Telephone Service), 158
Presence and Availability Management (PAM), 17
Presentation layer, 42
PRI (Primary Rate Interface), 166
Primary master nameserver, 140
Primary Rate Interface (PRI), 166
Private Branch Exchanges (PBX), 159
Private DTDs, 76
Procedural programming, 20—22
Proposed standards, 15
Protocol identifiers, 125
Protocols, 2, 31—39
   and APIs, 31—34
   definition of, 31—32
   and the Internet, 43—44
   text-based vs. binary, 34—39
PSTN (*see* Public Switched Telephone Network)
PSTN/Internet Internetworking (PINT), 12, 263—272
   architecture of, 265, 266
   features of, 267—268
   milestone services of, 264—265
   services of, 268—272
Public DTDs, 76
Public Switched Telephone Network (PSTN), 2, 3, 6—13, 39, 158—160
Pulse code modulation (PCM), 161—162

Q

Quality of Service (QoS), 193—197
Questions, 14

R

Raw sockets, 95
RBOCs (Regional Bell Operating Companies), 182
RDNs (*see* Relative distinguished names)
Real-Time Control Protocol (RTCP), 194—195
Real-Time UDP (RUDP), 213
RealtimeTransport Protocol (RTP), 195—196
Regional Bell Operating Companies (RBOCs), 182
REL (Release Message), 177
Relative distinguished names (RDNs), 147—149
Release Complete Message (RLC), 177
Release Message (REL), 177
Reliable channels, 201
Remote method calls, 93
Remote procedure calls (RPCs), 93, 96—101
Request messages (SIP), 227—229
Requests for Comments (RFCs), 15—16
Resolvers, 140
Resource records, 142—144
Resource Reservation Protocol (RSVP), 197
Response messages (SIP), 229—230

Index

RESULT components, 178
Reverse lookups, 142
RFCs (*see* Requests for Comments)
Rivest-Shamir-Adelman (RSA) public key cryptography, 42
RLC (Release Complete Message), 177
RMI registry, 110
Rollback, 93
Root elements, 74
Root node, 137
Routing, 40
RPCs (*see* Remote procedure calls)
RSA public key cryptography, 42
RSVP (Resource Reservation Protocol), 197
RTCP (*see* Real-Time Control Protocol)
RTP (*see* RealtimeTransport Protocol)
RUDP (Real-Time UDP), 213

S

S.100 standard, 316
SAX (*see* Simple API for XML)
SCCP (*see* Signaling Connection Control Part)
SCE (*see* Service Creation Environment)
Schemas, 78—84, 134
Schulzrinne, Henning, 223
SCM (Service Control Manager), 117
SCN (Switched Circuit Network), 158
SCPs (Service Control Points), 173
SCTP (*see* Stream Control Transport Protocol)
SDP (*see* Session Description Protocol)
Secondary master nameserver, 140
Semantic tags, 68
SEPs (signaling end points), 171
Server farm, 117
Server-server communication, 117
Service APIs, 384—385
Service Control Manager (SCM), 117
Service Control Points (SCPs), 173
Service Creation Environment (SCE), 185, 344—347
Service in the PSTN/IN Requesting Internet Service (SPIRITS), 12, 279—282
Service logic, 184, 286—287
Service Management System (SMS), 185
Service Provider Interface (SPI), 316—318
Service Switching Point (SSP), 172, 183
Servlets, 56—58
Session beans, 62
Session Description Protocol (SDP), 221, 222, 230—235
Session Initiation Protocol (SIP), 220—258, 260—283
    architecture of, 223—227
    call signaling with, 235—253
    and CFBL service, 277—279
    extensions of, 260—263
    forked invitations in, 249—253
    functions provided by, 222
    fundamental assumptions of, 222—223
    and H.323, 253—255
    headers, adding, 261—262
    interworking, 256—258
    JAIN SIP, 361—367
    methods, adding, 261
    music on hold, using, 273—277
    OPTIONS requests in, 262—263
    and PINT, 263—272
    proxy server, two-party call with, 246—248
    redirect server, two-party call with, 243—245
    registration in, 248—249
    request messages in, 227—229
    response messages in, 229—230
    and SDP, 230—235
    simple two-party call with, 235—243
    and SPIRITS, 279—282
Session layer, 42
SG (*see* Signaling Gateway)
SGCP (*see* Simple Gateway Control Protocol)
SGML (*see* Standard Generalized Markup Language)
Signaling, 163
Signaling Connection Control Part (SCCP), 174—175
Signaling end points (SEPs), 171
Signaling Gateway (SG), 204, 205, 207
Signaling links, 172
Signaling System 6 (SS6), 171
Signaling System 7 (SS7), 171—181
    network architecture of, 171—173
    protocols, 173—181
Signaling Transfer Points (STPs), 173
Signature, 100
Significant number, 165
Siloing, 6
Simple API for XML (SAX), 89, 90
Simple Gateway Control Protocol (SGCP), 207
Simple Mail Transfer Protocol (SMTP), 42
Simple Object Access Protocol (SOAP), 117—129
    and HTTP, 124—129
    and XML, 119—124
Simple type, 82
SIP (*see* Session Initiation Protocol)
SIP CGI, 293—298
    messages in, 296—297
    metavariables in, 295
    program output, processing of, 297—298
SIP servlets, 293, 298—303
SLAPD, 151—152
SMS (Service Management System), 185
SMTP (Simple Mail Transfer Protocol), 42
SOA records, 143
SOAP (*see* Simple Object Access Protocol)
Sockets, 94—95
Software, component-based, 29—31
Space-division switching, 161
SPC switches (*see* Stored program control switches)
Specialized resource functions (SRFs), 185
Speech encoding, 194—195
SpeechML, 303
SPI (*see* Service Provider Interface)
SPIRITS (*see* Service in the PSTN/IN Requesting Internet Service)

Index

SRFs (*see* Specialized resource functions)
SS6 (*see* Signaling System 6)
SS7 (*see* Signaling System 7)
SSP (*see* Service Switching Point)
SSPs (Service Control Points), 173
Standard Generalized Markup Language (SGML), 68, 73—75, 85
Standard Template Library, 53, 54
Standards, 15
Standards track RFCs, 15
Stateful proxies, 225
Stateless proxies, 225
Step-by-step switches, 160
Stored program control (SPC) switches, 161
STPs (Signaling Transfer Points), 173
Stream Control Transport Protocol (SCTP), 213, 214
Stream sockets, 94
Structure, 66
Stub compilers, 99—100
Study areas, 14
Study groups, 14
Subdomains, 138
Sun Microsystems, 48
Switch-based triggers, 185
Switch fabric, 160
Switch nodes, 291
Switched Circuit Network (SCN), 158
Switching, 160—162
Switching matrix, 160

T

T1, 16
TAC (Technical Advisory Council), 207
Tags, 67—77
Tandems, 172
TAPI (Telephony Applications Programming Interface), 315
TCAP (*see* Transaction Capabilities Application Part)
TCK (Technology Compatibility Kit), 353
TCP (*see* Transmission Control Protocol)
TCP/IP, 44
TDM (time division multiplexing), 162
TDPs (*see* Trigger detection points)
Technical Advisory Council (TAC), 207
Technology Compatibility Kit (TCK), 353
Telecommunication Industry Association (TIA), 16
Telecommunications and Internet Protocol Harmonization Over Networks (TIPHON), 16
Telephony, 158—188
  and analog vs. digital switching, 160—162
  and call processing, 163—165
  and Intelligent Network, 181—188
  and ISDN, 166—169
  and phone numbers, 165—166
  and SS7, 170—181
  and voice network, 158—160
  (*See also* IP telephony)
Telephony Applications Programming Interface (TAPI), 315
Telephony Routing Information Protocol (TRIP), 215—217
Telephony Services Applications Programming Interface (TSAPI), 315
Terminations, 212
Text-based protocols, 34—39
Text To Speech (TTS), 303
Thin client programming, 62
Third-party call control, 314
3rd Generation Partnership Project (3GPP), 16
TIA (Telecommunication Industry Association), 16
Time division multiplexing (TDM), 162
TIPHON (Telecommunications and Internet Protocol Harmonization Over Networks), 16
Transaction Capabilities Application Part (TCAP), 177—181
Transaction processing monitors, 93
Transactions, 93, 178, 211—212
Transmission Control Protocol (TCP), 42, 44
Transport, 11—12
Transport layer, 41—42, 43
Trigger detection points (TDPs), 184—185
TRIP (*see* Telephony Routing Information Protocol)
Trunking gateways, 204
TSAPI (Telephony Services Applications Programming Interface), 315
TTS (Text To Speech), 303

U

UAC (User Agent Client), 223
UAL (ULP adaptation layer), 214
UAS (*see* User Agent Server)
UDP (*see* User Datagram Protocol)
UDP/IP, 44
ULP adaptation layer (UAL), 214
ULP (upper layer protocol), 214
UML (Universal Modeling Language), 386
Unicast message, 44
Universal Modeling Language (UML), 386
Universal Resource Identifiers (URIs), 125, 216, 217
Universal Resource Locators (URLs), 125
Universally unique IDs, 115
Unmarshaling, 99
Unreliable channels, 201
Upper layer protocol (ULP), 214
URIs (see Universal Resource Identifiers)
URLs (Universal Resource Locators), 125
User Agent Client (UAC), 223
User Agent Server (UAS), 223—225
User Datagram Protocol (UDP), 42, 44, 94
User-to-server applications, 3—4
User-to-user applications, 4
`Uuids`, 115

Index

V

Values (tags), 74
Vertically integrated applications, 5
Visual J++, 55n
Visual programming languages, 344—346
Voice over IP (VoIP), 2, 10—11, 190
VoiceXML, 77, 303—309
VoIP (*see* Voice over IP)
VoxML, 303

W

W3C (*see* World Wide Web Consortium)
Wireless Application Protocol (WAP), 304
Working groups, 14
Working parties, 14
World Wide Web Consortium (W3C), 16, 71, 77
World Zones, 165

X

X.500, 144—149, 153—154
XML (*see* Extensible Markup Language)
XML parser, 88
XSL (*see* Extensible Stylesheet Language)

Z

Zone transfer, 140
Zones, 140—142

About the Author

Steve Mueller is Lead Member of Technical Staff for SBC Technology Resources, Inc., the applied research subsidiary of SBC Communications, Inc. He has more than fifteen years hands-on experience analyzing and programming communication network services.